FEMTOSECOND LASER
Techniques and Technology

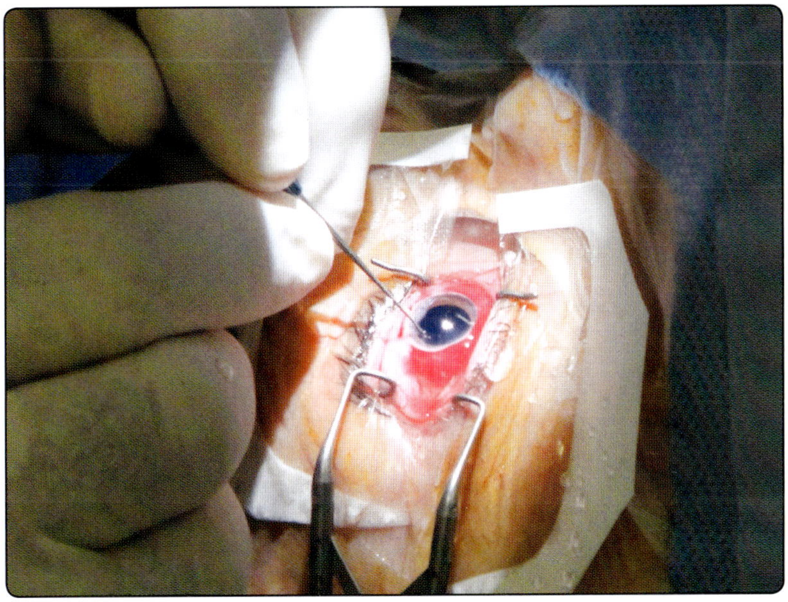

System requirement:

- **Windows XP or above**
- **DVD runs on web browser recommended—Google Chrome or Mozilla Firefox**
- **To view contents system requires JAVA—Plugins, if you not have, then install it by visiting on to troubleshoot section of the DVD**
- **To view videos system requires Flash Player**
- **For more details, you can visit troubleshoot section of the DVD**

Accompanying DVD-ROM is playable only in Computer and not in DVD player

Kindly wait for few seconds for DVD to autorun. If it does not autorun, then please do the following:

- Click on my computer
- Click the **CD/DVD drive** and after opening the drive, kindly double click the **launch.html**

DVD Contents

FEMTOSECOND LASER
Techniques and Technology

Editors-in-Chief

Ashok Garg MS PhD FRCS FIAO (Bel)
FRSM ADM FAIMS FICA
International and National Gold Medalist
Chairman and Medical Director
Garg Eye Institute and Research Centre
Dabra Chowk
Hisar, Haryana, India

Jorge L Alió MD PhD
Professor and Chairman of Ophthalmology
Medical Director
VISSUM, Instituto Oftalmologico de Alicante
Avda de Denia S/n 03016
Alicante, Spain

Guest Editor

Eric D Donnenfeld MD FACS
Clinical Professor
Ophthalmic Consultants of Long Island
Rockville Center, New York -11570
USA

JAYPEE - HIGHLIGHTS
MEDICAL PUBLISHERS, INC.

Honorary Editors

A John Kanellopoulos MD
Medical Director
Laservision gr Institute
17-Tsocha Street
Athens, Greece-11527

Mark Wevill MD
81 Grange Road
Dorridge
B-93 8QU
United Kingdom

Carlo F Lovisolo MD
Medical Director
QuattroElle Eye Center
Via Cusani, 7-9, 20121
Milano, Italy

Francisco Sanchez Leon MD
Director
Instituto Oftalmologico Novavision
Av Lomas Verdes 464
Naucalpan, Edo Mexico
Mexico CP53120

Aylin Kilic MD
Kudret Goz Hastanesi
Kennedy Caddesi No.71
Kavaklidere-Ankara
Turkey

Mahipal S Sachdev MD
Chairman and Medical Director
Centre for Sight
B-5/24, Safdarjung Enclave
New Delhi, India

Frederic Hehn MD
Chief and Medical Director
Centre de La Vision, Nations-Vision
23, Boulevarde de L'europe
54500, Vandoeuvre
France

Arun C Gulani MD
Director
Gulani Vision Institute
8075, Gate Parkway (W)
Suite 102
Jacksonville, Florida 32216
USA

Sharad Lakhotia MS
Director
Lakhotia Eye Centre Laser Institute
E-544, Greater Kailash Part-II
New Delhi, India

Bojan Pajic MD
Chairman and Medical Director
Swiss Eye Research Foundation
Eye Clinic Orasis,
Titlisstrasse 44
5734 Reinach AG
Switzerland

C Banu Cosar MD
Associate Professor of Ophthalmology
Sinpas Aqua City 1, Etap
H Block D: 13, Cekmekoy 34773
Istanbul, Turkey

Jaypee Brothers Medical Publishers (P) Ltd.

Headquarter
Jaypee Brothers Medical Publishers (P) Ltd
4838/24, Ansari Road, Daryaganj
New Delhi 110 002, India
Phone: +91-11-43574357
Fax: +91-11-43574314
Email: jaypee@jaypeebrothers.com

Overseas Offices
J.P. Medical Ltd.,
83 Victoria Street London
SW1H 0HW (UK)
Phone: +44-2031708910
Fax: +02-03-0086180
Email: info@jpmedpub.com

Jaypee-Highlights Medical Publishers Inc.
City of Knowledge, Bld. 237, Clayton
Panama City, Panama
Phone: + 507-301-0496
Fax: + 507- 301-0499
Email: cservice@jphmedical.com

Website: www.jaypeebrothers.com
Website: www.jaypeedigital.com

Inquiries for bulk sales may be solicited at: jaypee@jaypeebrothers.com

Femtosecond Laser: Techniques and Technology

First Edition: 2012

ISBN 978-93-5025-876-7

Printed at: Ajanta Offset & Packagings Ltd., New Delhi

DEDICATIONS

Dedicated to

- My respected Param Pujya Guru Sant Gurmeet Ram Rahim Singh Insaa Ji for his blessings and motivation
- My respected parents, teachers, my wife Dr Aruna Garg, son Abhishek and daughter Anshul for their constant support and patience during all these days of hard work
- My dear friend Dr Amar Agarwal, a renowned International Ophthalmologist for his constant support, guidance and expertise

Ashok Garg

Dedicated to

- To Mayca, Jorge, Fernando and Maria Lucia and most especially to my beloved wife Maria, for the hours that were taken from our family life to make this book possible.

Jorge L Alió

Contributors

AJ Kanellopoulos MD
Medical Director
Laser Vision Gr Institute
17, Tosocha Street, Athens, Greece-11521

Allon Barsan MD
Ophthalmic Consultants of Long Island
Rockville Center, New York -11570
USA

Arun C Gulani MD
Director
Gulani Vision Institute
8075 Gate Parkway (W)
Suite 102 and 103, Jacksonville
Florida-32216
USA

Ashok Garg MS PhD FRSM FRCS
Chairman and Medical Director
Garg Eye Institute and Research Centre
Dabra Chowk, Hisar, Haryana, India

Aylin Kilic MD
Kudret Goz Hastanesi
Kennedy Caddesi No.71
Kavaklidere-Ankara, Turkey

Bojan Pajic MD FEBO
Swiss Eye Research Foundation
Eye Clinic Orasis, Titlisstrasse 44, 5734, Reinach
Switzerland

Brigitte Pajic - Eggspuehler MD
Swiss Eye Research Foundation
Eye Clinic Orasis, Titlisstrasse 44, 5734, Reinach
Switzerland

Carlo Francesco Lovisolo MD
Medical Director
QuattroElle Eye Center, via Cusani, 7-9, 20121
Milano, Italy

C Banu Cosar MD
Associate Professor of Ophthalmology
Acibadem University
Department of Ophthalmology
Buyukdere Cad. No.:40, 34457 Maslak Istanbul
Turkey

Charu Khurana MS
Consultant Eye Surgeon
Centre for Sight, B-5/24, Safdarjung Enclave
New Delhi, India

Eric D Donnenfeld MD
Clinical Professor
Ophthalmic Consultants of Long Island
Rockville Center, New York -11570
USA

Eugenio Lipari PhD
QuattroElle Eye Center, via Cusani, 7-9, 20121
Milano, Italy

Farhad Hafezi MD PhD
Division of Ophthalmology
Department of Clinical Neurosciences
University Hospital of Geneva, Switzerland

Francisco Sánchez León MD
Instituto Nova Vision, Medical Director
Cornea, Refractive and Anterior Segment Clinic
Cd de Mexico, Acapulco, Mexico

Franz Fankhauser MD PhD
Swiss Eye Research Foundation
Eye Clinic Orasis, Titlisstrasse 44, 5734, Reinach
Switzerland

Frederic Hehn MD
Centre de La Vision
Nations - Vision
23, Boulevard de l'europe
54500, Vandoeuvre, France

Jaime Martiz MD
Instituto Nova Vision
Cornea, Refractive and Anterior Segment Clinic
Cd de Mexico
Acapulco, Mexico

Jorge L Alió MD PhD
Professor and Chairman
Instituto Oftalmologico De Alicante
Avda Denia 111, 03016
Edificio Vissum,Alicante
Spain

Joerg Muller MD
Swiss Eye Research Foundation
Eye Clinic Orasis
Titlisstrasse 44, 5734
Reinach, Switzerland

Mahipal S Sachdev MD
Chairman and Medical Director
Centre for Sight
B-5/24, Safdarjung Enclave
New Delhi, India

Mark Wevill MD
81, Grange Road
Dorridge, B93 8QU
United Kingdom

Maria Ott PhD
Institute of Physics
NMR Group, Faculty of Natural Sciences II
Martin Luther University
Halle-wittenberg, Betty-Heimann-
Str. 7, D-06120 Halle, Germany

Michael Mrochen PhD
Swiss Eye Research Foundation
Eye Clinic Orasis, Titlisstrasse 44, 5734, Reinach
Switzerland

Ming X Wang MD
Wang Vision Institute
Palmer Plaza Ste 1150, 1801 West End Avenue
Nashville, TN 37203, USA

Roberto Fernandez Buenaga MD
1, Vissum Corporacion
Alicante, Spain

Sharad Lakhotia MS
Director
Lakhotia Eye Centre Laser Institute
E-544, Greater Kailash Part-II, New Delhi, India

Slobodanka Latinovic MD
Swiss Eye Research Foundation
Eye Clinic Orasis, Titlisstrasse 44, 5734, Reinach
Switzerland

Foreword

Within the last decade, femtosecond laser technology has quickly blossomed from a nascent technology for flap creation during laser *in situ* keratomileusis to application in other field of cornea surgery, cataract surgery and most recently, diagnostic applications. The rapid incorporation of the femtosecond laser is due to the decreased trauma, greater precision, greater flexibility and increased safety compared to mechanical methods. The femtosecond laser seems to have applications for the outer in internal ocular structures.

In the cornea, the femtosecond laser can create incision architecture for keratoplasty for a variety of geometric patterns that are not necessarily repeatable even by a skilled surgeon. A more facile channel can performed for inserting corneal rings in the already compromised keratoconic cornea. A natural extension is use in corneal crosslinking. Application of the femtosecond for implantation of corneal inlays has been reported. Astigmatic keratotomy is yet another application can be performed by this laser.

Moving further into the eye, the femtosecond is currently being used for cataract surgery. In fact, the one of the most technically demanding aspects of the surgery can now be performed with the femtosecond. Although cataract surgery is already one of the most successful surgical procedures in medicine, safety outcomes may increase with the incorporation of "*femtosecond laser cataract surgery*". Dynamic light scattering and photon correlation spectroscopy allow the study of vitreoretinal changes to due to common pathology such as macular degeneration and diabetes.

As with any new technology, questions arise regarding economics, clinical flow and the nuances of incorporation into the daily routine. However, surgeon and patient experience over time will address these questions. Furthermore, any new surgical procedure has it is own set of complications, the femtosecond will be no different. However, the current focus is education of surgeons worldwide to provide guidance and mitigate adverse events and adverse experiences. This is the focus of this book.

The book highlights the surgical advances and diagnostic techniques outlined above from some of the most experienced surgeons worldwide. A host of topics are addressed that guide the reader through this rapidly evolving technology and the most current applications. The wide variation in application underlies the fact that there is an exciting future ahead for this technology.

I congratulate Professor Ashok Garg, Professor Jorge L Alió the Chief Editors and their dedicated team of editors and contributors who have made this book an excellent companion for refractive surgeons in their day-to-day clinical practice. I am sure, this is first fine book on Femtosecond Laser at an International level.

Alaa El-Danasoury FRCS
Professor of Ophthalmology
Magrabi Eye and Ear Hospital Jeddah
Khozam Street Kilo-3, Mekkah Road
PO Box 7344, Jeddah 21462
Kingdom of Saudi Arabia
E-mail: malaa@magrabi.com.sa

Preface

Refractive Surgery technology has undergone rapid advancements and innovations in last two decades. The latest innovation especially Femtosecond Laser offers new possibilities in the field of minimally invasive corneal surgery. Femtosecond Laser employs near infrared pulses to cut tissue with minimal collateral tissue damage. The highly localized tissue effect of low energy Femtosecond Laser shall expand the capabilities and precision of this technology in near future and may be used to create three-dimensional intrastromal resection with micron precision. The Femtosecond Laser has proved its versatility in Lamellar keratoplasty, customized trephination in penetrating keratoplasty, tunnel creation for intracorneal ring segments, astigmatic keratotomy for keratoprostheses, non-invasive trans-scleral glaucoma surgery, retinal imaging presbyopic surgery and cataract surgery. Advances in ultrafast laser technology continued to improve the surgical safety, efficiency, speed and versatility of Femtosecond Lasers in Ophthalmology.

Femtosecond Laser: Techniques and Technology, the first of its kind in world ophthalmology has been planned in such a way to provide complete insight of this technology in various ocular indications. The international experts of this field have covered the present and futuristic uses of this technology in ophthalmology in a step-by-step and lucid manner. Practically, the possible uses of Femtosecond Laser in anterior and posterior segment indications of ophthalmology have been explained for the benefits of ophthalmologists worldwide.

We are highly thankful to Shri Jitendar P Vij (Chairman and Managing Director), Mr Tarun Duneja (Director-Publishing) and all staff members of our publisher M/s Jaypee Brothers Medical Publishers Pvt Ltd, New Delhi, India, who took keen interest in this project and for speedy completion of this mega task.

We hope that the book shall meets the expectations of ophthalmologists and shall serve as Quick look Reference Book, for eye surgeons worldwide to sharpen their surgical skills.

Editors

Contents

Guest Editorial

Femtosecond Laser— An Introduction

Eric D Donnenfeld (USA)

As a practicing ophthalmologist for close to a quarter of a century, I have observed, and been part of, several landmark events in ophthalmology. During my residency training, I was able to watch surgery go from intracapsular to extracapsular this was followed by the advent of intraocular lenses. Later, I was able to take place in the revolution of small-incision surgery with the adoption of phacoemulsification. Finally, over the last 15 years, I have participated in the refractive corneal revolution, with the excimer laser, PRK, and LASIK. These landmark events have changed the course of ophthalmology, and have redefined our ability to treat patients and provide safe and effective visual rehabilitation from cataract surgery and refractive errors.

Today, we are on the verge of another important moment in ophthalmology; the emergence of femtosecond laser refractive cataract surgery. Only time will tell, if this becomes a landmark event. However, femtosecond laser refractive cataract surgery has the potential for being one of these rare moments where the course of ophthalmology changes dramatically due to the advent of new technology. In the past, intraocular lenses and phacoemulsification took years to become safe and effective. With femtosecond cataract surgery, we are moving from an analog to a digital treatment of cataracts, and I am hopeful that femtosecond refractive cataract surgery will level the playing field for all cataract surgeons. Early evidence suggests that this technology will provide better efficacy and safety right from day one.

My experience with femtosecond laser refractive cataract surgery has been one year and almost 1000 cases. As with any new technology, there is an increased chair time preoperatively. Now we discuss surgical options, such as phacoemulsification versus femtosecond laser cataract surgery, as well as having a conventional and premium intraocular lens discussion. In addition, there is an increased time in the operating room due to the laser and conventional cataract surgery. Most importantly, however, complications have been rare to date and the refractive and visual results have been superb. Overwhelmingly, the surgery has met my expectations for safety and efficacy. In my opinion, the greatest benefit of this technology has been a perfect capsulorhexis and the arcuate incisions. I have found that this technology is very helpful for mature lenses and weak zonule cases that were surgically demanding previously, and have now become easier with the advent of laser surgery.

In my experience, refractive outcomes have definitely improved. However, more data is necessary before we can determine the true efficacy of this procedure. One of the most interesting aspects of the surgery has been the way patients have overwhelmingly embraced laser refractive cataract surgery. As with any new technology, I am certain that there will be unexpected hurdles for all of us in the adoption of femtosecond laser refractive cataract surgery. However, if my early experience is any indicator, I believe that this technology is here to stay. The true test will be surgical outcomes and whether femtosecond refractive cataract surgery improves safety and efficacy for all surgeons and most importantly our patients. **I hope that five years from now we look back at femtosecond laser refractive cataract surgery as a landmark event in ophthalmology.**

1

Femtosecond Laser: Current Technology and Future Prospects

Ashok Garg (India)

Chapter Outline

- Future Prospects of Femtosecond Laser Technology

Femtosecond laser is the latest innovation in the field of ophthalmology with great future. The femtosecond laser was first introduced in 2001. It shall completely revolutionize the minimally invasive ophthalmic surgery in next five years with wider applications. Femtosecond laser (FS) shall offer new possibilities in the field of refractive surgery specially when using the laser as microkeratome. By generating microplasms inside corneal stroma with femtosecond pulses, it is possible to achieve a cutting effect inside tissue while leaving the anterior layer intact. The femtosecond laser employs near infrared pulses to cut tissue with minimal collateral tissue damage.

The highly localized tissue effect of low energy femtosecond laser may expand the capabilities and precision of this technology may be used to create three dimensional intrastromal resection with micron precision. The main technical specifications of femtosecond laser are:

1. Laser pulse repitition rate
2. Spot size, pulse energy and pulse pattern.

The use of femtosecond laser in LASIK is effective and safe procedure with less corneal higher-order aberrations resulting in better uncorrected visual acuity and contrast sensitivity.

Femtosecond laser can be also used to make channels for implant of intracorneal ring (Intacs) in the treatment of keratoconus and pellucid marginal corneal degeneration.

The use of femtosecond laser in the treatment of presbyopia is currently gaining acceptance. Presbyopia is by far the most common refractive error worldwide with no permanent therapeutic option available. All efforts to restore accommodation by the use of surgery have not led to a generally accepted therapy. However, there is evidence from recent research data that the use of femtosecond laser might influence the modulus of elasticity in the lens by creating intralenticular disruption.

Femtosecond laser photodisruption of the ocular lens yields a self-limited lesion with bubble that resolve with time. This suggests that the use of a low-energy femtosecond laser might be safe when modifying the lens for correction of presbyopia. A new procedure for presbyopia called intracor is being evolved with this platform.

Femtosecond laser technology offer a promising approach to minimally invasive anterior and posterior keratoplasty (Femto-PLAK) through small tunnel incision in corneal endothelial diseases. Femtosecond laser enables the surgeon to cut cornea none mechanically with cutting accuracy of ± 10 microns.

Penetrating keratoplasty can be performed with femtosecond laser safely and accurately. Femtosecond laser is highly useful in applanation free corneal surgery.

Recently FDA approved studies have showed that femtosecond laser corneal dissection provides an alternative to more challenging manual dissection methods for keratoprosthesis implantation.

Femtosecond laser allows highly precise and reproducible subsurface using long infrared wave lengths. One of the most important applications of this method is in glaucoma surgery. Permanent drainage channel may be created to reduce the elevated intraocular pressure.

Femtosecond laser enhance by indocyanine green can be used for photodynamic therapy because of effective closure of corneal neovascularization at a low energy level.

Femtosecond laser with specific dynamic light scattering and photon correlating spectroscopy shall be useful for diabetic and senile macular degeneration.

A number of commercially available femtosecond laser devices are in the market since the introduction of this new field by intralase. The various FS laser machines available as follows:

1. Intralase (Abbott Medical Optics).
2. Wavelight ultraflap LDV (Ziemer Ophthalmic System, Switzerland).
3. Femtec 20/10 lasers (Technolas Perfect Vision, Germany)
4. Visumax (Carl Zeiss)
5. Lens Sx (Alcon)
6. Lens AR
7. Optimedica FS system. (Santa Clara, USA). (Catalys precision laser system).
8. Wavelight FS 200 ultraflap laser system.
9. VICTUS femotsecond laser platform (Bausch and Lomb and Technolas).

New refinements are being systemactically introduced with improvements in pulserate, spot size, pulse energy and customized approaches for the use of the laser.

FUTURE PROSPECTS OF FEMTOSECOND LASER TECHNOLOGY

Femtosecond laser has evolved potentially over the past 10 years and further revolutionally advances are expected in next 5 years. With its ultra precise (minimal focal spots size), shorter laser pulse, 3D processing quality and photodisruption from mJ to nJ, it shall guide the major surgical changes in the ophthalmology. FS laser shall be used in a bigger way in following anterior and posterior segment indications with best diagnostic and therapeutic results.

- Refractive Surgery (cutting of laser *in situ* keratomileusis), intracor procedure based on controlled biomechanical manipulation for presbyopia and OCT controlled keratomileusis.
- Femtosecond laser cataract surgery is the major advancement in the near future. The new machines with a high resolution 3D confocal structured illumination shall help in well centered capsulotomy and minimal residual corneal astigmatism, excellent incision architecture and less energy used and better lens fragmentation. It shall also help in more precise refractive lens exchange and limbal relaxing incisions.
- OCT controlled microsurgery shall be more accurate in relation to vision and predictability.
- Femtosecond laser shall be highly useful in astigmatic correction. FS assisted astigmatic correction in higher degree astigmatism (laser astigmatic keratotomy) shows greater accuracy and lens complication advanced refinements are being introduced in new generation femtosecond laser devices.
- Femtosecond Laser lentotomy for presbyopia treatment. This shall allow increased flexibility and sliding of lens fibers that can partially restore the accommodative loss of the lens with aging.
- Nonrefractive corneal applications specially laser assisted anterior and posterior lamellar keratoplasty, cutting of donor buttons in endothelial keratoplasty, customised trephination in penetrating keratoplasty and keratolimbal autografts, Laser DSAEK and anterior lamellar keratoplasty (ALK).
- Cutting corneal pockets for insertion of biopolymer Keratoprosthesis.
- Femtosecond laser technology shall bring advances in nonrefractive surgery procedures like noninvasive transcleral glaucoma surgery. FS laser created trans-scleral channels.
- Femtosecond lasers with dynamic light scattering spectroscopy shall be highly useful in retinal imaging, vitreous analysis in diabetes, photodynamic therapy (PDT) and senile macular degeneration (ARMD), femtosecond Laser pulses shall be guiding force in sebum precision retinal surgery with adaptive optics.
- Femtosecond laser assisted applanation free corneal processing and in corneal biopsy.
- Femtosecond laser assisted tattooing shall serve to correct leukoria (cosmetic procedures).

Advances in femtosecond ultrafast laser technology continue to improve the surgical safety, efficiency, speed and versatility in ophthalmology.

New technologies in femtosecond laser specially complex polarization shaped pulse, stimmulated emission depletion (STED), ablation with different time delay, coherent control and algorithm 3D nanostructuring by photons polymerization, tissue engineering using 2 PP cross-linking shall change completely the scenario of modern ophthalmic surgery to actual minimal hands free noninvasive ophthalmic surgery.

BIBLIOGRAPHY

1. Bamba S, Rocha KM, Ramos-Esteban JC, Krueger RR. Incidence of rainbow glare after laser in situ keratomileusis flap

creation with a 60 kHz femtosecond laser. J. Cataract Refract. Surg. 2009;35(6):1082-6.

2. Blum M, Kunert K, Schroder M, Sekundo W. Femtosecond Lenticule extraction for the correction of myopia: preliminary 6-month results. Graefes Arch. Clin. Exp. Ophthalmol. 2010;248(7):1019-27.
3. Choi SK, Kim JH, Le D,. Oh SH. A new surgical technique : a femtosecond Laser-assisted Keratolimbal allograft procedure. Cornea 2010;29(8):924-49.
4. Hild M, Krause M, Riemann I, et al. Femtosecond Laser-assisted retinal imaging and ablation: experimental pilot study. Curr. Eye Res. 2008;33(4):351-63.
5. Krueger RR, Thoronton IL, Xu M, Bor Z, van den Berg TJ. Rainbow glare as an optical side effect of IntraLASIK. Ophthalmology 2008;115(7):1187-95.
6. Liu Y, Nakamura H, Witt TE, Edward DP, Gordon RJ. Femtosecond laser photodisruption of porcine anterior chamber angle; an ex vivo study. Ophthalmic Surg. Lasers Imaging. 2008;39(6):485-90.

2

Femtosecond LASIK: Practical Pearls

A John Kanellopoulos (Greece)

Chapter Outline

Throughout all these last six years of femto-LASIK, we have gained experience with several femtosecond lasers and I feel that there is certain intrinsic surgical pearls that one attains through using a femtosecond laser, that I think would be interesting even for the experienced LASIK surgeon with the microkeratome. So, I will try and summarize these in this short chapter.

1. In patient preoperative evaluation, corneal thickness is of the essence in any LASIK case, this should be reiterated in a femtosecond LASIK case, so we currently use two modes to evaluate corneal thickness. One is the Pentacam and the Pentacam II and obviously, we study, like most people, the anterior corneal surface, the posterior elevation, but more importantly I had found to spend most of my time evaluating the normality of corneal thickness, at looking at the corneal pachymetry map, that these tomographies can not give us. If the pachymetry map is round and has a symmetric thickness progression from the center to periphery, I put more value to that as a diagnostic tool than irregularity on the anterior curvature that may be the product of dryness or a transparency irregularity of the cornea. The example presented above is of normal anterior and posterior elevation and good thickness as noted (566 um), nevertheless, I consider this scan Abnormal as the thinnest point is inferonasal instead of central and the thickness progression is a distorted ellipse.

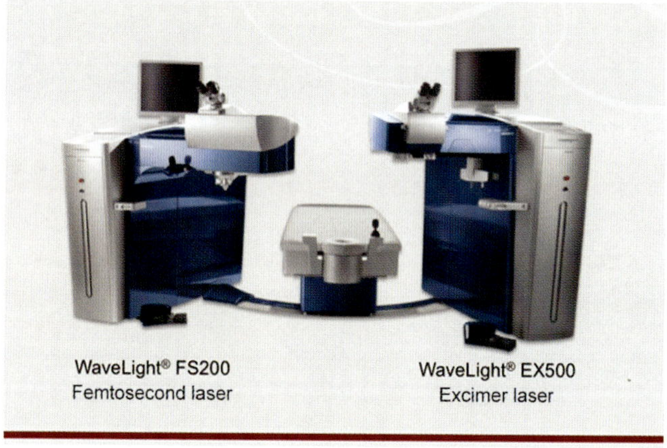

Fig. 2.1 WaveLight® refractive suite

We, additionally use, in cases where the thickness is borderline or there is a question about the normality of the cornea, a pachymetric map produced by our corneal OCT device, we use the Optovue OCT device for corneal imaging and the pachymetry map from the Optovue gives us accurate pachymetric measurements of the cornea as well as the pachymetry distribution of the cornea helps makes the diagnosis between a normal cornea or a cornea that is more suspicious for ectasia. So, pachymetry is of the essence.

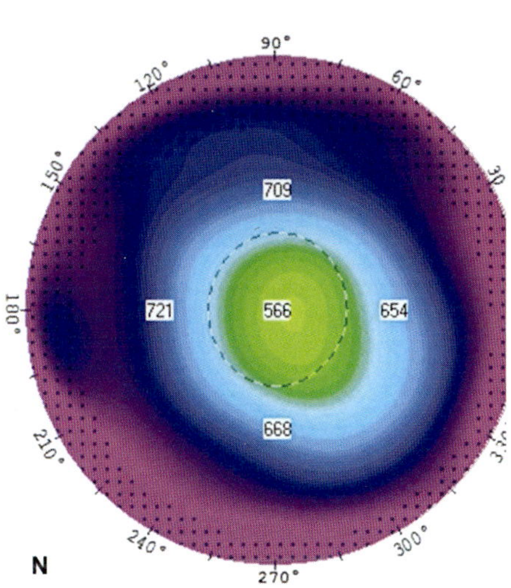

Fig. 2.2

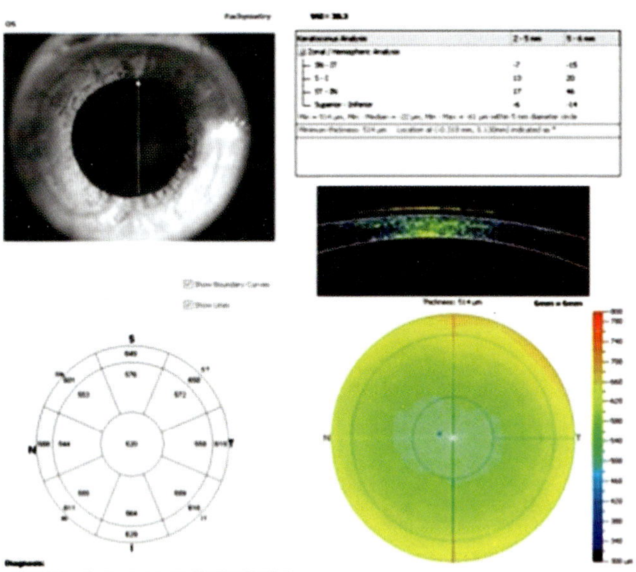

Fig. 2.3

2. On clinical evaluation, the difference in preoperative evaluation of femtosecond LASIK cases, is spending careful attention to any superficial corneal scars. It is common that patients who are contact lens wearers and they are the majority of people who decide to undergo LASIK for myopia, astigmatism and/or hyperopia, may have had a sterile infiltrate or even a bacterial keratitis that was not significant enough for them to remember. We know that this may scar Bowman's membrane and Bowman's membrane at that particular point may be absent, so in a regular microkeratome case this would not be important as the microkeratome blade cuts through the cornea without serious consequences from that corneal scar. In femtosecond LASIK, especially if we choose flaps under 120 microns in thickness, the significant pressure from the bubble air created within the lamellar cut of the flap, may find as a point of least resistance this Bowman's membrane scar and may "blow" the corneal coat through it and create a vertical gas break through or otherwise described as a femtosecond button hole. So, great attention is mandated on slit-lamp biomicroscopy before the femto-LASIK surgery such irregularities, especially in contact lens users. If this is omitted at the examination, most of these can be evident after the patient interface cone applanates the patient's cornea, especially in corneas that are overlined dark, iris is brown or dark brown in contrast to light blue or light green, some actual irregularity of the cornea can be seen at that point and, if the surgeon is quick enough, the thickness of the flap can be changed from let us say 100 microns to 120, which would be a more safe depth to perform a femtoflap and avoid a vertical gasp breakthrough.

3 Last, in preoperative evaluation, as we do with most LASIK cases, we access the actual ease with which the globe will be approached by the femtosecond laser patient interface cone. A lot of Southern Mediterranean men, a common patient that we see in Greece, have very deep set eyes and very prominent eye brows, the eye brow bone is very prominent making the distance between the tip of their nose and the actual surface of the cornea quite great and thus, a difficult approach with a femtosecond laser. We have had, through thousands of patients of the last five years, I had to exclude maybe a couple of patients from having surgery, because of that difficult access.

Fig. 2.4

INTRAOPERATIVE ISSUES (LASIKREFSUITE11)

As far as the intraoperative issues are concerned, I take special attention to explain hto the patient that the part of procedure that involves a femtosecond laser is a procedure that would probably be most uncomfortable for the patient, as they will feel pressure, especially if it is a deep set eye and we have to dilate the eyelids quite significantly with a led speculum, and there will be a black out time, the time that the suction will be placed and the flap will be created. The "blackout" time has great variability, depending on the femtosecond laser and the diameter of the flap created, as well as the spacing between each individual spots. Our experience with the FS60, was a flap time creation of about 30 seconds, which is quite significant if one considers that a microkeratome LASIK is under 10 seconds. Our latest femtosecond device, the Alcon/WaveLight FS200 takes us about 10 to 12 seconds from docking to flap creation, reducing this black out time significantly and increasing patient comfort.

I find, explaining this before the procedure carefully to the patient, very helpful as the patients tolerate this procedure more comfortably and we do employ countdown doing the flap creation, which also limits patient anticipation. If we are counting down, for instance, from 20 seconds to 0, I found this to give patients great comfort in anticipating those 20 seconds instead of nobody speaking those 20 seconds and having the patients wonder when is this black out time going to end and maybe moving towards the last few seconds, something that may complicate the flap creation process.

We have found, especially again in our Southern Mediterranean facial structures and especially males, to have to significantly tilt the patient's head towards their nose in order to get access to the cornea and avoiding the femtosecond device push on the nose, which number one is uncomfortable for the patient, number two creates a lot of resistance and possible loss of suction during the procedure. So, while applanating the eye with the patient interface, and I see directly on my screen how well the cornea is being applanated and how broad that area is, one of my assistants during the surgery looks directly at any contact between the cone and the nose, and if there is any contact that is verbalized to me from the assistant, then we redock with the head now tilted a little bit more nasaly, in order to avoid such contact. Obviously, after everything has been completed and we are ready to perform the procedure, the appropriate adjustment on the LASIK flap needs to be done on the laser computer software, as I have found clinically the need to decenter the flap that I see being planned on the computer, again nasaly, maybe sometimes all the way up to 1 mm. I found clinically otherwise, if the head is tilted nasaly and I do not perform the decentration on the

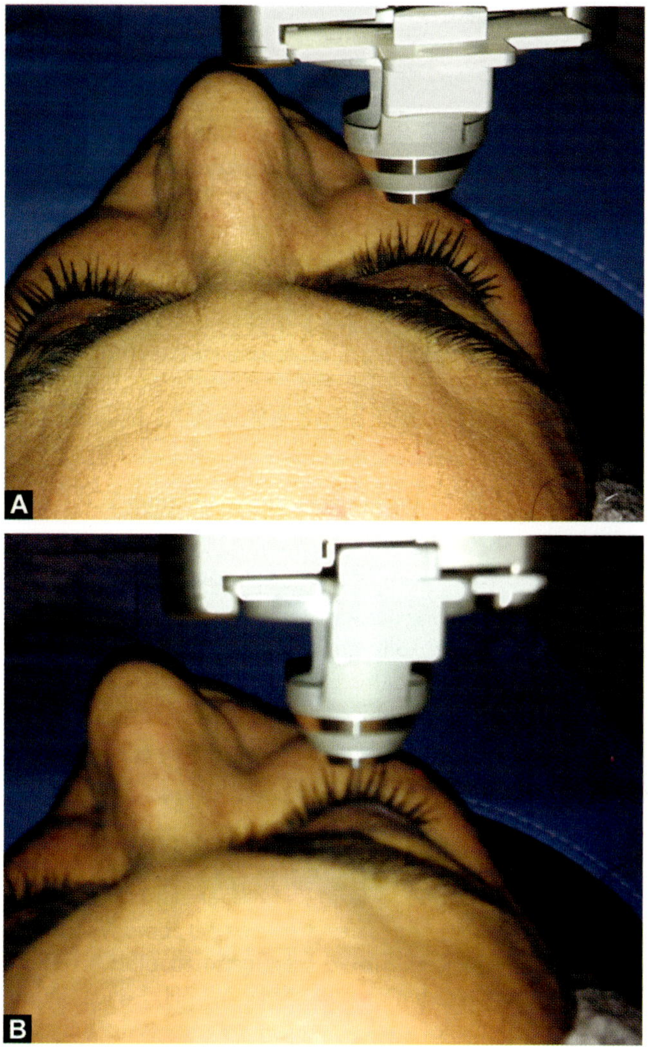

Figs 2.5A and B

plan of the procedure, my flaps to be deviated temporarily, which may be a problem.

So, although on the screen the flap may look well centered, according to the pupillary image behind of the cornea, if the head is significantly tilted to the side, then I find a need to plan my LASIK flap to be exaggerated nasaly, maybe a millimeter and I have found this a helpful peril in having my flaps more well-centered. Particular care on this point needs to be taken in hyperopic patients and I will discuss this more extensively in the chapter "looking into femtosecond laser in hyperopia".

Last, intraoperatively I find it extremely important to have a very reproducible protocol, how many drops or any material enters the surface of the cornea before that cornea is applanated with the patient interface. One has to consider that if we want to be extremely accurate in the flap thickness, we have to take into account the solution or the fluid that is going to be present between the epithelium of the cornea and the patient interface. For instance, if people used anesthetic gel before the procedure, that gel needs to be washed extremely well, otherwise this gel may interfere with the actual thickness we are going to attain. If we recall that thickness has been determined as the distance between the patient interface and the actual thickness planned. If between the patient interface and the corneal epithelium intervenes a layer of about 20 microns of lidocaine gel, then our flap is going to be cut 20 microns thinner than planned.

So, I always use a drop of alcaine and a drop of ofloxacin right before I start and obviously, no more and no less, and this has been built into our normogram, as we usually get 5 to 10 micron thinner flaps than we actually plan on our femtosecond laser protocol, and with thousands of cases now under our belt, both with the FS60 and last year with the FS200 by WaveLight, we feel comfortable that this is a very reproducible technique and does not interfere with the actual flap thickness and measurements afterwards.

Also, although LASIK can become a routine, the part where the femtosecond laser generates the lamellar cut is very important, because this is the time where quick reaction if irregularity, vertical gas breakthrough, a scar, any abnormality, patient movement can happen, the surgeon needs to step off the pedal and reconsider, redocking and completing the procedure or aborting the procedure. With the latest femtosecond lasers that reaction time has to be immediate, as for instance with the FS200 by Wavelight, the lamellar cut takes about 5 seconds, which is a very short time and very quick response time needs to be made, if any irregularity is noticed in lamellar cut.

We traditionally perform both the femtoflaps before we go on to the Excimer laser, as we have found it easier than doing flap first, laser second on one eye and then going back to the other eye to do flap first and Excimer laser. In our experience, that has been more uncomfortable to patients, because they invariably find the femto part being the toughest, and we feel that overcoming that part first is more convenient.

Also, key element here is to explain to the patient the focal point, both with the femto and the Excimer laser and obviously this is standard with all our LASIK protocol, recheck the patient's name, their date of birth, the eye they are treating, as in a day with multiple patients and a lot of confusion in the office, we have to always anticipate the possibility of having the wrong patient under the wrong data, a problem that could potentially happen in a busy laser center.

Then going on to the Excimer laser and both our Intralase FS60 platform, that is coupled with the 400 Hz WaveLight laser, as well as the new refractive suite by WaveLight Alcon, where the femtosecond is the FS200 and the Excimer laser is the EX500 Concerto, by WaveLight. The bed connecting the femto and the laser is the same, so the patient is automatically transferred on to the Excimer laser. I wait to lift the flap after the femtosecond laser has finished and the data on the Excimer laser are ready to go. I use an aspirating speculum because, right before I lift the flap, I try to irrigate the eye squeezing of the eye, the pressure from the procedure can have meibomian gland secretions present in the corneal surface and if the eyes are not irrigated and that fluid not evacuated from the surface, it can find its way within the flap, the under side of the flap or the surface that is going to be treated with the laser as an irregular ablation. So, irrigating or rather, aspirating speculum irrigating the surface well and then lifting the flap, only after the data on the Excimer are ready to go and we have checked that this is the correct eye, the correct patient that is represented on the Excimer laser panel.

The online pachymetry present in the EX500 is an excellent tool of double checking the correct flap thickness and if there is enough corneal space on the stroma to perform the ablation planned. In our older platform, the FS60 by Intralase and the 400 Hz WaveLight laser, we use a Sonomed pachymetry measurement in order to attain this number. With the EX500 this is done automatically by the laser, optically, and is present on the screen, but right after the flap is lifted. So, then the procedures take place and then the flap repositioned, we use copious irrigation, at least half a bottle of BSS, with a disposable, single use Hersh Visitec LASIK Cannula, to irrigate under the flap and the surface, and again this is where the aspirating speculum is been activated, in order to evacuate all the solution and not have debris from the laser room air or the meibomian secretions re-enter the flap.

The flap is then repositioned, if it is a femtoflap I do not use a wet, weckcell sponge to "fine tune it" in place, because I have found that LASIK flaps with the femtosecond laser have less slippage on the surface and if I manipulate the flap with a wet spear-sponge I may create striae. So I try to lay the flap down correctly, I use a drop of milky Pred Forte solution, in order to drop it on the surface of the eye and help that flap, and that drop helps me delineate the flap edges and make sure the flap is in correct position.

And finally, I use a Johnston applanator which is a nice tool to applanate the flap on to the stromal bed and take out a potential microstriae from high myopic corrections. The device is available by Rhein, and it is basically a very large, 20 mm lens, that is very flat about 30 diopters and by pushing the lens in the center of the cornea, we make sure that the actual central part of the flap is applanated on to the laser bed that has just been ablated.

Following that, a few drops of ofloxacin and then, finishing, we always use an Acuvue daily contact lens of zero power to lay on the cornea, extra antibiotic drops of ofloxacin and Pred Forte on to the contact lens, and then we release the LASIK speculum carefully and we leave the contact lens in from the evening of the procedure, we perform our LASIK procedures usually in the early evening, until the next morning in order to minimize the friction between the actual eyelids of the patient and the LASIK flap. We have found this measure, performed the last three years in our practice, to almost zero the chance of flap slippage and flap striae, although it is, I have to admit, sometimes uncomfortable for some patients in the morning as they wake up and they have had this lens in for several hours.

So, this is basically in review, our LASIK protocol for femtosecond laser. It is illustrated in other video on our website: www.brilliantvision.com.

The basic advantages are:

1. Exact parameters of flap diameter/thickness/position in the cornea.
2. The flap has not began until the side cut is performed. An irregular flap can be aborted with no consequences.
3. Thicknesses of 100 microns can reproducibly be achieved.
4. The flap architecture is planar, avoiding more biomechanical destabilization of the cornea.
5. Fewer high order aberrations here are some results with the Wavelight Refractive Suite over the last 1½ year.

HYPEROPIC FEMTO-LASIK

I have come to realize that hyperopia is not just the opposite of myopia. Hyperopic patients have several distinct differences:

1. They are almost always undercorrected. This is due to variable degree of accommodating spasm, resulting in common "regression" following LASIK and 2nd and 3rd treatments to be necessary.
2. They have large amounts of angle kappa. As we will expand herein the visual axis of hyperopes is usually 0.5 to 1.5 mm off the geometric center of the pupil. This is in stark contrast than myopes and needs special attention during laser correction.
3. In our population of southern Mediterranean PRK appears to be a poor option for hyperopic correction, even of small amounts, due to the increased regression percentages (up to 35%) and the invariable scarring associated with it, despite of MMC use. LASIK, therefore, becomes the only laser correction option in my opinion in our patient population.

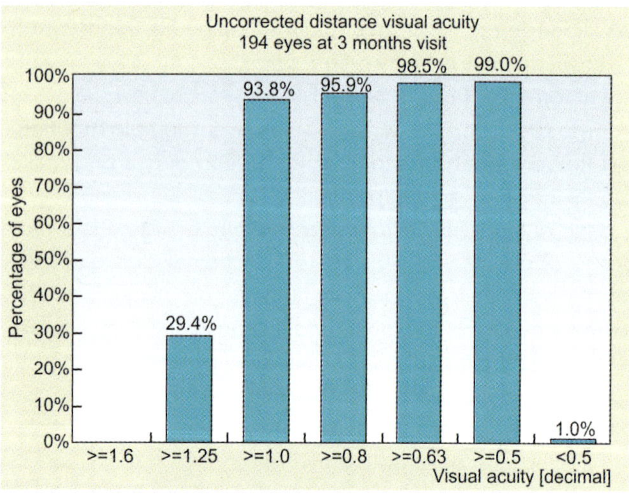

Fig. 2.6 Myopic LASIK, 3 months

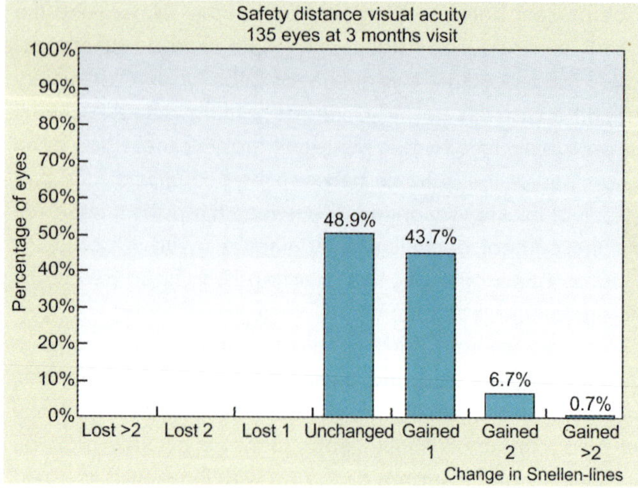

Fig. 2.7 Myopic LASIK, 3 months safety data

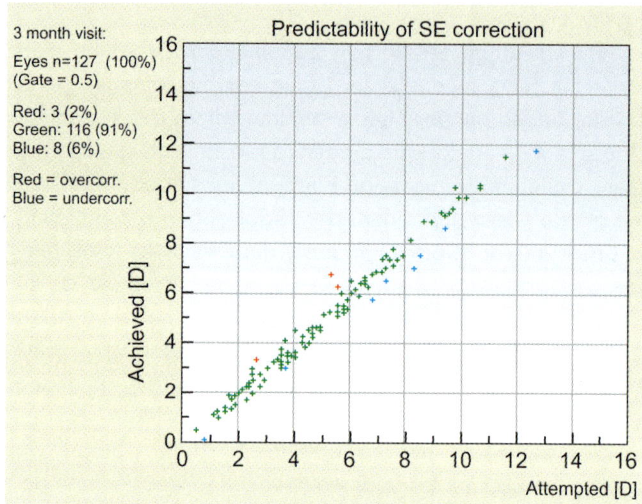

Fig. 2.8 Myopic LASIK, 3 months predictability

4. The mean age of the hyperope corrected in our practice in near or over the presbyopic age making overcorrection in the nondominant eye a serious consideration in our practice to be discussed and demonstrated by contact lenses in all hyperopes as will be discussed below.

These are the basic reasons of why, I feel hyperopic laser corrections with LASIK deserve a special chapter.

While the lasers and software that drives them become more sophisticated, there are still alignment errors during photoablation that could lead to decentration and nonhomogenous ablation patterns.

Decentered ablations can lead to negative visual effects including irregular astigmatism, reduced best corrected visual acuity, glare etc.[1,2] Thus, proper centration of the ablation zone during refractive procedures is an essential goal. However, a large angle kappa, which is defined as the difference between the primary line of sight and the pupillary axis, may cause alignment errors during photoablation. This issue is more important in hyperopic patients that tend to have larger angle kappa. Because of this, some refractive surgeons prefer to alter the location of the ablation to take into account a large angle kappa. However, there is little data in the literature concerning the normative values of angle kappa in healthy subjects.

Angle Kappa

Angle kappa is determined by the intersection between the line of sight and the pupillary axis. Because the fovea lies slightly temporal to the point at which the pupillary axis intersects with the posterior pole of the globe, the normal angle kappa is slightly positive. Evaluation of angle kappa before refractive surgery has gained significant importance because ablation zone centration during refractive surgery is a critical step. Uozato and Guyton asserted that centering on the pupil is the proper method of centration because the photoreceptors are aimed toward the center of a normal pupil; their method has since become the standard practice. However, Pande and Hillman stated that the optimal centration is the corneal intercept of the visual axis because this is the line joining the fovea to the fixation point. They have concluded that coaxially sighted corneal reflex was nearest the corneal intercept of the visual axis. Recently Nepomuceno et al. performed hyperopic LASIK with the ablation centered on the coaxially sighted corneal light reflex. They have concluded that traditional centering method based on the entrance of the pupil could lead to decentration in the presence of a large angle kappa, especially in the hyperopic patients.There are various methods, which measure the amount of angle kappa. Synoptophore is one of the most widely used methods in the clinical practice. It measures angle kappa depending on the corneal reflection methods. Recently, newer devices such as Orbscan II (Orbtek Inc. Bausch and Lomb, Rochester, USA) have become commercially available and can be used in the evaluation of angle kappa. But, there is no published data evaluating the reliability of Orbscan in measuring angle kappa. We have reported in the significance and magnitude of angle kappa when studying normal hyperopic population. We have also previously reported our experience in hyperopic LASIK–utilizing a mechanical microkeratome (Moria M_2) and the Wavelight Allegretto Excimer laser.

Hikmet Basmak, Afsun Sahin, Nilgun Yildirim, Thanos D Papakostas, and A John Kanellopoulos, Measurement of angle kappa with synoptophore and Orbscan II in a normal population, J Refract Surg 2007.

We found that there is a significant correlation between positive refractive errors and large positive angle kappa values. Refractive surgeons must take into account angle kappa especially in hyperopic patients in order to avoid complications related to decent ration of ablation zone.

The figures below depict the same planned Excimer profile for the correction of hyperopic astigmatism on the left: centered on the pupillary center and on the right: adjusted by topography to take into consideration and adjust for angle kappa.

Figures on a hyperopic LASIK flap (Figs 2.11 and 2.12) created with the M_2 microkeratome by Moria showing the gross nasal decentration of the flap in order to accommodate the applicable ablation decentration in reference to the pupillary center.

We have presented our experience in comparing hyperopic standart to topography-guided LASIK (AAO 2007):

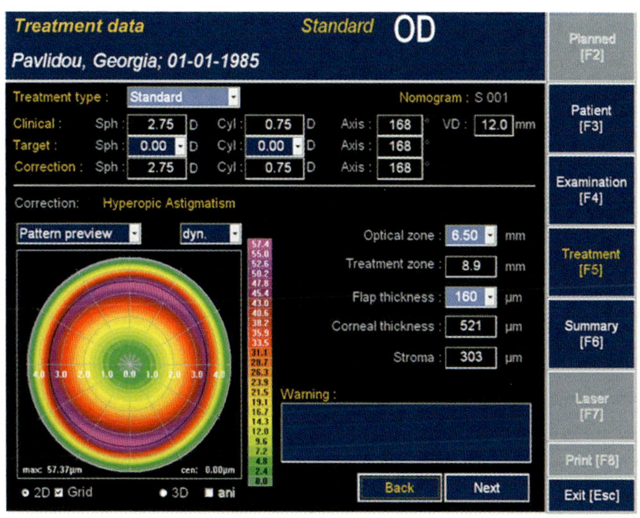

Fig. 2.9

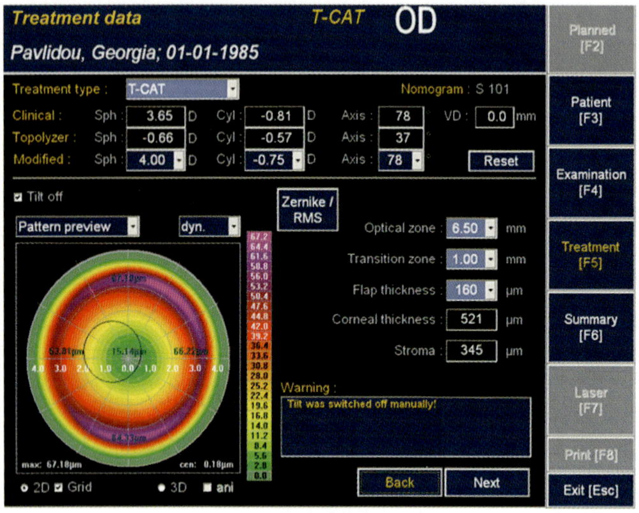

Fig. 2.10 LASIK flap needs to be decentered as well to accommodate challenging for surgeon

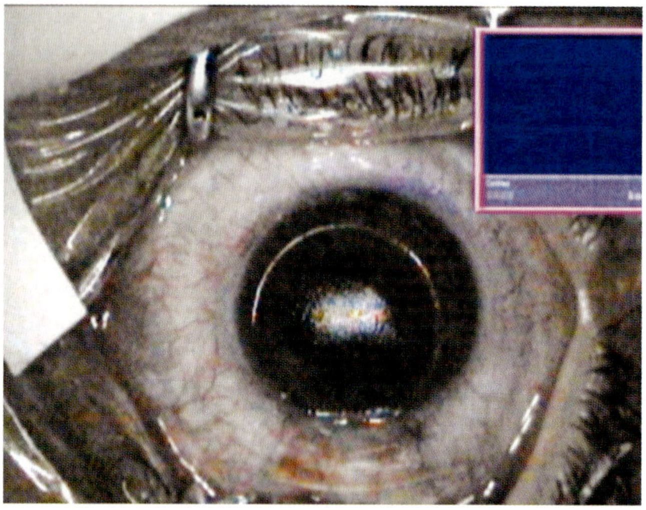

Fig. 2.12

Topography-guided femto-LASIK and Standard LASIK appear to be both safe and effective for hyperopia. Topography-guided femto-LASIK appears be superior in regard to regression, residual astigmatism, contrast sensitivity and estimated ablation decentration in regard to the visual axis. Hyperopic LASIK utilizing the topography-guided WaveLight ALLEGRETTO Excimer laser appears to be safe and effective in the correction of low, moderate and high hyperopia and/or hyperopic astigmatism. The results appear to be safe and predictable for the low and moderate hyperopia groups; they appear to carry similar promise in the high hyperopia/high cylinder group. The topography guided platform appears to better center the ablation to the visual axis vs the papillary center with high level of reproducibility in all cases, and compared to our previously published results with standard hyperopic treatments with this laser.

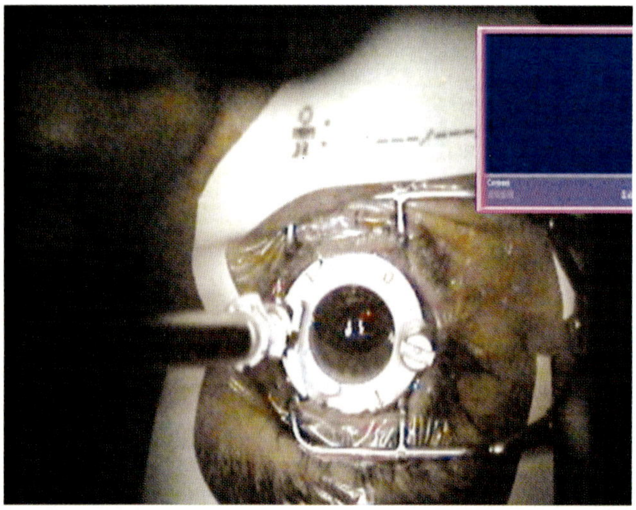

Fig. 2.11

Measurements

We currently use as our treatment values the dilated (with one drop of mydriacyl 1%) pupil measuremnts and not cycloplegic measurements as we feel that they do not represent a physiologic condition. We make it an effort to trial contact lenses on all hyperopic patients prior to their LASIK in order to appreciate the proposed correction better. This is the point that we showcase to them overcorrecting the nondominant eye by 0.5D and how well they can tolerate this. This is well tolerated by the great majority and offers a great near advantage up to ages of 55!

We use Placido disk topography along with the Pentacam, a tool that better illustrates the degree of angle kappa. As with most hyperopic measurements the dilated pupil measurement is usually 0.5 to 1 diopter more hyperopic. We use an optical zone of 6.5 mm (the treatment zone is 6.5 to 8.5 mm) and use the upper limit of 6 diopters of spherical equivalent-meaning up to +6 of hyperopia or up to +6 in spherical equivalent for mixed astigmatism (it is possible to treat a +7 -2.5D astigmatism as this would be the equivalent of +4.5 +2.5D of astigmatism). We treat with these parameters, but discuss with patients prior to their surgery the fact that they may be myopic 0.5 to 1D 1-3 weeks following the femto-LASIK, until their accommodative spasm relaxes. We have a 3 percent re-treatment rate with this protocol, a fact that greatly reinforces this approach. We have to share though the fact the hyperopes become extremely anxious during this 1-3

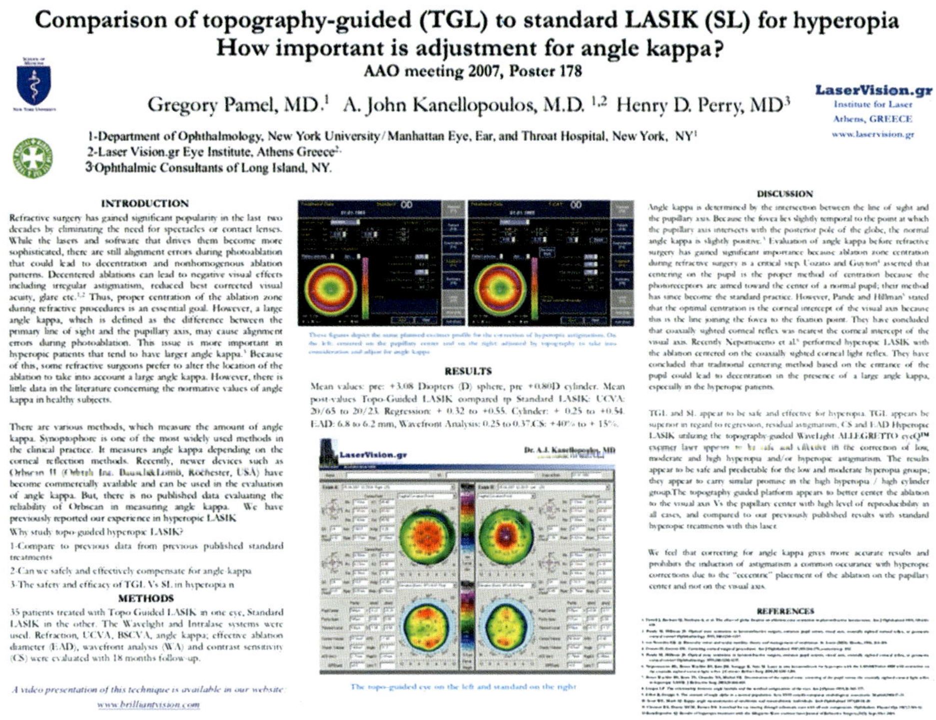

Fig. 2.13

week myopic interval and they require ample reassurance and plenty of discussion time.

We feel that correcting for angle kappa gives more accurate results and prohibits the induction of astigmatism a common occurrence with hyperopic corrections due to the "eccentric" placement of the ablation on the papillary center and not on the visual axis.

We have had the opportunity to enhance our hyperopic femto-LASIK treatments these last 12 months with the new Alcon/Wavelight refractive suite platform. This platform utilizes the FS200 femtosecond laser with distinct advantages for hyperopes the:

1-very large applanation are on the cornea allowing for comfortable "nasal" decentration of the flap in order to accommodate the angle kappa-corrected ablation center.

In the two figures (Fig.s 2.13 and 2.14) the reader can appreciate the Excimer treatment plan (the refractive suite utilizes the ultra-fast EX500 Excimer that is linked to the Pentacam HD-called the Oculyzer II as a platform device) in reference to the pupil and clearly understand the necessity of an equally nassaly placed ablation. The actual 9.5 mm flap created by the FS200 femtolaser is noted in the second picture. As an additional safety feature the refractive suite is inter-connected via its network not only between the 2 lasers but additionally to the Pentacam device (the Oculyzer) for enhanced safety in

Fig. 2.14

data processing. The actual Pentacam scans of each eye of the patient are fed seamlessly to the EX500 and processed by surgeon and staff. The treatment plan is viewed on the EX500 screen, but also superim posed over the femto flap

Patient file 14.02.2011		OD	WaveLight
Created by Lasik	Ultraflap Treatments Performed		Page 2 of 3 pages

Date: 03.02.2011 15:26:42 Treatment Type: Standard Status: Finished

Treatment Parameters (Standard)

Ablation

Abl. Zone	Max. Depth	Min. Pachy	Res. Stroma
9.0 mm	103 um	492 µm	249 µm

Flap

Diameter	Thickness	Side Cut Angle	Canal Width	Canal Length Offset
9.5 mm	140 µm	70°	1.3 mm	0 mm

Hinge

Position	Length	Angle	Width
90°	3.6 mm	45°	0.4 mm

Laser pulse energy (measured)

Bed Cut	Side Cut
0.7 µJ	0.9 µJ

Laser separations

Bed Cut		Side Cut	
Spot Separations	Line Separations	Spot Separations	Line Separations
8.0 µm	8.0 µm	5.0 µm	3.0 µm

Treatment Screenshot (Standard)

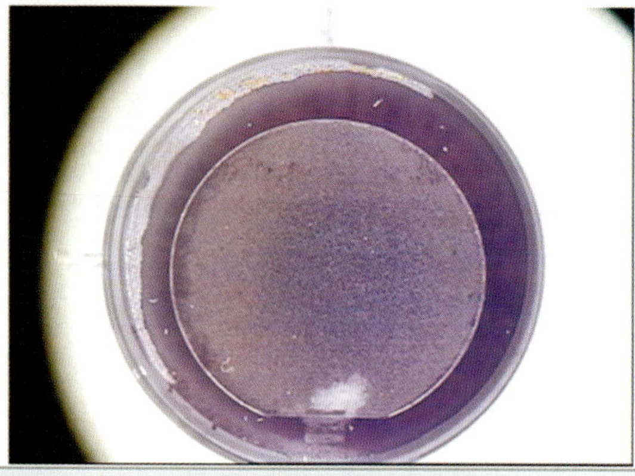

Comments

Fig. 2.15

planned on the relative flap-making process that precedes on the FS200 femtosecond laser. This is an extremely helpful tool for the learning curve of performing topography-guided hyperopic LASIK.

Figure 2.15 represents the final FS200 report documenting all flap parametes used, as well as the actual flap position at the termination of the sede cut. Please note the nassaly (to the left) "decentration of the flap in regard to the pupil.

On last consideration for hyperopes, especially the over 40 age group is narrow angle. We caution clinicians that hyperopes of over 3 diopters may have significantly narrow angles, a fact that needs to be evaluated thouroughly especially when considering that the progressive cataract formation in these patients will further narrow the angle and potentially lead to angle closure attack(s). It is reasonable to consider premature cataract extraction in these patients and/or laser peripheral iridotomy as this approach may eliminate both the hyperopic correction and the phacomorphic danger of glaucoma.

BIBLIOGRAPHY

1. 3D finite element model of aqueous outflow to predict the effect of femtosecond laser created partial thickness drainage channels. Chai D, Chaudhary G, Mikula E, Sun H, Juhasz T. Lasers Surg Med. 2008 Mar;40(3):188-95. PMID: 18366082 [PubMed - indexed for MEDLINE]
2. Agarwal A, Brubaker JW, Mamalis N, Kumar DA, Jacob S, Chinnamuthu S, Nair V, Prakash G, Meduri A, Agarwal A. Femtosecond-assisted lamellar keratoplasty in atypical Avellino corneal dystrophy of Indian origin. Eye Contact Lens. 2009;35(5):272-4. PMID: 19657278 [PubMed - indexed for MEDLINE]
3. Anterior chamber gas bubble formation during femtosecond laser flap creation for LASIK. Srinivasan S, Rootman DS. J Refract Surg. 2007;23(8):828-30. PMID: 17985804 [PubMed - indexed for MEDLINE]
4. Bahar I, Kaiserman I, Lange AP, Levinger E, Sansanayudh W, Singal N, Slomovic AR, Rootman DS. Femtosecond laser versus manual dissection for top hat penetrating keratoplasty. Br J Ophthalmol. 2009;93(1):73-8. Epub 2008 Oct 16. PMID: 18927225 [PubMed - indexed for MEDLINE]
5. Bahar I, Kaiserman I, Mashor RS, McAllum P, Slomovic A, Rootman D. Femtosecond LASIK combined with astigmatic keratotomy for the correction of refractive errors after penetrating keratoplasty. Ophthalmic Surg Lasers Imaging. 2010;41(2):242-9. doi: 10.3928/15428877-20100303-14. PMID: 20307044 [PubMed - in process].
6. Bindewald-Wittich A, Han M, Schmitz-Valckenberg S, Snyder SR, Giese G, Bille JF, Holz FG. Two-photon-excited fluorescence imaging of human RPE cells with a femtosecond Ti:Sapphire laser. Invest Ophthalmol Vis Sci. 2006;47(10):4553-7. PMID: 17003452 [PubMed - indexed for MEDLINE]Free Article
7. Blum M, Kunert K, Gille A, Sekundo W. LASIK for myopia using the Zeiss VisuMax femtosecond laser and MEL 80 excimer laser. J Refract Surg. 2009;25(4):350-6. PMID: 19431925 [PubMed - indexed for MEDLINE]

8. Blum M, Kunert K, Schröder M, Sekundo W. Femtosecond lenticule extraction for the correction of myopia: preliminary 6-month results. Graefes Arch Clin Exp Ophthalmol. 2010. [Epub ahead of print] PMID: 20130899 [PubMed - as supplied by publisher] Related citations.
9. Boxer Wachler BS, Korn TS, Chandra NS, Michel FK. Decentration of the optical zone: centering of the pupil versus the coaxially sighted corneal light reflex in hyperopic LASIK. J Refractive Surg 2003;19:464-5. 8. Looper LP . The relationship between angle lambda and the residual astigmatism of the eyes. Am J Optom 1959;36:365-77.
10. Brown JS, Wang D, Li X, Baluyot F, Iliakis B, Lindquist TD, Shirakawa R, Shen TT, Li X. In situ ultrahigh-resolution optical coherence tomography characterization of eye bank corneal tissue processed for lamellar keratoplasty. Cornea. 2008;27(7):802-10. PMID: 18650667 [PubMed - indexed for MEDLINE]
11. Buzzonetti L, Petrocelli G, Laborante A, Mazzilli E, Gaspari M, Valente P. Arcuate keratotomy for high postoperative keratoplasty astigmatism performed with the intralase femtosecond laser. J Refract Surg. 2009;25(8):709-14. PMID: 19714795 [PubMed - indexed for MEDLINE]
12. Calvo R, McLaren JW, Hodge DO, Bourne WM, Patel SV. Am J Ophthalmol. Corneal Aberrations and Visual Acuity After Laser In Situ Keratomileusis: Femtosecond Laser Versus Mechanical Microkeratome. 2010;12. [Epub ahead of print] PMID: 20227675 [PubMed - as supplied by publisher].
13. Case report: femtosecond laser-assisted small incision deep lamellar endothelial keratoplasty. Lee DH, Chung TY, Chung ES, Azar DT. Korean J Ophthalmol. 2008 Mar;22(1):43-8. PMID: 18323705 [PubMed - indexed for MEDLINE]Free PMC ArticleFree text
14. Chan CC, Ritenour RJ, Kumar NL, Sansanayudh W, Rootman DS. Cornea. Femtosecond laser-assisted mushroom configuration deep anterior lamellar keratoplasty. 2010;29(3):290-5. PMID: 20098303 [PubMed - in process]
15. Chang JS, Lau S. Intraoperative flap re-cut after vertical gas breakthrough during femtosecond laser keratectomy. J Cataract Refract Surg. 2010;36(1):173-7. PMID: 20117723 [PubMed - indexed for MEDLINE]
16. Chastang P, Hoang-Xuan T J Fr Ophtalmol. Femtosecond laser-assisted keratoplasties. 2008;31(9):921-35. French. PMID: 19107065 [PubMed - indexed for MEDLINE]Free Article
17. Cheng YY, Hendrikse F, Pels E, Wijdh RJ, van Cleynenbreugel H, Eggink CA, van Rij G, Rijneveld WJ, Nuijts RM. Preliminary results of femtosecond laser-assisted descemet stripping endothelial keratoplasty. Arch Ophthalmol. 2008;126(10):1351-6. PMID: 18852412 [PubMed - indexed for MEDLINE]Free Article
18. Cheng YY, Kang SJ, Grossniklaus HE, Pels E, Duimel HJ, Frederik PM, Hendrikse F, Nuijts RM. Histologic evaluation of human posterior lamellar discs for femtosecond laser Descemet's stripping endothelial keratoplasty. Cornea. 2009;28(1):73-9. PMID: 19092410 [PubMed - indexed for MEDLINE]
19. Cheng YY, Pels E, Nuijts RM. Femtosecond-laser-assisted Descemet's stripping endothelial keratoplasty. J Cataract Refract urg. 2007;33(1):152-5. PMID: 17189814 [PubMed - indexed for MEDLINE]
20. Cheng YY, Schouten JS, Tahzib NG, Wijdh RJ, Pels E, van Cleynenbreugel H, Eggink CA, Rijneveld WJ, Nuijts RM. Efficacy and safety of femtosecond laser-assisted corneal endothelial keratoplasty: a randomized multicenter clinical trial. Transplantation. 2009;88(11):1294-302. PMID: 19996929 [PubMed - indexed for MEDLINE]
21. Cheng YY, Tahzib NG, van Rij G, van Cleynenbreugel H, Pels E, Hendrikse F, Nuijts R. Femtosecond laser-assisted inverted mushroom keratoplasty. Cornea. 2008;27(6):679-85. PMID: 18580260 [PubMed - indexed for MEDLINE]
22. Choi SK, Kim JH, Lee D, Oh SH, Lee JH, Ahn MS. Creation of an extremely thin flap using IntraLase femtosecond laser. J Cataract Refract Surg. 2008;34(5):864-7. PMID: 18471648 [PubMed - indexed for MEDLINE]
23. Choi SK, Kim JH, Lee D. The effect of femtosecond laser lamellar dissection at various depths on corneal endothelium in the recipient bed of the porcine eye. Ophthalmic Surg Lasers Imaging. 2010;41(2):255-60. doi: 10.3928/15428877-20100303-16. PMID: 20307046 [PubMed - in process].
24. Chung JL, Seo KY, Yong DE, Mah FS, Kim TI, Kim EK, Kim JK. Antibiotic susceptibility of conjunctival bacterial isolates from refractive surgery patients. Ophthalmology. 2009;116(6):1067-74. Epub 2009. PMID: 19395038 [PubMed - indexed for MEDLINE]
25. Chung SH, Roh MI, Park MS, Kong YT, Lee HK, Kim EK. Mycobacterium abscessus keratitis after LASIK with IntraLase femtosecond laser. Ophthalmologica. 2006;220(4):277-80. PMID: 16785761 [PubMed - indexed for MEDLINE]
26. Clement RA, Dunne MCM, Barnes DA. A method for ray tracing through schematic eyes with off-axis components. Ophthalmic Physiol Opt 1987;7:149-52.
27. Comparison of corneal aberration changes after laser in situ keratomileusis performed with mechanical microkeratome and IntraLase femtosecond laser: 1-year follow-up. Buzzonetti L, Petrocelli G, Valente P, Tamburrelli C, Mosca L, Laborante A, Balestrazzi E. Cornea. 2008;27(2):174-9. PMID: 18216572 [PubMed - indexed for MEDLINE]
28. Comparison of the femtosecond laser (IntraLase) versus manual microkeratome (Moria ALTK) in dissection of the donor in endothelial keratoplasty: initial study in eye bank eyes. Jones YJ, Goins KM, Sutphin JE, Mullins R, Skeie JM. Cornea. 2008;27(1):88-93. PMID: 18245973 [PubMed - indexed for MEDLINE]
29. Comparison of the femtosecond laser and mechanical keratome for laser in situ keratomileusis. Chan A, Ou J, Manche EE. Arch Ophthalmol. 2008;126(11):1484-90. PMID: 19001213 [PubMed - indexed for MEDLINE]Free Article
30. Complications of sub-Bowman's keratomileusis with a femtosecond laser in 3009 eyes. Chang JS. J Refract Surg. 2008;24(1):S97-101. PMID: 18269158 [PubMed - indexed for MEDLINE]

31. Corneal endothelial viability after femtosecond laser preparation of posterior lamellar discs for Descemet-stripping endothelial keratoplasty. Cheng YY, Pels E, Cleutjens JP, van Suylen RJ, Hendrikse F, Nuijts RM. Cornea. 2007;26(9):1118-22. PMID: 17893547 [PubMed - indexed for MEDLINE]
32. Dhaliwal DK, Mather R. New developments in corneal and external disease–LASIK. Ophthalmol Clin North Am. 2003 Mar;16(1):119-25. Review. PMID: 12683255 [PubMed - indexed for MEDLINE]
33. Effect of hinge position on corneal sensation and dry eye after laser in situ keratomileusis using a femtosecond laser. Mian SI, Shtein RM, Nelson A, Musch DC. J Cataract Refract Surg. 2007;33(7):1190-4. PMID: 17586374 [PubMed - indexed for MEDLINE]
34. Effert R, Gruppe S. The amount of angle alpha in a normal population. Acta XVII concilii europaeae strabologicae associeatis. Madrid;1988:17-21.
35. Fankhauser F, Niederer PF, Kwasniewska S, van der Zypen E. Supernormal vision, high-resolution retinal imaging, multiphoton imaging and nanosurgery of the cornea–a review. Technol Health Care. 2004;12(6):443-53. Review. PMID: 15671599 [PubMed - indexed for MEDLINE]
36. Farid M, Steinert RF, Gaster RN, Chamberlain W, Lin A. Comparison of penetrating keratoplasty performed with a femtosecond laser zig-zag incision versus conventional blade trephination. Ophthalmology. 2009 Sep;116(9):1638-43. Epub 2009. PMID: 19646760 [PubMed - indexed for MEDLINE]
37. Farid M, Steinert RF. Deep anterior lamellar keratoplasty performed with the femtosecond laser zigzag incision for the treatment of stromal corneal pathology and ectatic disease. J Cataract Refract Surg. 2009;35(5):809-13.PMID: 19393878 [PubMed - indexed for MEDLINE] Related citations
38. Femtosecond laser corneal ablation threshold: dependence on tissue depth and laser pulse width. Sun H, Han M, Niemz MH, Bille JF. Lasers Surg Med. 2007;39(8):654-8. PMID: 17886278 [PubMed - indexed for MEDLINE]
39. Femtosecond laser-assisted corneal surgery. Mian SI, Shtein RM. Curr Opin Ophthalmol. 2007;18(4):295-9. Review. PMID: 17568205 [PubMed - indexed for MEDLINE]
40. Femtosecond laser-assisted penetrating keratoplasty: stability evaluation of different wound configurations. Bahar I, Kaiserman I, McAllum P, Rootman D. Cornea. 2008;27(2):209-11. PMID: 18216578 [PubMed - indexed for MEDLINE]
41. Femtosecond laser-assisted retinal imaging and ablation: experimental pilot study. Hild M, Krause M, Riemann I, Mestres P, Toropygin S, Löw U, Brückner K, Seitz B, Jonescu-Cuypers C, König K. Curr Eye Res. 2008 Apr;33(4):351-63. PMID: 18398710 [PubMed - indexed for MEDLINE]
42. Femtosecond laser-assisted sutureless anterior lamellar keratoplasty. Yoo SH, Kymionis GD, Koreishi A, Ide T, Goldman D, Karp CL, O'Brien TP, Culbertson WW, Alfonso EC. Ophthalmology. 2008;115(8):1303-7, 1307.e1. Epub 2008 Jan 2. PMID: 18171586 [PubMed - indexed for MEDLINE]
43. Femtosecond-assisted diagnostic corneal biopsy (FAB) in keratitis. Yoo SH, Kymionis GD, O'Brien TP, Ide T, Culbertson W, Alfonso EC. Graefes Arch Clin Exp Ophthalmol. 2008;246(5):759-62. Epub 2008 Mar 6. PMID: 18322693 [PubMed - indexed for MEDLINE]
44. Flap thickness reproducibility in laser in situ keratomileusis with a femtosecond laser: optical coherence tomography measurement. Kim JH, Lee D, Rhee KI. J Cataract Refract Surg. 2008;34(1):132-6. PMID: 18165093 [PubMed - indexed for MEDLINE]
45. Ghanem RC, Azar DT. Femtosecond-laser arcuate wedge-shaped resection to correct high residual astigmatism after penetrating keratoplasty. J Cataract Refract Surg. 2006;32(9):1415-9. PMID: 16931248 [PubMed - indexed for MEDLINE]
46. Ghanem RC, de la Cruz J, Tobaigy FM, Ang LP, Azar DT. LASIK in the presbyopic age group: safety, efficacy, and predictability in 40- to 69-year-old patients. Ophthalmology. 2007;114(7):1303-10. Epub 2007 Mar 26. PMID: 17382397 [PubMed - indexed for MEDLINE]
47. Goggin M. Femtosecond keratotomy. Ophthalmology. 2010;117(4):847-8; author reply 848. No abstract available. PMID: 20346822 [PubMed - indexed for MEDLINE].
48. Hainline BC, Price MO, Choi DM, Price FW Jr. Central flap necrosis after LASIK with microkeratome and femtosecond laser created flaps. J Refract Surg. 2007;23(3):233-42. PMID: 17385288 [PubMed - indexed for MEDLINE]
49. Harissi-Dagher M, Azar DT. Can J Ophthalmol. Femtosecond laser astigmatic keratotomy for postkeratoplasty astigmatism. 2008;43(3):367-9. PMID: 18443607 [PubMed - indexed for MEDLINE]Free Article
50. Hoffart L, Proust H, Matonti F, Conrath J, Ridings B. Am J Ophthalmol. Correction of postkeratoplasty astigmatism by femtosecond laser compared with mechanized astigmatic keratotomy. 2009;147(5):779-87, 787.e1. Epub 2009 Feb 20. PMID: 19232560 [PubMed - indexed for MEDLINE]
51. Hoffart L, Proust H, Matonti F, Ridings B, Conrath J. Am J Ophthalmol. Short-term results of penetrating keratoplasty performed with the Femtec femtosecond laser. 2008;146(1):50-55. Epub 2008 Apr 24. PMID: 18439558 [PubMed - indexed for MEDLINE]
52. Holzer MP, Mannsfeld A, Ehmer A, Auffarth GU. Early outcomes of INTRACOR femtosecond laser treatment for presbyopia. J Refract Surg. 2009;25(10):855-61. doi: 10.3928/1081597X-20090917-06. PMID: 19835325 [PubMed - indexed for MEDLINE]
53. Holzer MP, Rabsilber TM, Auffarth GU. Am J Ophthalmol. Penetrating keratoplasty using femtosecond laser. 2007;143(3):524-6. Epub 2006 Dec 18. PMID: 17317405 [PubMed - indexed for MEDLINE]
54. Holzer MP, Rabsilber TM, Auffarth GU. Femtosecond laser-assisted corneal flap cuts: morphology, accuracy, and histopathology. Invest Ophthalmol Vis Sci. 2006;47(7):2828-31. PMID: 16799021 [PubMed - indexed for MEDLINE]Free Article

55. Hu MY, McCulley JP, Cavanagh HD, Bowman RW, Verity SM, Mootha VV, Petroll WM. Comparison of the corneal response to laser in situ keratomileusis with flap creation using the FS15 and FS30 femtosecond lasers: clinical and confocal microscopy findings. J Cataract Refract Surg. 2007;33(4):673-81. PMID: 17397742 [PubMed - indexed for MEDLINE]
56. Huang SC, Chen HC. Overview of laser refractive surgery. Chang Gung Med J. 2008;31(3):237-52. Review. PMID: 18782946 [PubMed - indexed for MEDLINE]Free Article
57. Ide T, Yoo SH, Kymionis GD, Haft P, O'Brien TP. Second femtosecond laser pass for incomplete laser in situ keratomileusis flaps caused by suction loss. J Cataract Refract Surg. 2009;35(1):153-7. PMID: 19101438 [PubMed - indexed for MEDLINE]
58. Ide T, Yoo SH, Kymionis GD, Leng T, Marini C, Stanciu NA, O'Brien TP. Descemet stripping automated endothelial keratoplasty tissue preparation with femtosecond laser and contact lens. Cornea. 2010;29(1):93-8. PMID: 19907309 [PubMed - indexed for MEDLINE]
59. Ignacio TS, Nguyen TB, Chuck RS, Kurtz RM, Sarayba MA. Top hat wound configuration for penetrating keratoplasty using the femtosecond laser: a laboratory model. Cornea. 2006;25(3):336-40. PMID: 16633036 [PubMed - indexed for MEDLINE]
60. In vitro noncontact intravascular femtosecond laser surgery in models of branch retinal vein occlusion. Toropygin S, Krause M, Riemann I, Hild M, Mestres P, Seitz B, Khurieva E, Ruprecht KW, Löw U, Gatzioufas Z, König K. Curr Eye Res. 2008;33(3):277-83. PMID: 18350439 [PubMed - indexed for MEDLINE]
61. Incidence, possible risk factors, and potential effects of an opaque bubble layer created by a femtosecond laser. Kaiserman I, Maresky HS, Bahar I, Rootman DS. J Cataract Refract Surg. 2008;34(3):417-23. PMID: 18299066 [PubMed - indexed for MEDLINE]
62. Intacs for keratoconus. Rabinowitz YS. Curr Opin Ophthalmol. 2007;18(4):279-83. Review. PMID: 17568203 [PubMed - indexed for MEDLINE]
63. Jonas JB, Vossmerbaeumer U. Femtosecond laser LASIK flap preparation with conical incision and positional spikes. J Cataract Refract Surg. 2004;30(5):1107-8. PMID: 15130651 [PubMed - indexed for MEDLINE]
64. Kaluzny BJ, Gora M, Karnowski K, Grulkowski I, Kowalczyk A, Wojtkowski M. Br J Ophthalmol. Imaging of the lens capsule with an ultrahigh-resolution spectral optical coherence tomography prototype based on a femtosecond laser. 2010;94(3):275-7. PMID: 20215371 [PubMed - in process].
65. Kanellopoulos AJ: Results of hyperopia treatment with the Allegretto-Wave excimer laser: Journal of Refractive Surgery;21(5) Sept/Oct 2005.
66. Keratitis after implantation of intrastromal corneal ring segments (Intacs) aided by femtosecond laser for keratoconus correction: case report and description of the literature. Levy J, Lifshit T. Eur J Ophthalmol. 2010. pii: 540A63D2-DD1F-494D-8B62-1C25DE6BAD41. [Epub ahead of print] PMID: 20155711 [PubMed - as supplied by publisher]
67. Kessel L, Eskildsen L, van der Poel M, Larsen M. Non-invasive bleaching of the human lens by femtosecond laser photolysis. PLoS One. 2010;16;5(3):e9711. PMID: 20300521 [PubMed - in process]Free PMC ArticleFree text.
68. Kim JH, Choi SK, Lee D. The comparison of femtosecond laser-assisted penetrating keratoplasty with conventional surgery in terms of endothelial safety: ex vivo study using porcine eyes. Cornea. 2009;28(7):812-6. PMID: 19574902 [PubMed - indexed for MEDLINE]
69. Kim JH, Lee D, Hahn TW, Choi SK. New surgical strategy for corneal tattooing using a femtosecond laser. Cornea. 2009;28(1):80-4. PMID: 19092411 [PubMed - indexed for MEDLINE]
70. Kim JY, Kim MJ, Kim TI, Choi HJ, Pak JH, Tchah H. A femtosecond laser creates a stronger flap than a mechanical microkeratome. Invest Ophthalmol Vis Sci. 2006;47(2):599-604. PMID: 16431956 [PubMed - indexed for MEDLINE]Free Article
71. Kitzmann AS, Bourne WM, Patel SV. Confocal microscopy of a femtosecond laser LASIK flap before separation. Am J Ophthalmol. 2007;143(4):691-3. Epub 2006 Dec 8. PMID: 17386281 [PubMed - indexed for MEDLINE]Free PMC ArticleFree text
72. Ko TH, Fujimoto JG, Duker JS, Paunescu LA, Drexler W, Baumal CR, Puliafito CA, Reichel E, Rogers AH, Schuman JS. Comparison of ultrahigh- and standard-resolution optical coherence tomography for imaging macular hole pathology and repair. Ophthalmology. 2004 Nov;111(11):2033-43. PMID: 15522369 [PubMed - indexed for MEDLINE]Free PMC ArticleFree text
73. Kook D, Derhartunian V, Bug R, Kohnen T. Cornea. Top-hat shaped corneal trephination for penetrating keratoplasty using the femtosecond laser: a histomorphological study. 2009;28(7):795-800. PMID: 19574905 [PubMed - indexed for MEDLINE]
74. Krueger RR, Juhasz T, Gualano A, Marchi V. The picosecond laser for nonmechanical laser in situ keratomileusis. J Refract Surg. 1998;14(4):467-9. PMID: 9699173 [PubMed - indexed for MEDLINE]
75. Kucumen RB, Dinc UA, Yenerel NM, Gorgun E, Alimgil ML. Immediate evaluation of the flaps created by femtosecond laser using anterior segment optical coherence tomography. Ophthalmic Surg Lasers Imaging. 2009;40(3):251-4. PMID: 19485288 [PubMed - indexed for MEDLINE]
76. Kucumen RB, Yenerel NM, Gorgun E, Alimgil ML. Management of postkeratoplasty ametropia: IntraLASIK after penetrating keratoplasty. Eur J Ophthalmol. 2008;18(6):877-85. PMID: 18988156 [PubMed - indexed for MEDLINE]
77. Kumar NL, Kaiserman I, Shehadeh-Mashor R, Sansanayudh W, Ritenour R, Rootman DS. IntraLase-Enabled Astigmatic Keratotomy for Post-Keratoplasty Astigmatism: On-Axis Vec-

tor Analysis. Ophthalmology. 2010. [Epub ahead of print] PMID: 20163860 [PubMed - as supplied by publisher]
78. Kunert KS, Blum M, Reich M, Dick M, Russmann C. Effect of a suction device for femtosecond laser on anterior chamber depth and crystalline lens position measured by OCT. J Refract Surg. 2009;25(11):1005-11. doi: 10.3928/1081597X-20091016-06. Epub 2009 Nov 13. PMID: 19921769 [PubMed - indexed for MEDLINE]
79. Kurtz RM, Horvath C, Liu HH, Krueger RR, Juhasz T. Lamellar refractive surgery with scanned intrastromal picosecond and femtosecond laser pulses in animal eyes. J Refract Surg. 1998;14(5):541-8. PMID: 9791821 [PubMed - indexed for MEDLINE]
80. Lee D, Kim JH, Oh SH, Choi SK, Kim JK. Femtosecond laser lamellar keratoplasty to aid visualization for cataract surgery. J Refract Surg. 2009;25(10):902-4. doi: 10.3928/1081597X-20090617-03. Epub 2009 Oct 12. PMID: 19835332 [PubMed - indexed for MEDLINE]
81. Lee J, Winokur J, Hallak J, Azar DT. Femtosecond dovetail penetrating keratoplasty: surgical technique and case report. Br J Ophthalmol. 2009;93(7):861-3 PMID: 19553511 [PubMed - indexed for MEDLINE]t
82. Li Y, Netto MV, Shekhar R, Krueger RR, Huang D. A longitudinal study of LASIK flap and stromal thickness with high-speed optical coherence tomography. Ophthalmology. 2007;114(6):1124-32. Epub 2007 Feb 23. PMID: 17320959 [PubMed - indexed for MEDLINE]
83. Lifshitz T, Levy J, Klemperer I, Levinger S. Anterior chamber gas bubbles after corneal flap creation with a femtosecond laser. J Cataract Refract Surg. 2005;31(11):2227-9. PMID: 16412944 [PubMed - indexed for MEDLINE]
84. Lifshitz T, Levy J, Mahler O, Levinger S. Peripheral sterile corneal infiltrates after refractive surgery. J Cataract Refract Surg. 2005;31(7):1392-5. PMID: 16105612 [PubMed - indexed for MEDLINE]
85. Lim T, Yang S, Kim M, Tchah H. Am J Ophthalmol. Comparison of the IntraLase femtosecond laser and mechanical microkeratome for laser in situ keratomileusis. 2006;141(5):833-9. PMID: 16678504 [PubMed - indexed for MEDLINE]
86. Lin A, Gaster RN. Suction loss during femtosecond laser incision for penetrating keratoplasty. Cornea. 2009;28(3):362-4. PMID: 19387246 [PubMed - indexed for MEDLINE] Related citations
87. Malta JB, Soong HK, Shtein R, Banitt M, Musch DC, Sugar A, Mian SI. Femtosecond laser-assisted keratoplasty: laboratory studies in eye bank eyes. Curr Eye Res. 2009;34(1):18-25. PMID: 19172466 [PubMed - indexed for MEDLINE]
88. Masters BR, So PT. Antecedents of two-photon excitation laser scanning microscopy. Microsc Res Tech. 2004;63(1):3-11. PMID: 14677127 [PubMed - indexed for MEDLINE]
89. McAllum P, Kaiserman I, Bahar I, Rootman D. Femtosecond laser top hat penetrating keratoplasty: wound burst pressures of incomplete cuts. Arch Ophthalmol. 2008;126(6):822-5. PMID: 18541847 [PubMed - indexed for MEDLINE]Free Article
90. Meltendorf C, Burbach GJ, Bühren J, Bug R, Ohrloff C, Deller T. Corneal femtosecond laser keratotomy results in isolated stromal injury and favorable wound-healing response. Invest Ophthalmol Vis Sci. 2007;48(5):2068-75. PMID: 17460262 [PubMed-indexed for MEDLINE]Free Article
91. Meltendorf C, Burbach GJ, Ohrloff C, Ghebremedhin E, Deller T. Intrastromal keratotomy with femtosecond laser avoids profibrotic TGF-beta1 induction. Invest Ophthalmol Vis Sci. 2009;50(8):3688-95. Epub 2009. PMID: 19387066 [PubMed - indexed for MEDLINE Related citations
92. Meltendorf C, Schroeter J, Bug R, Kohnen T, Deller T. Corneal trephination with the femtosecond laser. Cornea. 2006;25(9):1090-2. PMID: 17133060 [PubMed - indexed for MEDLINE]
93. Mian SI, Li AY, Dutta S, Musch DC, Shtein RM. Dry eyes and corneal sensation after laser in situ keratomileusis with femtosecond laser flap creation Effect of hinge position, hinge angle, and flap thickness. J Cataract Refract Surg. 2009;35(12):2092-8. PMID: 19969213 [PubMed - indexed for MEDLINE]
94. Mian SI, Soong HK, Patel SV, Ignacio T, Juhasz T. In vivo femtosecond laser-assisted posterior lamellar keratoplasty in rabbits. Cornea. 2006;25(10):1205-9. PMID: 17172899 [PubMed - indexed for MEDLINE]
95. Montés-Micó R, Rodríguez-Galietero A, Alió JL. Femtosecond laser versus mechanical keratome LASIK for myopia. Ophthalmology. 2007;114(1):62-8. Epub 2006 Oct 27. PMID: 17070593 [PubMed - indexed for MEDLINE]
96. Morishige N, Yamada N, Teranishi S, Chikama T, Nishida T, Takahara A. Detection of subepithelial fibrosis associated with corneal stromal edema by second harmonic generation imaging microscopy. Invest Ophthalmol Vis Sci. 2009;50(7):3145-50. Epub 2009 Feb 21. PMID: 19234355 [PubMed - indexed for MEDLINE]
97. Muftuoglu O, Prasher P, Chu C, Mootha VV, Verity SM, Cavanagh HD, Bowman RW, McCulley JP. Laser in situ keratomileusis for residual refractive errors after apodized diffractive multifocal intraocular lens implantation. J Cataract Refract Surg. 2009;35(6):1063-71. PMID: 19465293 [PubMed - indexed for MEDLINE]
98. Nagy Z, Takacs A, Filkorn T, Sarayba M. Initial clinical evaluation of an intraocular femtosecond laser in cataract surgery. J Refract Surg. 2009;25(12):1053-60. doi: 10.3928/1081597X-20091117-04. PMID: 20000286 [PubMed - indexed for MEDLINE]
99. Nakamura H, Liu Y, Witt TE, Gordon RJ, Edward DP. Femtosecond laser photodisruption of primate trabecular meshwork: an ex vivo study. Invest Ophthalmol Vis Sci. 2009;50(3):1198-204. Epub 2008 Oct 3. PMID: 18836174 [PubMed - indexed for MEDLINE]Free Article
100. Nepomuceno RL, Boxer Wachler BS, Kim JM, Scruggs R, Sato M. Laser in situ keratomileusis for hyperopia with the LADARVision 4000 with centration on the coaxially sighted corneal light reflex. J Cataract Refract Surg 2004;30:1281-6.

101. Non-penetrating deep sclerectomy for glaucoma surgery using the femtosecond laser: a laboratory model. Bahar I, Kaiserman I, Trope GE, Rootman D. Br J Ophthalmol. 2007;91(12):1713-4. No abstract available. PMID: 18024818 [PubMed - indexed for MEDLINE]Free PMC ArticleFree text
102. Novel technique of corneal biopsy by using a femtosecond laser in infectious ulcers. Kim JH, Yum JH, Lee D, Oh SH.Cornea. 2008;27(3):363-5. PMID: 18362669 [PubMed - indexed for MEDLINE]
103. Nubile M, Carpineto P, Lanzini M, Calienno R, Agnifili L, Ciancaglini M, Mastropasqua L. Femtosecond laser arcuate keratotomy for the correction of high astigmatism after keratoplasty. Ophthalmology. 2009;116(6):1083-92. Epub 2009. PMID: 19395035 [PubMed - indexed for MEDLINE]
104. Panday VA, Reilly CD. Refractive surgery in the United States Air Force. Curr Opin Ophthalmol. 2009;20(4):242-6. Review. PMID: 19398912 [PubMed - indexed for MEDLINE]
105. Pande M, Hillman JS. Optical zone centration in keratorefractive surgery; entrance pupil center, visual axis, coaxially sighted corneal reflex, or geometric corneal center? Ophthalmology 1993;100:1230–7.
106. Pande M, Hillman JS. Optical zone centration in keratorefractive surgery; entrance pupil centers, visual axis, coaxially sighted corneal reflex, or geometric corneal center? Ophthalmology 1993;100:1230-7.
107. Patel SV, Maguire LJ, McLaren JW, Hodge DO, Bourne WM. Femtosecond laser versus mechanical microkeratome for LASIK: a randomized controlled study. Ophthalmology. 2007;114(8):1482-90. Epub 2007 Mar 13. PMID: 17350688 [PubMed - indexed for MEDLINE]
108. Petroll WM, Bowman RW, Cavanagh HD, Verity SM, Mootha VV, McCulley JP. Assessment of keratocyte activation following LASIK with flap creation using the IntraLase FS60 laser. J Refract Surg. 2008;24(8):847-9. PMID: 18856242 [PubMed - indexed for MEDLINE]Free PMC ArticleFree text
109. Pietilä J, Huhtala A, Jääskeläinen M, Jylli J, Mäkinen P, Uusitalo H. LASIK flap creation with the Ziemer femtosecond laser in 787 consecutive eyes. J Refract Surg. 2010;26(1):7-16. doi: 10.3928/1081597X-20101215-02. Epub 2010 Jan 11. PMID: 20199007 [PubMed - indexed for MEDLINE]
110. Piñero DP, Alio JL, El Kady B, Coskunseven E, Morbelli H, Uceda-Montanes A, Maldonado MJ, Cuevas D, Pascual I. Refractive and aberrometric outcomes of intracorneal ring segments for keratoconus: mechanical versus femtosecond-assisted procedures. Ophthalmology. 2009;116(9):1675-87. Epub 2009 Jul 29. PMID: 19643498 [PubMed - indexed for MEDLINE]
111. Piñero DP, Alio JL, Morbelli H, Uceda-Montanes A, El Kady B, Coskunseven E, Pascual I. Refractive and corneal aberrometric changes after intracorneal ring implantation in corneas with pellucid marginal degeneration. Ophthalmology. 2009;116(9):1656-64. Epub 2009 Jul 29. PMID: 19643482 [PubMed - indexed for MEDLINE]
112. Piñero DP, Alio JL, Uceda-Montanes A, El Kady B, Pascual I. Intracorneal ring segment implantation in corneas with post-laser in situ keratomileusis keratectasia. Ophthalmology. 2009;116(9):1665-74. Epub 2009. PMID: 19643485 [PubMed - indexed for MEDLINE]
113. Quantitative assessment of corneal wound healing following IntraLASIK using in vivo confocal microscopy. McCulley JP, Petroll WM. Trans Am Ophthalmol Soc. 2008;106:84-90; discussion 90-2. PMID: 19277224 [PubMed - indexed for MEDLINE]Free PMC ArticleFree text
114. Regularity of human corneal flaps prepared by femtosecond laser technology. Vossmerbaeumer U, Jonas JB. J Refract Surg. 2008;24(6):645-8. PMID: 18581793 [PubMed - indexed for MEDLINE]
115. Results of penetrating keratoplasty performed with a femtosecond laser zigzag incision initial report. Farid M, Kim M, Steinert RF. Ophthalmology. 2007;114(12):2208-12. PMID: 18054639 [PubMed - indexed for MEDLINE]
116. Rosa AM, Neto Murta J, Quadrado MJ, Tavares C, Lobo C, Van Velze R, Castanheira-Dinis A. Femtosecond laser versus mechanical microkeratomes for flap creation in laser in situ keratomileusis and effect of postoperative measurement interval on estimated femtosecond flap thickness. J Cataract Refract Surg. 2009;35(5):833-8. PMID: 19393881 [PubMed - indexed for MEDLINE]
117. Sansanayudh W, Bahar I, Kumar NL, Shehadeh-Mashour R, Ritenour R, Singal N, Rootman DS. Intrastromal corneal ring segment SK implantation for moderate to severe keratoconus. J Cataract Refract Surg. 2010;36(1):110-3. PMID: 20117713 [PubMed - indexed for MEDLINE]
118. Sarayba MA, Juhasz T, Chuck RS, Ignacio TS, Nguyen TB, Sweet P, Kurtz RM. Femtosecond laser posterior lamellar keratoplasty: a laboratory model. Cornea. 2005;24(3):328-33. PMID: 15778607 [PubMed - indexed for MEDLINE]
119. Sarayba MA, Kurtz RM, Nguyen TT, Ignacio T, Mansoori M, Sweet PM, Chuck RS. Femtosecond laser-assisted intracorneal keratoprosthesis implantation: a laboratory model. Cornea. 2005;24(8):1010-4. PMID: 16227853 [PubMed - indexed for MEDLINE]
120. Sawa M, Awazu K, Takahashi T, Sakaguchi H, Horiike H, Ohji M, Tano Y. Br J Ophthalmol. Application of femtosecond ultrashort pulse laser to photodynamic therapy mediated by indocyanine green. 2004;88(6):826-31. PMID: 15148220 [PubMed - indexed for MEDLINE]Free PMC ArticleFree text
121. Schallhorn SC, Tanzer DJ, Kaupp SE, Brown M, Malady SE. Comparison of night driving performance after wavefront-guided and conventional LASIK for moderate myopia. Ophthalmology. 2009;116(4):702-9. PMID: 19344822 [PubMed - indexed for MEDLINE]
122. Scott WE, Mash AJ. Kappa angle measurements of strabismic nad nonstrabismic individuals. Arch Ophthalmol 1973;89:18-20.
123. Second-harmonic imaging of cornea after intrastromal femtosecond laser ablation. Han M, Zickler L, Giese G, Walter M, Loesel FH, Bille JF. J Biomed Opt. 2004;9(4):760-6. PMID: 15250763 [PubMed - indexed for MEDLINE]

124. Seitz B, Brünner H, Viestenz A, Hofmann-Rummelt C, Schlötzer-Schrehardt U, Naumann GO, Langenbucher A. Am J Ophthalmol. Inverse mushroom-shaped nonmechanical penetrating keratoplasty using a femtosecond laser. 2005;139(5):941-4. PMID: 15860317 [PubMed - indexed for MEDLINE]
125. Seitz B, Langenbucher A, Hofmann-Rummelt C, Schlötzer-Schrehardt U, Naumann GO. Nonmechanical posterior lamellar keratoplasty using the femtosecond laser (femto-plak) for corneal endothelial decompensation. Am J Ophthalmol. 2003;136(4):769-72. PMID: 14516834 [PubMed - indexed for MEDLINE]
126. Sekundo W, Kunert K, Russmann C, Gille A, Bissmann W, Stobrawa G, Sticker M, Bischoff M, Blum M. First efficacy and safety study of femtosecond lenticule extraction for the correction of myopia: six-month results. J Cataract Refract Surg. 2008;34(9):1513-20. Erratum in: J Cataract Refract Surg. 2008 Nov;34(11):1819. PMID: 18721712 [PubMed - indexed for MEDLINE]
127. Shabayek MH, Alió JL. Intrastromal corneal ring segment implantation by femtosecond laser for keratoconus correction. Ophthalmology. 2007;114(9):1643-52. Epub 2007 Apr 2. PMID: 17400293 [PubMed - indexed for MEDLINE]
128. Slade SG, Durrie DS, Binder PS. A prospective, contralateral eye study comparing thin-flap LASIK (sub-Bowman keratomileusis) with photorefractive keratectomy. Ophthalmology. 2009;116(6):1075-82.
129. Sletten KR, Yen KG, Sayegh S, Loesel F, Eckhoff C, Horvath C, Meunier M, Juhasz T, Kurtz RM. An in vivo model of femtosecond laser intrastromal refractive surgery. Ophthalmic Surg Lasers. 1999;30(9):742-9. PMID: 10574496 [PubMed - indexed for MEDLINE]
130. Sonigo B, Iordanidou V, Chong-Sit D, Auclin F, Ancel JM, Labbé A, Baudouin C. In vivo corneal confocal microscopy comparison of intralase femtosecond laser and mechanical microkeratome for laser in situ keratomileusis. Invest Ophthalmol Vis Sci. 2006;47(7):2803-11. PMID: 16799017 [PubMed - indexed for MEDLINE]Free Article
131. Soong HK, Malta JB. Femtosecond lasers in ophthalmology. Am J Ophthalmol. 2009;147(2):189-197.e2. Epub 2008 Oct 18. Review. PMID: 18930447 [PubMed - indexed for MEDLINE]
132. Soong HK, Mian S, Abbasi O, Juhasz T. Femtosecond laser-assisted posterior lamellar keratoplasty: initial studies of surgical technique in eye bank eyes. Ophthalmology. 2005;112(1):44-9. PMID: 15629819 [PubMed - indexed for MEDLINE]
133. Stachs O, Schumacher S, Hovakimyan M, Fromm M, Heisterkamp A, Lubatschowski H, Guthoff R. Visualization of femtosecond laser pulse-induced microincisions inside crystalline lens tissue. J Cataract Refract Surg. 2009;35(11):1979-83. PMID: 19878832 [PubMed - indexed for MEDLINE]
134. Steinert RF, Ignacio TS, Sarayba MA. Am J Ophthalmol. "Top hat"-shaped penetrating keratoplasty using the femtosecond laser. 007;143(4):689-91. Epub 2006 Dec 22. PMID: 17386280 [PubMed - indexed for MEDLINE]
135. Stern D, Lin WZ, Puliafito CA, Fujimoto JG. Femtosecond optical ranging of corneal incision depth. Invest Ophthalmol Vis Sci. 1989;30(1):99-104. PMID: 2912917 [PubMed - indexed for MEDLINE]Free Article
136. Sub-epithelial gas breakthrough during femtosecond laser flap creation for LASIK. Srinivasan S, Herzig S. Br J Ophthalmol. 2007;91(10):1373. PMID: 17895418 [PubMed - indexed for MEDLINE]
137. Sutton G, Hodge C. Accuracy and precision of LASIK flap thickness using the IntraLase femtosecond laser in 1000 consecutive cases. J Refract Surg. 2008;24(8):802-6. PMID: 18856234 [PubMed - indexed for MEDLINE]
138. Tanter M, Touboul D, Gennisson JL, Bercoff J, Fink M. High-resolution quantitative imaging of cornea elasticity using supersonic shear imaging. IEEE Trans Med Imaging. 2009;28(12):1881-93. Epub 2009 May 5. PMID: 19423431 [PubMed - indexed for MEDLINE]
139. Terrell J, Bechara SJ, Nesburn A, et al. The effect of globe fixation on ablation zone centration in photorefractive keratectomy. Am J Ophthalmol 1995;119:612–9.
140. The LADAR6000: results in highly myopic and highly astigmatic eyes. Lee VW. J Refract Surg. 2006;22(9):S980-2.
141. Toropygin SG, Krause M, Riemann I, Seitz B, Mestres P, Ruprecht KW, König K. In vitro femtosecond laser-assisted nanosurgery of porcine posterior capsule. J Cataract Refract Surg. 2008;34(12):2128-32. PMID: 19027571 [PubMed - indexed for MEDLINE]
142. Toyran S, Liu Y, Singha S, Shan S, Cho MR, Gordon RJ, Edward DP. Femtosecond laser photodisruption of human trabecular meshwork: an in vitro study. Exp Eye Res. 2005;81(3):298-305. PMID: 16129097 [PubMed - indexed for MEDLINE]
143. Uozato H, Guyton DL. Centering corneal surgical procedures. Am J Ophthalmol 1987;103:264-75; correction p. 852.
144. von Jagow B, Kohnen T. Corneal architecture of femtosecond laser and microkeratome flaps imaged by anterior segment optical coherence tomography. J Cataract Refract Surg. 2009;35(1):35-41. PMID: 19101422 [PubMed - indexed for MEDLINE]
145. von Noorden GK. (): Binocular vision and ocular motility: theory and management of strabismus. St. Louis (MO): Mosby; 1996;163-205.
146. Vossmerbaeumer U, Jonas JB. Structure of intracorneal femtosecond laser pulse effects in conical incision profiles. Graefes Arch Clin Exp Ophthalmol. 2008;246(7):1017-20. Epub 2008 Apr 30. PMID: 18446357 [PubMed - indexed for MEDLINE]
147. Wei AJ, Hong JX, Xu JJ. Zhonghua Yan Ke Za Zhi. Advances in penetrating keratoplasty technique. 2010;46(1):85-7. Chinese. PMID: 20388326 [PubMed - in process]
148. Yang X, Jiang F, Song Y, Peng C, Sheng S. Accidental macular injury from prolonged viewing of a plasma flash produced by a femtosecond laser. Li X. Ophthalmology. 2010;2. [Epub ahead of print] PMID: 20202686 [PubMed - as supplied by publisher].

3

The Benefits of Femtosecond Laser Applications

Mark Wevill (United Kingdom)

Chapter Outline

- Femtosecond LASIK
- Femtosecond Astigmatic Keratotomies
- Intracor
- Other Femtosecond Laser Applications

INTRODUCTION

Femtosecond lasers are changing eye surgery. The list of applications is growing and the benefits in terms of improved safety and efficacy is making femtosecond technology the gold standard in some surgical techniques. The author has been using femtosecond lasers for the last 5 years for LASIK, axial keratotomy and presbyopia procedures. Some procedures such as the Intracor procedure would not be possible without a femtosecond laser and the femtosecond laser has made LASIK and axial keratotomies more accurate and safe.

FEMTOSECOND LASIK

Ultralase reviewed the results of LASIK in 23, 496 eyes comparing the results obtained with the femtosecond laser and the Moria One Use Plus (OUP) microkeratome (130 µ head). The surgeons were all experienced microkeratome users with collectively tens of thousands of procedures experience with microkeratomes. Previous audits had demonstrated that the Moria OUP was a good microkeratome, better than the original Hansatome used previously. There were better results and fewer complications with the Moria OUP. The same surgeons were all new to the femtosecond laser (Intralase FS60, AMO Abbott); therefore the eyes studied were operated on during the early phase of the surgeons' learning curve. All eyes were operated on within the same 12-month period. The OUP group included 6,920 eyes, and femtosecond (FS) group included of 14,457 eyes. The mean ages of the patients, mean refractive errors, range of refractive errors and percentage of hyperopes and myopes were similar. The UCDVA, spherical equivalents, defocus equivalents, percentage of patients seeing 6/6 and within ± 0.5D at the patient's last visit were all statistically significantly better with the femtosecond flap group. The UCDVA was 0.07 LogMAR (±0.17) for the OUP group and -0.02 (±0.12) for the FS group ($p<0.01$). The mean postoperative spherical equivalent for the OUP eyes was -0.22 (±0.47) and for the FS group was -0.07 (±0.38). The defocus equivalents for the 2 groups were 0.47 (±0.48) and 0.32 (±0.40) respectively. The percentage of patients seeing 6/6 in the OUP group was 76.4 percent and in the FS group 82.3 percent ($p<0.05$). The percentage of patients within ± 0.5D were 71.6 percent and 81.6 percent respectively ($p<0.05$). The femtosecond laser was also safer with more eyes in the femtosecond group gaining lines of vision and fewer losing lines (Fig. 3.1).

These results demonstrated the improved efficacy and safety of femtosecond LASIK (using the Intralase FS60) compared to microkeratome LASIK (using the Moria OUP).[1] Other studies published in the peer review literature have confirmed the safety and efficacy of femtosecond LASIK and demonstrated reduced higher order aberrations, less backscatter and improved contrast sensitivity with FS LASIK.

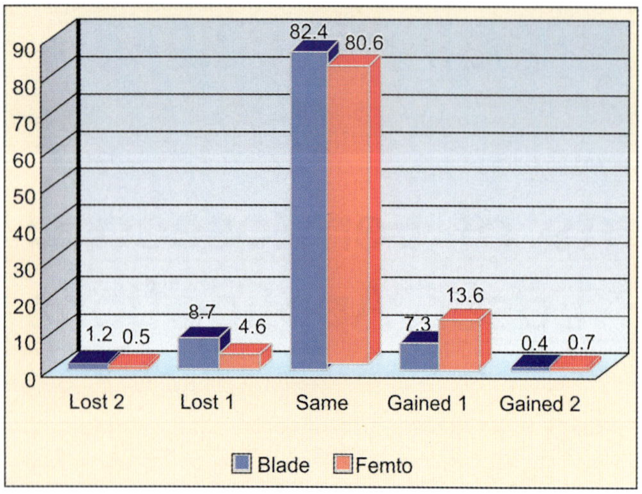

Fig. 3.1 Safety: Lines lost and gained in microkeratome and femtosecond LASIK

The complications produced by the two flap making technologies were also assessed. Another cohort of over 25,000 eyes was retrospectively reviewed over a different time period. OUP LASIK was done in 2,874 eyes and 23,011 had FS LASIK. Any complication which resulted in a change in or a delay in the operative treatment plan was included in the study. The OUP complication rate was 0.6 percent and the femtosecond rate was 0.3 percent, which was similar to the rate reported by Chang in 2008.[2] Most of the complications were minor and consisted of suction breaks (both groups), abrasions (0.17% OUP group only), vertical gas breakthrough (Femtosecond group only (0.07%), incomplete flaps (0.14% OUP only) and free caps (0.07% OUP only).[3]

However, the implications for the patient are more severe with the microkeratome complications (abrasions, incomplete flaps and free caps) than with vertical gas breakthrough with the femtosecond laser. Therefore another study was done assessing the outcomes of the different flap complications. 120 consecutive flap complications were analysed. Microkeratome complications in 80 eyes and femtosecond laser complications in 40 eyes were analyzed. The mean time between the development of the flap complication and subsequent excimer laser refractive correction was 76 days in the microkeratome group and 22 days in the femtosecond group. This is one of the most valuable aspects of femtosecond lasers to the surgeon. For both the patient and the surgeon, it is a great advantage to be able to complete the planned treatment as soon as possible after a complication. With the femtosecond laser, the surgeon has to cope with a dissatisfied patient for less than half the time as a microkeratome complication patient. As discussed previously, the surgeons were all experienced microkeratome users with a good knowledge of management of microkeratome flap complications, but were less experienced femtosecond users. With more experience, these surgeons are now treating femtosecond flap complications even earlier with good results.[4]

Further analysis of the cause and management and outcomes of twenty sequential femtosecond laser flap complications were reviewed retrospectively. There were repeated suction losses in 15 eyes, incomplete laser flaps in 3 eyes and vertical gas breakthrough in 2 eyes. Seven of the eyes had LASEK on the same day as the femtosecond flap complication. Nine had LASEK on a subsequent day. Four had LASIK with a deeper flap on a subsequent day. At discharge or 1 year later, all the patients had 6/12 uncorrected visual acuities or better, 84 percent had 6/6 or better and no eye lost 2 lines of best corrected acuity. This data demonstrated the efficacy and safety of excimer laser refractive correction of eyes with previous femtosecond laser flap complications.[5]

FEMTOSECOND ASTIGMATIC KERATOTOMIES

The femtosecond laser is useful for other refractive surgery applications. Many refractive surgeons would not do LASIK on a healthy eye (no features of keratoconus or media opacities) with a refractive error of, for example +6.00 / -7.00 x 180. Refractive correction with glasses or contact lenses is difficult. Despite this, patients are sometimes advised to avoid surgery or may be offered a toric lens implant. With a refractive lens exchange procedure, a young patient then faces "iatrogenic presbyopia" or the side effects of multifocal intraocular lenses (Fig. 3.2).

Guell described a technique of combining astigmatic keratotomies (created with a blade) with LASIK to correct high astigmatism.[6] At the Femtosecond Laser Forum meeting in Paris in 2010, the results of 16 eyes with high hyperopic or mixed astigmatism which had femtosecond laser axial keratotomies followed by LASIK for their high refractive errors were presented. The eyes had a mean spherical equivalent of +0.27 (Range +3.00 to -2.00), mean astigmatism was 5.56 (Range +4.00 to +7.00) and the best corrected visual acuity was 0.01 LogMAR (0.2 to -0.2). Femtosecond astigmatic keratotomies were done on the eyes.

At least 3 months later they had LASIK for the residual refractive error. After at least 1 month of post LASIK follow up, the mean spherical equivalent was -0.30 LogMAR, mean astigmatism of 0.69D (Range 0.00 to -2.25), a mean vector analyzed cylinder change of 5.3D (range 3.35 to 8.00), mean LogMAR best corrected acuity of 0.02 (0.2 to -0.2) and a mean uncorrected LogMAR acuity of 0.08 (range 0.22 to -0.16). No eyes lost 2 or more lines of best corrected acuity.[7] The scattergrams at each stage (Figs 3.2 to 3.5) of the procedure illustrate the effectiveness of the procedure:

Satisfaction surveys indicated that all 16 patients were "satisfied" with the results, 7 were "very pleased", and there were no patients who "regretted having it done" or "hoped for better". Therefore for these difficult to correct eyes, the procedure seems to be safe and effective, and studies are ongoing.

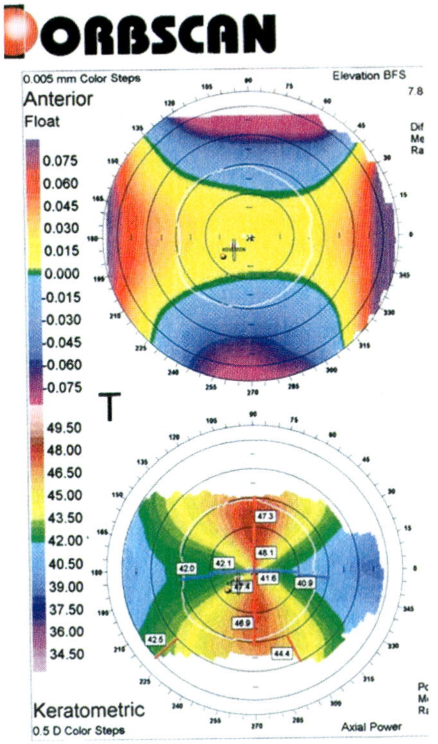

Fig. 3.2

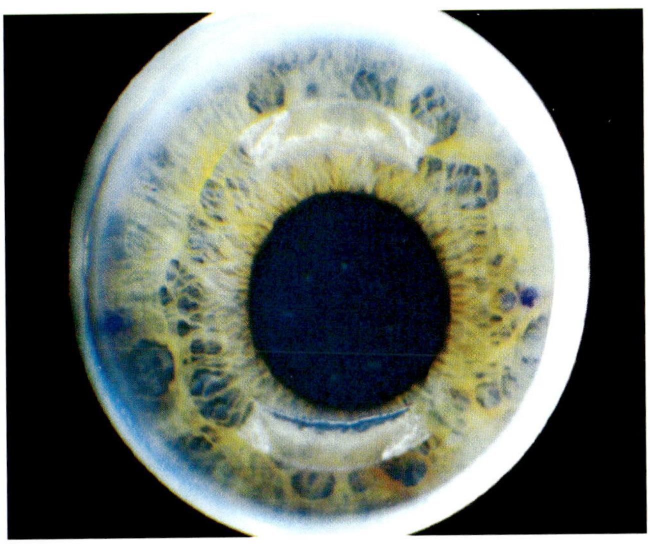

Fig. 3.3 Scattergram 1: Preoperative astigmatism

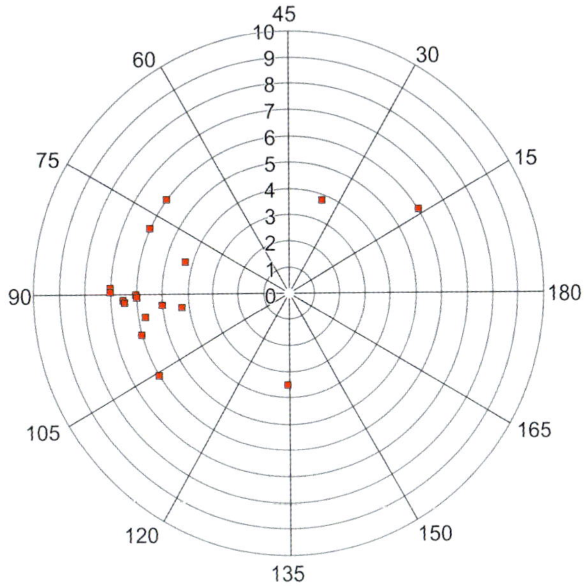

Fig. 3.4 Scattergram 2: The post arcuate incisions, pre LASIK

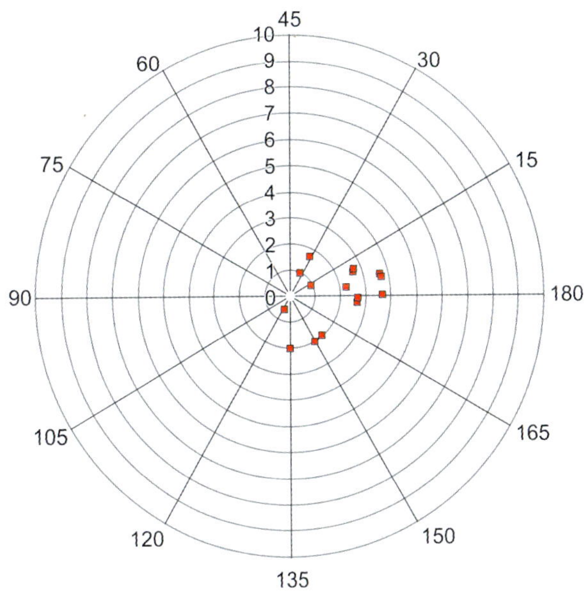

Fig. 3.5 Scattergram 3: The post LASIK scattergram

INTRACOR (FIGS 3.6 to 3.8)

Another application for femtosecond laser technology is the Intracor procedure. Intrastromal rings are cut into the cornea of a non-dominant emmetropic presbyopic eye of a

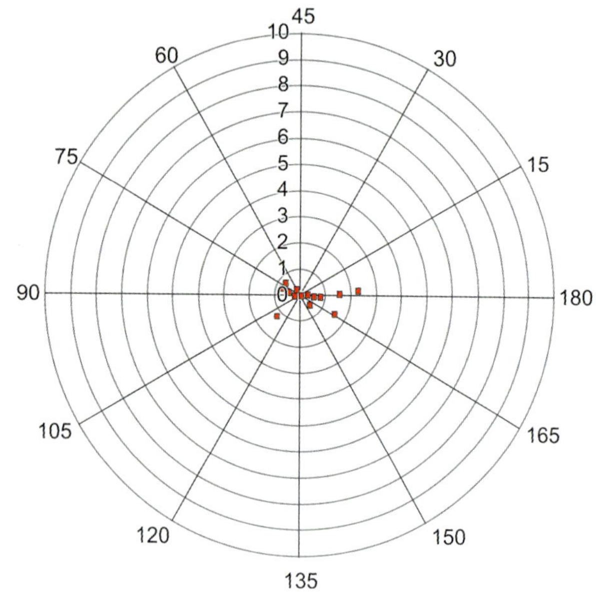

Fig. 3.6 No eyes lost 2 or more lines of best corrected acuity

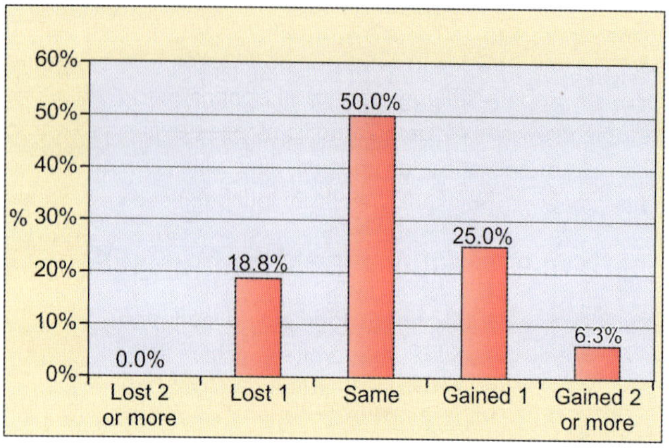

Fig. 3.7

binocular patient to produce central steepening of the cornea. The results of this procedure have been reported by Holzer and others.[8] Our results were similar with near vision improving and preserving as much distance vision as possible. The results in 44 eyes at 6 months after Intracor were reported at the Geneva Presbymania 2011 meeting.[9] Most eyes became slightly myopic with a reduction of spherical equivalent from +0.32 (-0.25 to +1.00) to -0.34 (-1.50 to +0.75) and a reduction of UCDVA from 1.01 to 0.97. However there was an improvement of uncorrected near visual acuity in the treated eye of J12 (J8 to J15) to J4 (J2 to J8) with 84 percent of eyes seeing between J2 and J5. 92.6

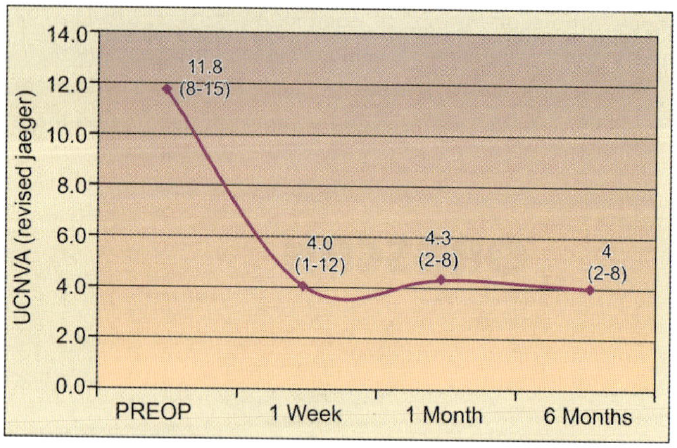

Fig. 3.8 Mean uncorrected near visual acuity

percent of patients were satisfied to extremely satisfied with the result.

OTHER FEMTOSECOND LASER APPLICATIONS

At Ultralase we are implanting presbyopic corneal inlays under femtosecond laser flaps for presbyopia treatments and have cut channels for Intacs implants with the femtosecond laser for the treatment of keratoconus. Other presbyopic inlays are implanted in pockets cut by the femtosecond laser. Femtosecond only refractive surgery is done using the Flex and SMILE techniques. Femtosecond full thickness and lameller corneal grafts are improving the safety and efficacy of keratoplasties. Femtosecond cataract surgery is in its infancy but is already producing consistently accurate capsulorrhexis diameters and reducing the phaco power transmitted into the eye in cataract surgery. Work is being done to increase the flexibility of the presbyopic crystalline lens with femtosecond lasers. And the list of eye surgical applications for femtosecond lasers will continue to grow. We have entered a new and exciting era of ophthalmic surgery with this precision instrument.

REFERENCES

1. H Eleftheriades, J Dermott. Comparison of the results of femtosecond and microkeratome LASIK in over 20,000 eyes, poster presentation. Winter ESCRS Barcelona, 2008.
2. J. Chang JS. Complications of sub-Bowman's keratomileusis with a femtosecond laser in 3,009 eyes. J Refract Surg. 2008;24(1):97-101.
3. J Dermott, Flap complications in femtosecond and microkeratome LASIK in over 25,000 eyes. Free paper presentation ESCRS Berlin, 2008.

4. J Dermott. Outcomes of flap complications with a microkeratome and the femtosecond laser. Free paper presentation ESCRS Barcelona, 2009.
5. M Wevill, J Dermott. Visual outcomes of eyes with Intralase flap complications. Free paper presentation ESCRS Paris, 2010.
6. Güell JL, Vazquez M. Correction of high astigmatism with astigmatic keratotomy combined with laser in situ keratomileusis. J Cataract Refract Surg. 2000;26(7):960-6.
7. M Wevill, J Dermott. Treatment of high astigmatism with femtosecond laser incisions. Presentation at the Femtosecond Laser Users Forum, Paris 2010.
8. Holzer MP, et al. Early outcomes of INTRACOR femtosecond laser treatment for presbyopia. J Refract Surg. 2009;25(10):855-61.
9. M Wevill, J Dermott, Intracor for Presbyopia: Early experience, presentation at Presbymania Geneva 2011.

4

Femtosecond–assisted Keratoplasty Techniques

A John Kanellopoulos (Greece)

Chapter Outline

- Lamellar Keratoplasty
- Keratoprosthesis
- Video KeraKlear

INTRODUCTION

The femto-second laser, originally introduced to replace mechanical microkeratomes for LASIK surgery, soon became a very eloquent scalpel for the cornea, with its ability to split cornea tissues in a very precise manner. It was inevitable that femtosecond lasers would follow that path of eloquent keratoplasty techniques. We have utilized femtosecond lasers in keratoplasty since their first introduction; it was the IEK IntraLase Enabled Keratoplasty software in 2006, with the FS-60 IntraLase femtosecond laser. That particular software enabled the preparation on a donor cornea secured in the artificial chamber, to prepare a lamellar graft of pre-calculated diameter depth, as well as to create a very deep lamellar graft in order to employ, in deep lamellar keratoplasty.

We found interest at that time in utilizing the femtosecond laser in DSEAK surgery, which now, with the use of femtosecond laser is not anymore defined as a keratome-related procedure. We did present the first DSEAK case utilizing a femtosecond laser at the winter ESCRS meeting in 2007, in Athens. A video of this procedure is included in this presentation.

Video: iDSEK

Under this technique, the donor cornea is secured in an artificial chamber and then placed under the femtosecond laser where a predetermined 420 to 450 micron depth and 9 mm diameter of button is prepared in a similar fashion where a lamellar keratoplasty button would be prepared.

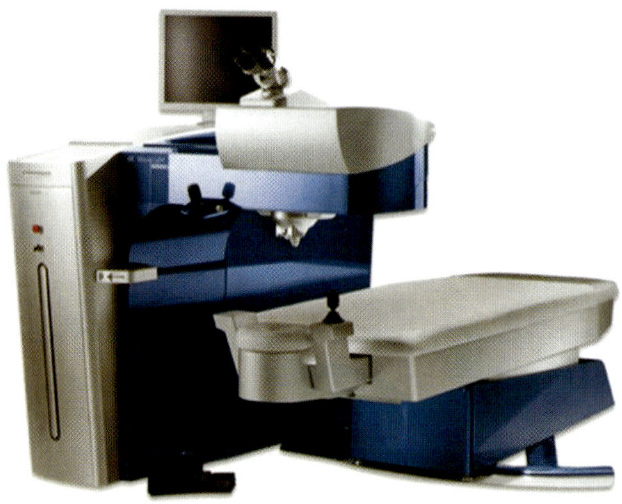

Fig. 4.1

Now, the strategy behind this is to obtain the donor button that will have just the posterior lamellae of stroma, about 100 microns, and obviously Descemet's membrane and the endothelial cells which are placed on a standard donor trephination block and the preferred diameter DSEAK graft is being prepared.

The advantages of the femtosecond laser used in DSEAK surgery is the avoidance of creating a very thick DSEAK graft, as almost invariably the lamellar part of the preparation secures at least 400 micron deep superficial stroma being removed. Other advantages are the fact that when the DSEAK graft is being prepared with microkeratome, it's usually a lenticular graft, much thicker in the periphery than in the center and this may induce more hyperopia in the recipient.

We have presented a poster with a series of cases with relatively good results with this technique:

The AAO poster presentation at the 2009 AAO on a small case series with femto-assisted DSEK:

LAMELLAR KERATOPLASTY

Lamellar keratoplasty is probably the most simple femto-assisted of the ones described herein;

The technique is simple:

Select the diameter, depth for the host and then again for the donor.

It requires a relatively clear host in order for the laser to penetrate and dissect the lamellar part to be removed. It may not require even suturing as the graft may adhere in a fashion similar to a LASIK flap.

KERATOPROSTHESIS

The next category of keratoplasty procedures that may be facilitated with a femtosecond laser is keratoprosthesis surgery. We have employed both in Dohlman (Boston) keratoprosthesis and the KeraClear inlay keratoprosthesis. We have been able to refine the donor vehicle cornea with the femtosecond laser in order to be able to attach to it the Dohlman keratoprosthesis. In a Dohlman keratoprosthesis a 9.5 mm central donor cornea is being prepared and then a central trephination to this graft of 3 mm is performed in order for the keratoprosthesis to go through it. The two parts of the prosthesis get locked anteriorly and posteriorly to the donor graft and the vehicle graft and the prosthesis with the vehicle graft are then transplanted onto the cornea that has a severe external disease, in a fashion similar to a penetrating keratoplasty.

The advantages of using the femtosecond laser here, again, is precision, a predetermined site cut that could be straight, angled or even zigzag in appearance and a video of this will also be displayed in this chapter.

VIDEO KERAKLEAR

Last, penetrating keratoplasty that used to be the most common transplantation procedure is now under 50 percent of procedures performed in the United States. Still, the original hope was that using a femtosecond laser, both on the donor and the host corneas, we would be able to reduce the amount of post-operative astigmatism, that is one of the major problems in postoperative rehabilitation after penetrating keratoplasty.

The initial experience with penetrating keratoplasty is relatively promising. As far as getting a very good opposition of donor and host tissue, the most popular technique is the zigzag technique both on the donor tissue and on the host tissue, which enables a larger area of contact between graft-host interface and thus the possibility of a stronger adherence of the graft to the host cornea. Nevertheless, post-operative astigmatism appears to be not significantly reduced with the use of the femtosecond laser. More techniques and proce-

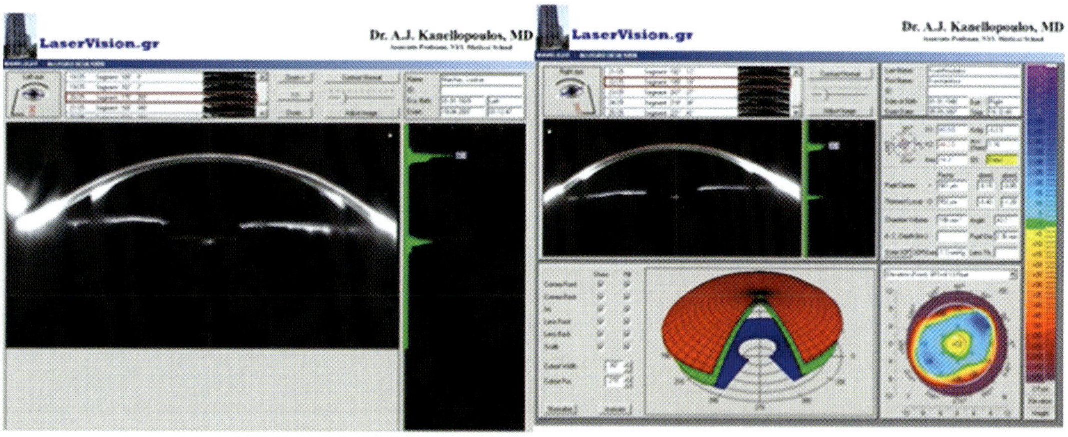

Fig. 4.2 A microkeratome DSEAK case on the left and an Intralase assisted DSEK case on the right studied with the pentacam (thicker edges in the left case)

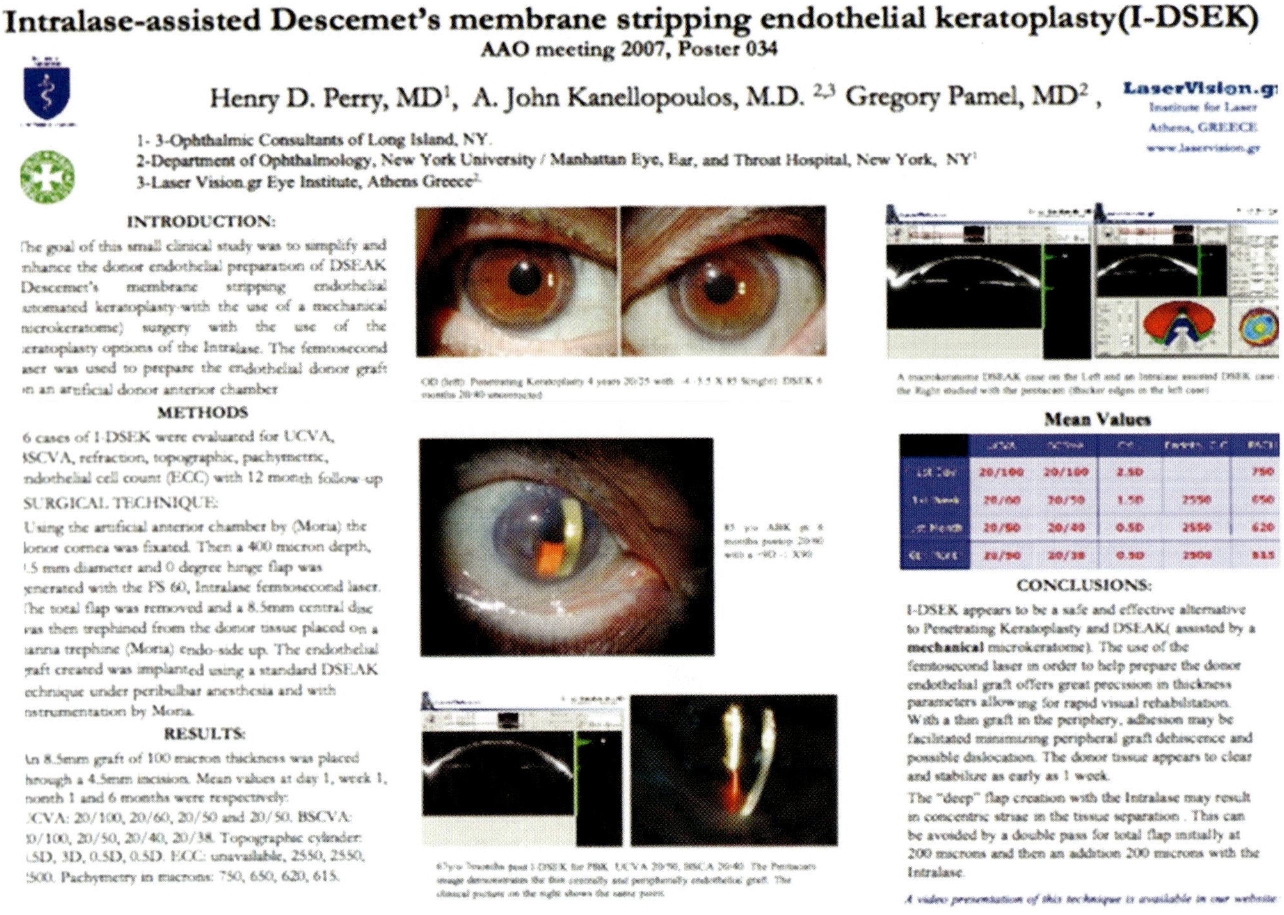

Fig. 4.3

dures are being introduced into facilitating penetrating keratoplasty with the use of femtosecond laser.

We are currently using the WaveLight FS200 femtosecond laser which has the ability of performing any type of side-cut incisions to a depth of 800 microns and with a maximum diameter of 10 mm, thus being able to perform a completely bladeless penetrating keratoplasty. Examples of this will be shown in the video as well.

BIBLIOGRAPHY

1. [Advances in penetrating keratoplasty technique.] Wei AJ, Hong JX, Xu JJ. Zhonghua Yan Ke Za Zhi. 2010 Jan;46(1):85-7. Chinese. PMID: 20388326 [PubMed - in process]
2. Accidental Macular Injury from Prolonged Viewing of a Plasma Flash Produced by a Femtosecond Laser. Yang X, Jiang F, Song Y, Peng C, Sheng S, Li X. Ophthalmology. 2010 Mar 2. [Epub ahead of print] PMID: 20202686 [PubMed - as supplied by publisher]
3. Corneal Aberrations and Visual Acuity After Laser In Situ Keratomileusis: Femtosecond Laser Versus Mechanical Microkeratome. Calvo R, McLaren JW, Hodge DO, Bourne WM, Patel SV. Am J Ophthalmol. 2010 Mar 12. [Epub ahead of print] PMID: 20227675 [PubMed - as supplied by publisher]
4. Femtosecond keratotomy. Goggin M. Ophthalmology. 2010 Apr;117(4):847-8; author reply 848. No abstract available. PMID: 20346822 [PubMed - indexed for MEDLINE]
5. Femtosecond LASIK combined with astigmatic keratotomy for the correction of refractive errors after penetrating keratoplasty. Bahar I, Kaiserman I, Mashor RS, McAllum P, Slomovic A, Rootman D. Ophthalmic Surg Lasers Imaging. 2010 Mar-Apr;41(2):242-9. doi: 10.3928/15428877-20100303-14. PMID: 20307044 [PubMed - in process]
6. Imaging of the lens capsule with an ultrahigh-resolution spectral optical coherence tomography prototype based on a femtosecond laser. Kaluzny BJ, Gora M, Karnowski K, Grulkowski I, Kowalczyk A, Wojtkowski M. Br J Ophthalmol. 2010 Mar;94(3):275-7. PMID: 20215371 [PubMed - in process]
7. IntraLase-Enabled Astigmatic Keratotomy for Post-Keratoplasty Astigmatism: On-Axis Vector Analysis. Kumar NL, Kaiserman I, Shehadeh-Mashor R, Sansanayudh W, Ritenour R,

Rootman DS. Ophthalmology. 2010 Feb 15. [Epub ahead of print] PMID: 20163860 [PubMed - as supplied by publisher]

8. Keratitis after implantation of intrastromal corneal ring segments (Intacs) aided by femtosecond laser for keratoconus correction: case report and description of the literature. Levy J, Lifshit T. Eur J Ophthalmol. 2010 Jan 25. pii: 540A63D2-DD1F-494D-8B62-1C25DE6BAD41. [Epub ahead of print]
9. LASIK flap creation with the Ziemer femtosecond laser in 787 consecutive eyes. Pietilä J, Huhtala A, Jääskeläinen M, Jylli J, Mäkinen P, Uusitalo H. J Refract Surg. 2010 Jan;26(1):7-16. doi: 10.3928/1081597X-20101215-02. Epub 2010 Jan 11. PMID: 20199007 [PubMed - indexed for MEDLINE]
10. Non-invasive bleaching of the human lens by femtosecond laser photolysis. Kessel L, Eskildsen L, van der Poel M, Larsen M. PLoS One. 2010 Mar 16;5(3):e9711. PMID: 20300521 [PubMed - in process]Free PMC ArticleFree text
11. The effect of femtosecond laser lamellar dissection at various depths on corneal endothelium in the recipient bed of the porcine eye. Choi SK, Kim JH, Lee D. Ophthalmic Surg Lasers Imaging. 2010 Mar-Apr;41(2):255-60. doi: 10.3928/15428877-20100303-16. PMID: 20307046 [PubMed - in process]

5

Femtosecond Lasers in Ophthalmology

Mahipal S Sachdev, Charu Khurana (India)

Chapter Outline

- Steps of Femto-cataract/Blade-free Cataract Surgery

We stand at the threshold of a revolution that will possibly completely change the face of Ophthalmology, a revolution heralded, guided and fired by the FEMTOSECOND LASER. With its ever-expanding repertoire of uses, this laser will soon be used in almost every aspect of ophthalmic science. From its beginning in LASIK, the femtosecond laser has gone on to Intra-Corneal Ring Segments and keratoplasties, both full thickness and partial thickness. With multiple research teams having successfully used it for various corneal incisions, the femtosecond laser is now going to treat cataracts...the very core of ophthalmology. From capsulotomy to nuclear softening and chopping as well as creation of side ports and corneal incisions, everything is now possible with the femtosecond laser.

A femtosecond laser uses an infrared beam of light to precisely separate tissue through a process called photo-disruption by generating pulses as short as one-quadrillionth of a second (10^{-15} = femtosecond). It has a wavelength of 1053 nm and is based upon the technology whereby focused laser pulses divide material at the molecular level without transfer of heat or impact to the surrounding tissue (Figs 5.1 and 5.2).

The most popular use of the femtosecond laser lies in the creation of a corneal flap in a lasik procedure eliminating the use and risk of a micro-keratome and blade and increasing the overall safety, precision and accuracy. The laser beam is focused on a pre-programmed depth and position within the cornea with each pulse forming a microscopic bubble. As the laser moves painlessly back and forth, the bubbles connect to form a flap with no trauma to adjacent tissue, the entire process taking around 10-20 seconds. The surgeon then lifts the flap to allow treatment by excimer laser (Figs 5.3 to 5.14). Laser specifications which can be modified to meet individual patient's needs include flap diameter, depth, hinge location and width and side-cut architecture. The laser also creates

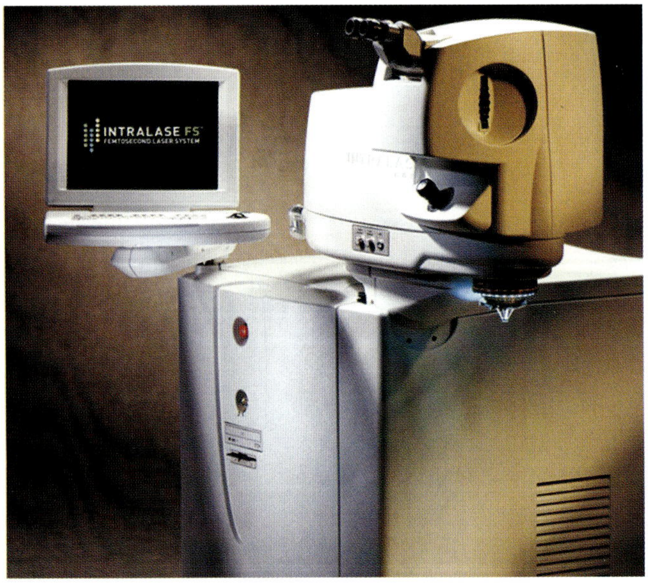

Fig. 5.1 The IntraLase femtosecond laser

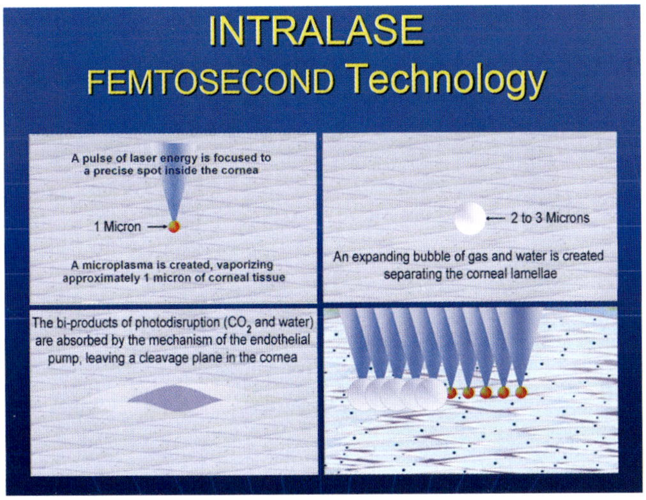

Fig. 5.2 Separation of corneal lamellae by focused laser energy

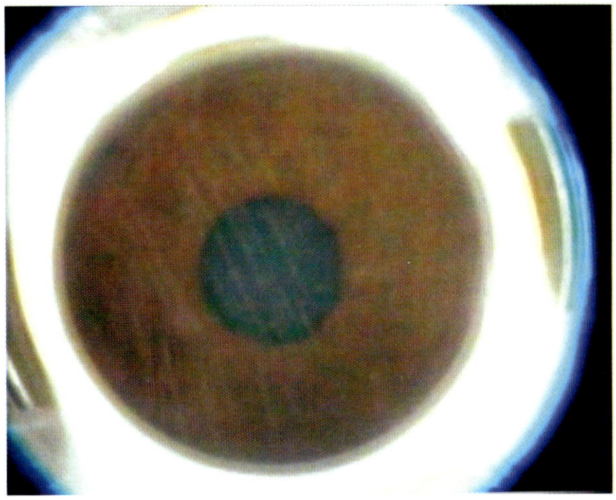

Fig. 5.3 Applanation of the corneal surface

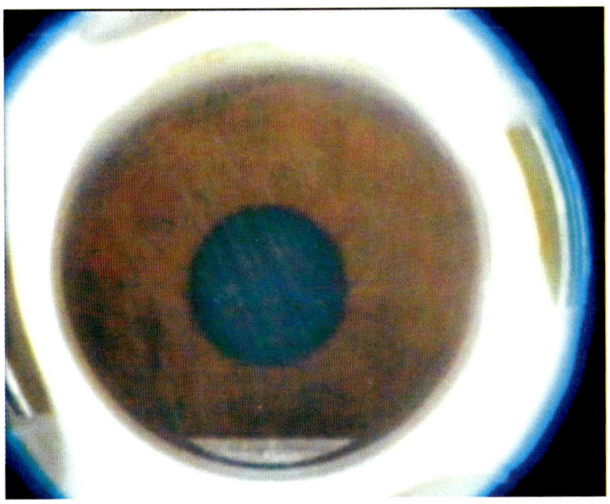

Fig. 5.4 Creation of laser pocket at a deeper depth

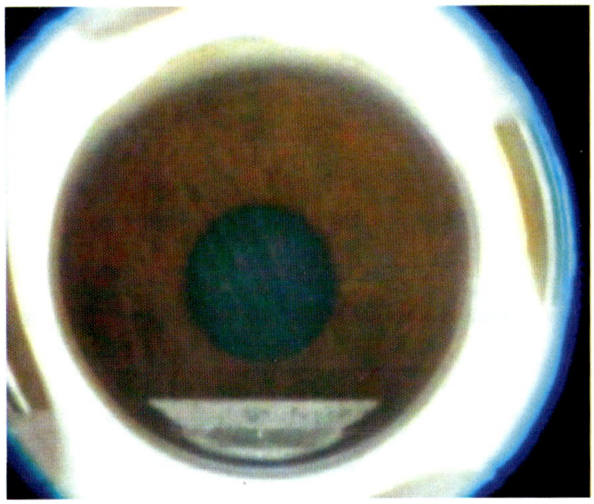

Fig. 5.5 Progression of the femtolaser procedure

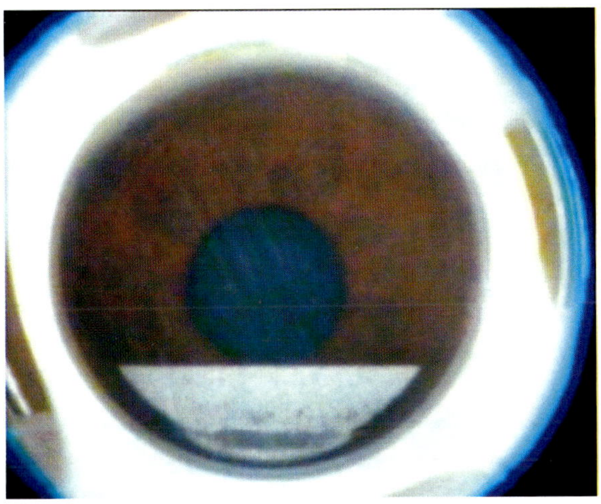

Fig. 5.6 Laser beam traveling forward creating a tissue plane

Fig. 5.7 Femtolaser assisted flap more than 50 percent complete

Fig. 5.8 Femtolaser assisted flap almost 80 percent complete

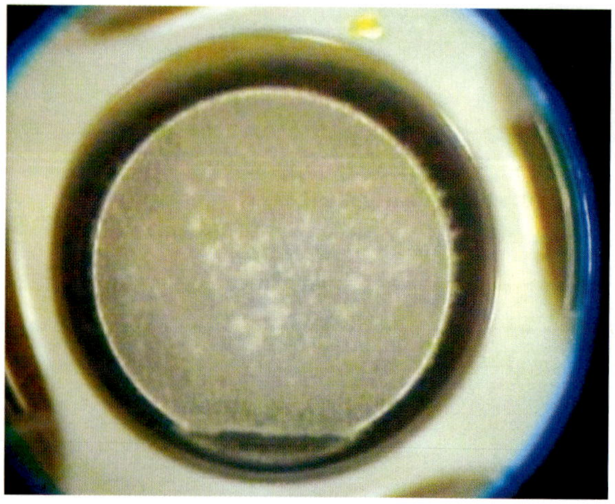

Fig. 5.9 Flap complete, side cut being delivered

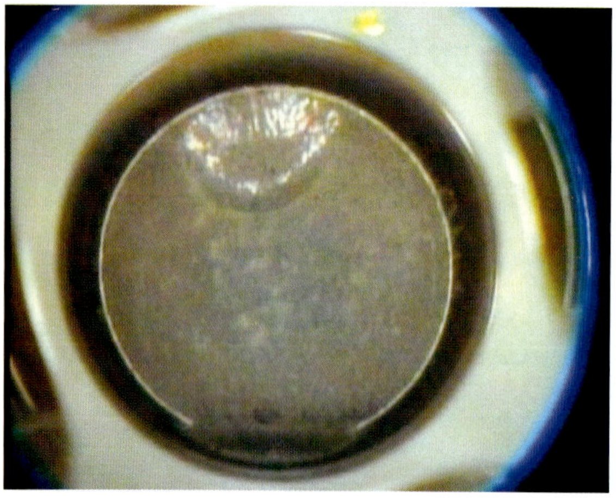

Fig. 5.10 'Burp' where a condensed water bubble appears at the cleavage plane

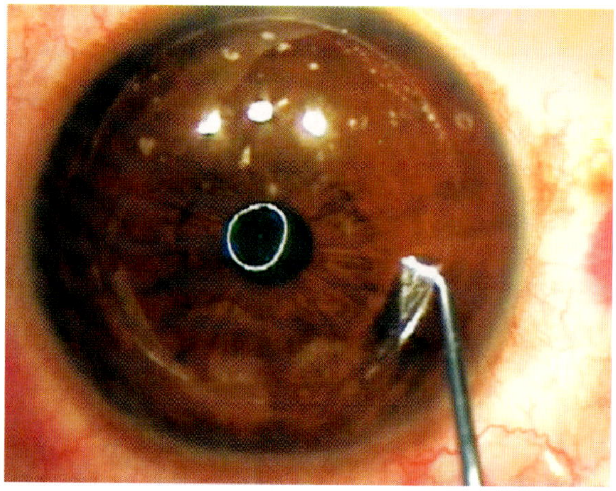

Fig. 5.11 Lifting the flap by delineating the edge

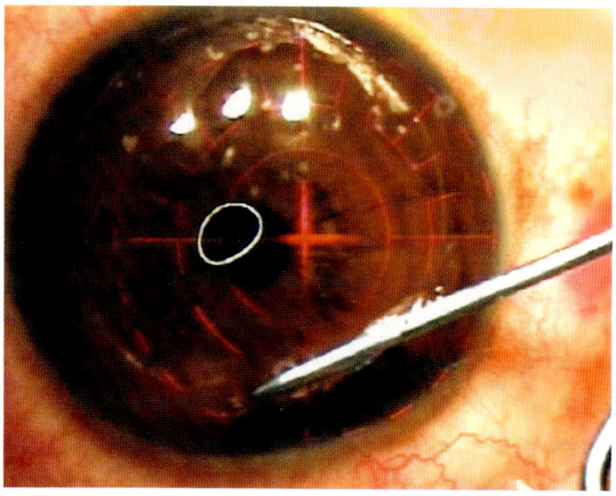

Fig. 5.12 Flap lifted up from the stromal bed

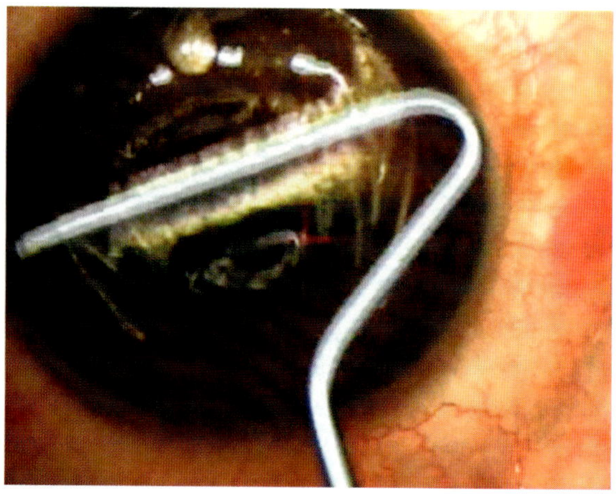

Fig. 5.13 Flap completely separated

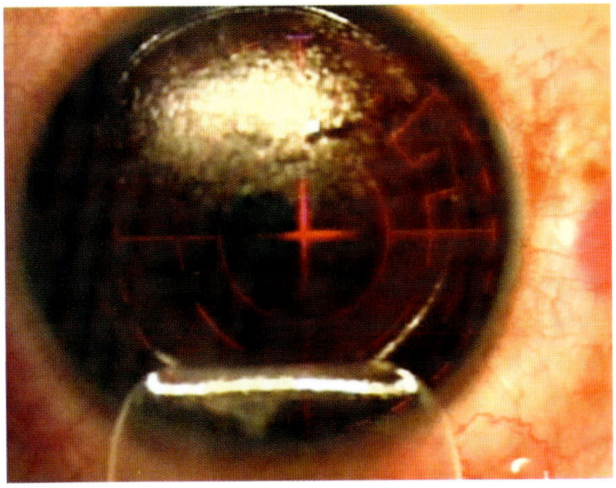

Fig. 5.14 Excimer ablation being performed on the stromal bed

a distinctive beveled edge flap which allows for precise repositioning and alignment after Lasik is completed.

The precision of the femtosecond laser helps to create flaps of exact size, shape and depth and markedly reduces the risk of blade-related flap complications such as free caps, button holes, incomplete or decentered flaps. Not only is the visual acuity better but the incidence of post-op dry eye symptoms is reduced. It also creates fewer high and low-order aberrations which may cause glare and haloes at night.

The femtosecond laser has also revolutionized other forms of corneal surgery such as corneal transplantation and Intrastromal ring implantation. The femtosecond laser enables surgeons to create straight, angled and arcuate incisions which allow faster healing and improved visual recovery in penetrating keratoplasty. The mushroom-shaped incision preserves more host endothelium than the traditional trephine approach while the top-hat incision allows for transplantation of large endothelial surfaces. The zig-zag incision approach provides a smooth transition between host and donor tissue and allows for a hermetic wound seal. Intralase enabled keratoplasty (IEK) establishes secure grafts requiring less suture tension and reduced incidence of astigmatism leading to a faster and better visual recovery.

The Femtosecond laser is also now being used for insertion of intra-corneal ring segments (ICRS) in the treatment of keratoconus and corneal ectasia especially those occurring post lasik. Instead of using a mechanical spreader, the femtosecond laser is used to create channels at a pre-determined depth with a high degree of accuracy. The laser parameters which need to be specified include channel depth, entry incision length and width and inner and outer diameters of the channels. The narrower the channel the greater the effect, however a narrower channel also results in increased difficulty in inserting ICRS. A compromise between maximal outcome and ease of insertion is achieved. The entry wound and channel creation takes 8-10 seconds after which the entry wound is opened with a Sinskey hook and the ring segment is carefully pushed forward into the channel till the edge lies within 1 mm from the entry wound (Figs 5.15 to 5.17). Various studies have reported better results with femtosecond laser as compared to mechanical group with a significant reduction in average keratometry, spherical equivalent refraction and surface asymmetry index and improvement in BSCVA, UCVA and spectacle and contact lens tolerance.

Continued innovation in the femtosecond laser technology has led to the introduction of the newer generations of femtosecond lasers which have a higher frequency giving the surgeon greater precision, control and safety and reducing the time spent during the laser procedure. Additional features include variation in the flap shape and diameters to allow for elliptical flaps for better results in hyperopic patients. Another facet is the change in angulation of the flap edge to achieve a better flap apposition and more secure flap healing with greater stability. The other new innovation is the 'all femtosecond laser keratomileusis' which eliminates the excimer laser completely and performs the entire procedure using only the femtosecond laser: the FLEX and SMILE technologies.

The latest breakthrough in the use of femtosecond technology is introduction of 'laser cataract surgery' wherein the femtosecond laser is able to create all required surgical incisions with perfect dimensions and architecture, a perfectly centered and sized capsulotomy and help to liquefy and chop the lens. Along with this, precise corneal incisions can be created to treat any pre-existing astigmatism resulting in near perfect outcomes.

The US FDA cleared femto-second lasers for cataract surgery in 2010 and 4 companies OptiMedica, LensAR, LenSx and Bausch and Lomb are developing this technology for widespread use. The clinical trials of femto-second cataract surgery in India have been completed and this laser has now become commercially available in India.

The laser causes 43 percent reduction of phaco-emulsification energy and 51 percent reduction in time which improves the safety profile. In addition the well centered, precise continuous capsulorhexis done by the laser is an important consideration for premium IOLs and when combined with limbal relaxing incisions provide a real refractive cataract surgery experience.

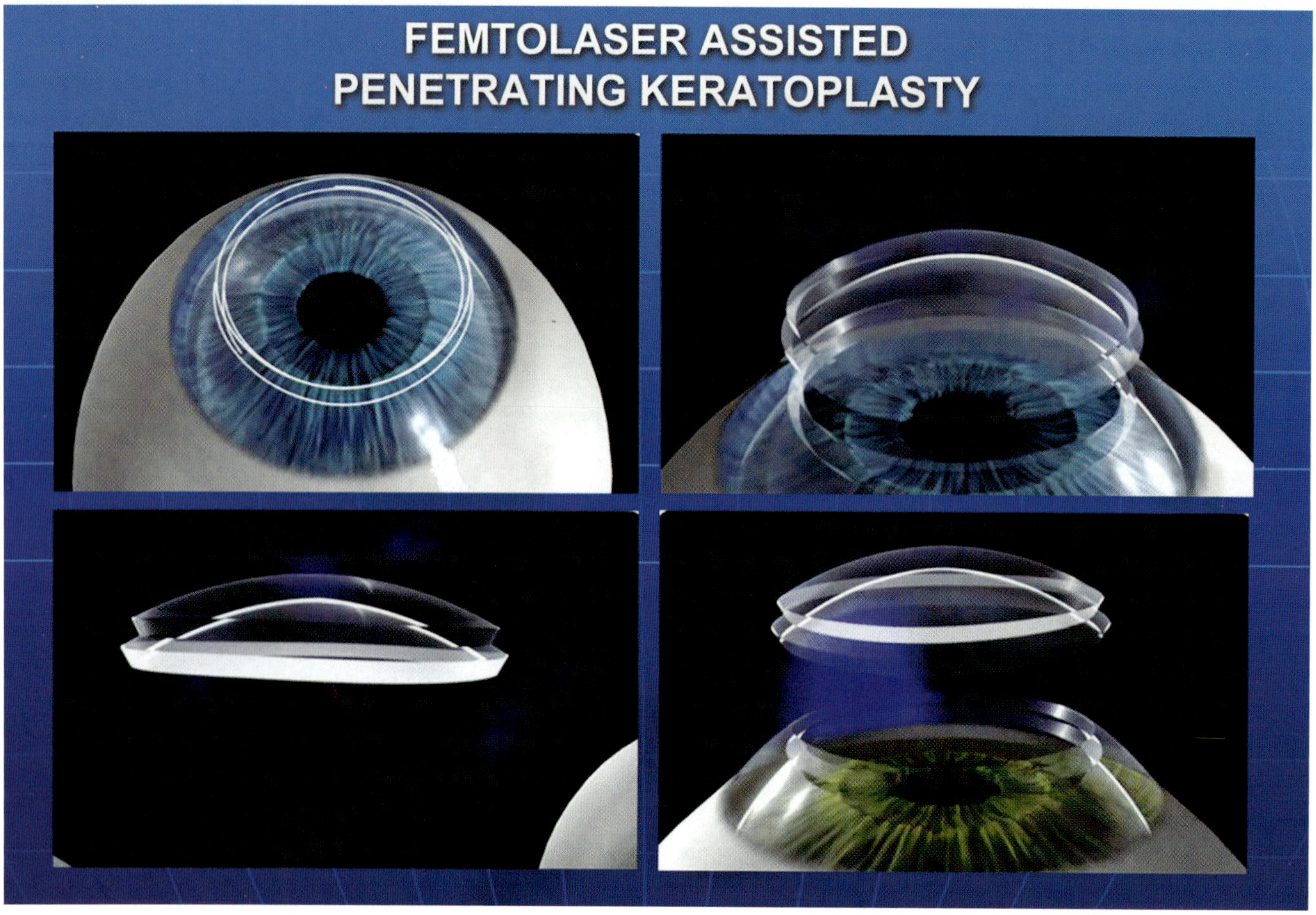

Fig. 5.15 Femtolaser assisted penetrating keratoplasty

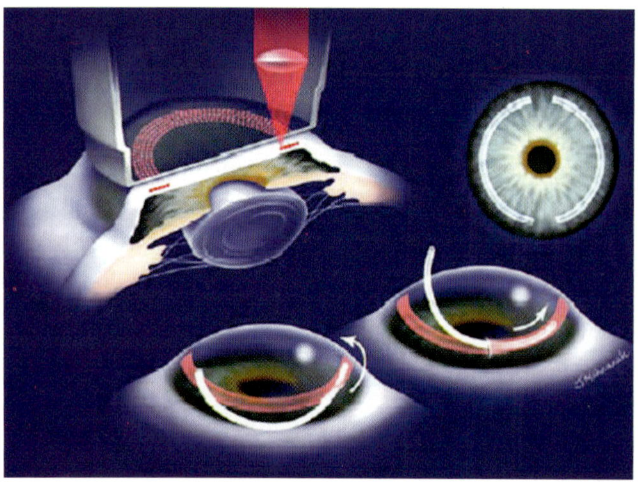

Fig. 5.16 Intra-stromal ring implants (ICRS) placement using the femtosecond laser

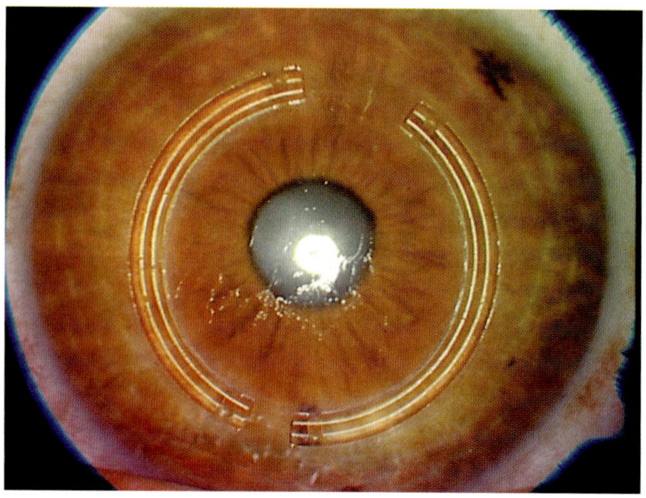

Fig. 5.17 Intra-stromal ring implants (postoperative)

STEPS OF FEMTO-CATARACT/ BLADE-FREE CATARACT SURGERY

1. Using anesthetic eye drops, the eye is numbed.
2. Wire speculum is placed.
3. Cornea is applanated with a disposable patient interface cone.
4. A high resolution video-microscope combined with a real time OCT captures images of the anterior segment of the patient's eye (Fig. 5.18).
5. The images are analyzed and each step is planned and customized according to the patients, requirement and surgeons' preference.
6. Size, shape and location of the incisions is mapped and marked.
7. The laser foot pedal is depressed to begin the laser procedure which includes the following:
   - Capsulorhexis is created at a depth of 500 microns in the anterior capsule
   - Lens fragmentation incisions are performed staying 1000 microns above the posterior capsule
   - Primary and secondary corneal incisions are carried out
   - Arcuate incisions are created to tackle any co-existing astigmatism
8. The patient is then wheeled under an operating microscope and the surgery completed with removal of the capsulorhexis flap, aspiration of the lens fragments and cortical clean-up. The IOL is implanted and the procedure completed.

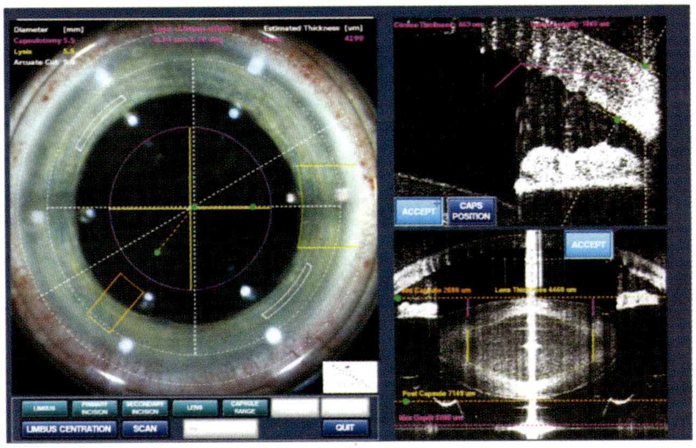

Fig. 5.18 The integrated OCT in a femtosecond cataract laser projects a real time image of the cornea, lens, iris and capsule on a video-microscope. The surgeon selects the corneal incisions and lens treatment and overlays them on the live OCT image

Various studies have shown that Optical coherence tomography (OCT)-guided nuclear fragmentation improves refractive results, reduces the risk of complications such as posterior capsule rupture, endophthalmitis and endothelial cell density loss. The first clinical results show that femtosecond laser incisions achieved perfect diameter accuracy in 100 percent of cases, while only 10 percent of surgeons' manual cuts were within 0.25 mm error. With more than 6,000 procedures performed world wide, this is definitely the cataract surgery of the future. The complication reported so far is Capsular Blockage Syndrome (CBS) or Femtocataract Intracapsular Gas Syndrome in which accumulation of fluid after hydrodissection in the capsular bag led to capsular rupture with nucleus drop. Advanced grade of cataract was the common risk factor along with old age. While only 2 cases have been published so far, further experience will help to tackle this problem in femtocataract cases.

Apart from this femtosecond lasers are being evaluated for correction of presbyopia and may be used not only for creation of concentric corneal rings to change the corneal curvature but also for modifying the lens to improve accommodation. Creation of permanent drainage channels for glaucoma surgery, retinal imaging and photo-disruption and treatment of corneal neo-vascularization are the other vistas of application of femtosecond laser which are undergoing research and evaluation.

T he turnover rate for both medical knowledge and technology is about 50 percent every 4 to 5 years. Femtosecond lasers maybe the one technological breakthrough which may change the course of ophthalmic care forever. Let us stay aware and willing to understand and adapt to the changing face of science and technology and grow into the future wiser and smarter.

6

Corneal Crosslinking Treatment and Femtosecond Laser Pocket

Aylin Kilic (Turkey), A John Kanellopoulos (Greece)

Chapter Outline

- Video: ICXL KCN

Riboflavin 0.1 percent UVA CXL technique has traditionally been introduced with administration of riboflavin from the ocular surface.

It typically (according to the initial Dresden protocol) requires the removal of the epithelium and saturation of the corneal stroma with riboflavin large molecule, which diffuses through the cornea into the anterior chamber. The riboflavin molecule collects along side the collagen to be crosslinked and is photosensitized by the exposed UVA light in the presence of oxygen to create a reactive singlet oxygen species that induces collagen crosslinking.[1]

Kanellopoulos introduced at the ISRS/AAO meeting in New Orleans in 2007, the administration of riboflavin directly within the stroma via a femtosecond laser created pocket.

The thought process in this alternative technique, was that the creation of a femtosecond laser pocket with a small 10 degree side cut, would not weaken the cornea biomechanically since it is shown that the side cut may induce cornea weakening. Further the selection of the exact depth that the riboflavin would be introduced would make the CXL procedure more customised instead of applying UV light to the whole cornea and potentially the cornea endothelium. Last the absence of riboflavin from the surface of the cornea would facilitate the permeability to UV light through the intact epithelium, Bowman's and to the part of the stroma that has been selectively soaked with riboflavin and possibly attaining more effective CXL.

The downside of intracorneal administration would be the time limitation of substantial riboflavin present in the cornea after a single application. This was calculated to be about 10 minutes. Thus the employment of higher fluence UV light was introduced by Kanellopoulos along with this technique.

The initial cases treated were early progressive keratoconus with cornea thickness over 500 microns. A 100 micron pocket was created with the FS60 Intralase femtolaser, of 7 mm diameter and a 10 degree side cut placed at 12 o'clock.

After instilling 0.1 ml of 0.1 percent riboflavin sodium phosphate solution the yellow riboflavin color was visualized within the cornea stroma.

Immedietely after the conrea was exposed to 10 mW/cm^2 UV light fluence for 10 minutes.

VIDEO: ICXL KCN

The clinical results published in 2009 at the JRS showed comparable stability and cone flattening to the standard Dresden protocol. The advantages were significant; no postoperative pain and next day ability to full activities involving vision.

At the same year's IROC CXL meeting in Zurich (2007) a similar technique of 2 consecutive pockets at 350 microns and then 150 microns was introduced by Kanellopoulos as a palliative measure for symtomatic bullous keratopathy. The main concept here was to CXL as much as stroma as possible in order to acchive stroma deturgecence and possible reduction of cornea edema in bullous keratopathy. The procedure appeared to be effective in temporizing these patients for 6-12 months and possibly improving anterior chamber visualization for the performance of a DSEAK procedure video ICXLpbkV12).

This was work of staging the intrastromal riboflavin administration in bullous keratopathy. This was published by Kruger, EE and Kanellopoulos.[2] The benefit of this staging process is that delivery of riboflavin does not require the removal of epithelium (Fig. 6.1).

The benefit of this staged CXL technique in reducing the corneal thickness of our experimental bullous keratopathy model was shown to be effective in the 5 treated corneas which underwent statistically significant reductions of 256 microns (P:0.0002) in ultrasound central cornea thickness.

It was theorialized[3] that by instilling the riboflavin solution paracentrally within the cornea, with minimal epithelial surface disturbance, and by using higher fluence UVA light source the following could be targeted:

1. Faster re-epithelization and visual rehabilitation.
2. Faster diffusion/soak period as the riboflavin is directly injected at depth via the intrastromal pocket, providing greater shielding near the endothelium.
3. Selective CXL of the anterior two-thirds of the cornea.

The intrastromal pocket CXL for a small series of keratoconus cases was subsequently published by Kanellopoulos in 2009.[3] As noted above Femtosecond laser was used to create an intrastromal pocket with a nasal hinge at 100 microns corneal depth, 7 mm diameter centered at the pupil, and a 10 degree temporal side-cut. A 0.1 ml dose of 0.1 percent riboflavin solution was administered twice with a 25 gauge cannula. In this study, no adverse effects were noted in any of cases studied (Figs 6.2 and 6.3).

This novel epithelium sparing, rapid soak-and-treat method of intrastromal riboflavin instillation and higher fluence UVA light for CXL appeared to be clinically safe and effective. No negative effect (ectasia, epithelial ingrowth) was noted due to femtosecond laser-created pocket. Because minimal epithelial injury occurs using this technique, postoperative pain appears to be significantly reduced.

The currrent intrastromal CXL technique involves the FS200 Wavelight femtosecond laser (video: ICXLFS200). The unique characteristic of this femtolaser is that the venting "chimmney" created before the pocket serves as the passage for the pocket preparation and instillation of the riboflavin, thus avoiding the older side cut at the pocket perimeter as shown with the Intralase femtolaser. The unique advantage of this simple difference is the reduction of inadvertent tear produced with the older pockets that enlarged the side cut width and potentially produced a larger biomechanical change in these compromised corneas.

Intrastromal Ring and Crosslinking Combined Treatment with Riboflavin Injection Into the Stromal Femtolaser Pocket

Corneal crosslinking can stabilize the cornea biomechanical properties and prevent progression in keratoconus. To date, epithelial debridement has been described as a crucial step in the CXL technique prior to riboflavin application and UVA exposure. However, epithelial removal for riboflavin application results in postoperative pain, which makes the procedure unpleasant for patient. To avoid this disadvantage, a new technique for riboflavin application without epithelial removal was recently proposed by Kilic.[4]

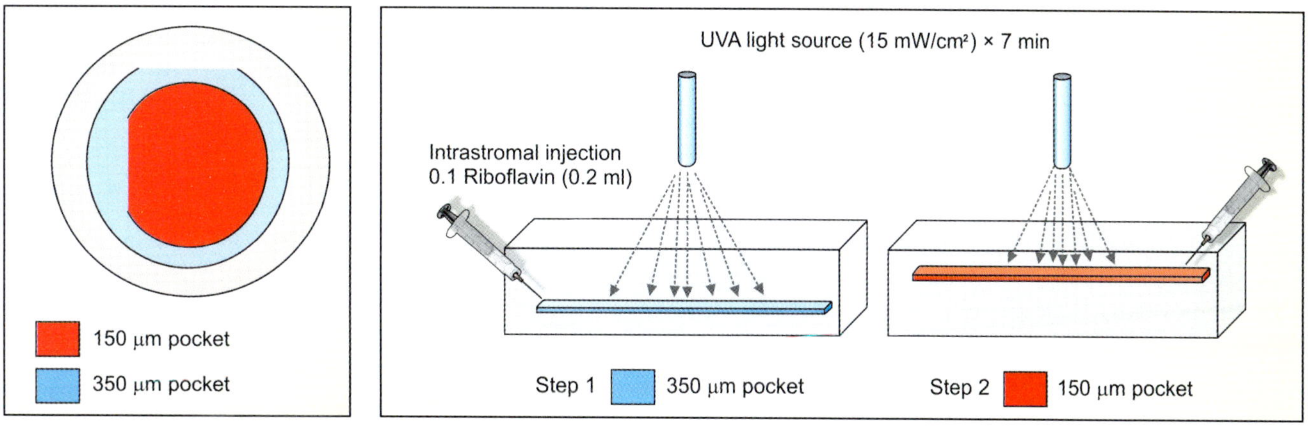

Fig. 6.1 Intrastromal pocket for bullous keratopathy treatment (Kruger RR, Ramos-Esteban JC, Kanellopoulos AJ. Staged intrastromal delivery of riboflavin with UVA cross-linking in advanced bullous keratopathy: laboratory investigation and first clinical case. J Refract Surg 2008; 24:730-6.)

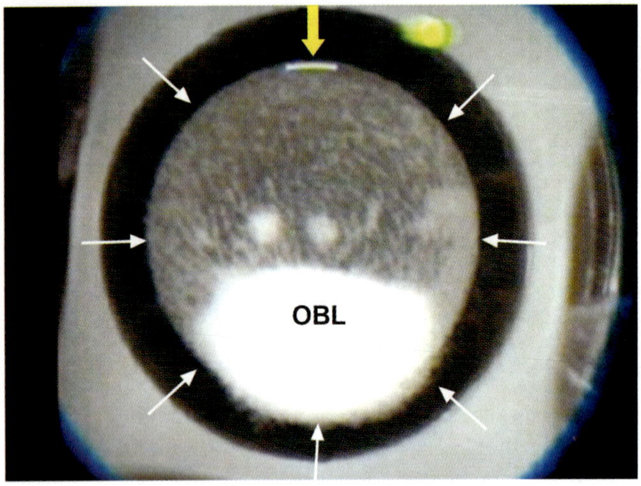

Fig. 6.2 Clinical photograph demonstrating the intracorneal "pocket" created with the Intralase FS60 femtosecond laser (white arrows). Lamellar dissection within the cornea was planned at 100 μm depth, in a 7 mm diameter centered on the pupil. The side-cut was 10° which can be seen at the top of the picture (yellow arrow), depicting the actual vertical separation that will make it possible to enter the pocket from the corneal surface. (OBL = Opaque bubble layer)

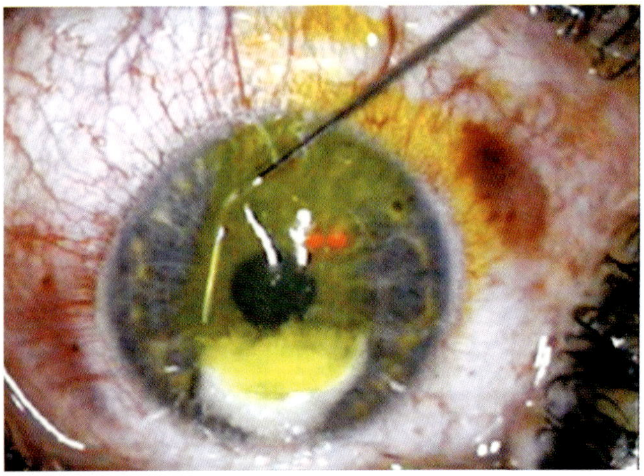

Fig. 6.3 Clinical photograph demonstrating the infusion of 0.1 ml of 0.1% riboflavin solution within the femtosecond laser created intrastromal pocket to facilitate intrastromal cross-linking

An intrastromal pocket was created by means of femtosecond laser system at a corneal depth ranging from 70 to 90 microns with a 70 degree side cut and 7 mm diameter in eyes with previously implanted corneal ring segment. Riboflavin 0.1 percent in dextrose was injected directly into the corneal pocket. Significant corneal haze was observed in all cases in early period, and resolved over time. No postoperative pain was observed reported in any case operated with the pocket technique.[4] In this study all patients enrolled had undergone previous corneal surgery (intracorneal ring segment), recently we reported noncomparative retrospective case series 131 eyes of 105 keratoconic patients, underwent combined CXL and intracorneal ring segment (ICRS) implantation treatment. All eyes had combined surgery ICRS and simultaneous transepithelial CXL with 20 percent alcohol application and riboflavin injection into the corneal channel.[5]

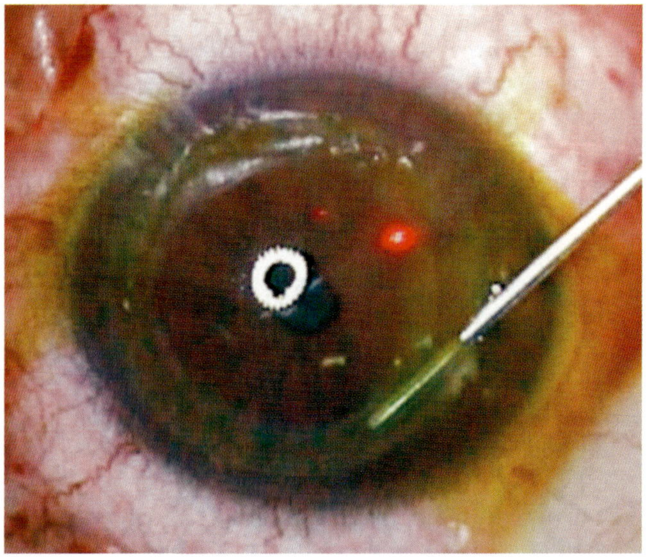

Fig. 6.4 Intrastromal riboflavin injection into the corneal channel

Intacs SK (Addition technology Information, AAO, Las Vegas) segments were inserted to 80 percent depth of cornea in all eyes using Intralase (60 Hz, Intralase Corporation, Irvine CA). The pulse duration was 600 femtoseconds, with the inner to outer diameter of the Intacs tunnel set from 6.0 mm to 7.3 mm. Spot size was 1 micron and the energy was 1.5 microjoules. 20 percent alcohol was applicated with 8.5 mm trephine (same procedure with LASEK) to disturb epithelial tight-junctions for transepithelial CXL to increase riboflavin penetration. Death epithelium without epithelial tight junctions was confirmed by using confocal microscopy after alcohol. Riboflavin solution was injected with 25 G cannula into the corneal femtochannel until riboflavin pass completely 360 degree and fill all channel size (Fig. 6.3). Then, immediately Intacs segments were implanted inferiorly and superiorly based on patients' preoperative spherical equivalent and the location of the cone.

Combined Intacs and CXL treatment is effective in keratoconic eyes with intracorneal riboflavin injection (Fig. 6.4). Intracorneal riboflavin injection into the tunnel is safe, may provide more penetration without remove epithelium.[5]

Alio et al[4] compared the clinical and confocal microscopic outcomes achieved with two different procedures for CXL in eyes with keratoconus: CXL with epithelial debridement and transepithelial CXL with intrastromal pocket. They reported CXL with intrastromal riboflavin injection seemed to be as effective for corneal and refractive changes as classic CXL.

Further studies and experience will provide additional data on the safety and efficacy of femto-assisted CXL.

A natural evolution of these techniques is the prophylactic use of CXL in LASIK cases.

It was introduced at the AAO meeting in 2009 by Kanellopoulos, as a means of reinforcing lost biomechanical stability even in normal routine LASIK cases. It has recently been refined with the introduction of the KXL CXL device from Avedro (Waltham, MA, USA). In this technique at the completion of a routine femto-LASIK case a drop of riboflavin solution is placed onto the bear stroma surface for 60 seconds. The solution rapidly diffuses into the underlying stroma. Then the flap is repositioned and no further riboflavin is used. Immediately after the cornea is exposed to 30 mW/cm^2 fluence for 90 seconds. This enables UV light to go through the flap and CXL the flap to the underlying stroma as well as reinforce the residual stroma bed biomechanically. The technique has become extremely popular in Greece, and Japan and has recently (Jan 2012) been CE marked for the European union. Kanellopoulos et al will be presenting at the 2012 ASCRS meeting in Chicago data on LASIK Xtra in hyperopes. It appears to provide higher stability and less regreassion pressumed by halting the biomechanical shift intrinsic to heperopic LASIK that may result in regression of the effect.

REFERENCES

1. Wollensak G, Spoerl E, Seiler T. Riboflavin/ultraviolet-a-induced collagen crosslinking for the treatment of keratoconus. Am J Ophthalmol 2003;135:620-7.
2. Kruger RR, Ramos-Esteban JC, Kanellopoulos AJ. Staged intrastromal delivery of riboflavin with UVA cross-linking in advanced bullous keratopathy: laboratory investigation and first clinical case. J Refract Surg 2008;24:730-6.
3. Kanellopoulos AJ. Collagen cross-linking in early keratoconus with riboflavin in a femtosecond laser-created pocket: initial clinical results. J Refract Surg 2009;25:1034-7.
4. Alio JL, Toffaha BT, Pinero DP, Klonowski P, Javaloy J. cross-linking in progressive keratoconus using an epithelial debridement or intrastromal pocket technique after previous corneal ring segment implantation. J Refract Surg 2011;27(10):737-43.
5. Kiliç A, Kamburoglu G, Akinci A. Riboflavin injection into the corneal channel for crosslinking and intacs combined treatment. J Cataract Refract Surg, 2012 (in press).

7

Corneal Inlays for Presbyopia and Femtosecond Laser Implantation

Francisco Sánchez León (México)

Chapter Outline

- Synthetic Keratophakia History
- Features and Surgical Technique
- AcuFocus Inlay Features
- Surgical IntraLase Surgical Technique Step-by-Step
- Invue Inlay Features and Surgical Technique
- The Presbylens™ Corneal Inlay: An Adjustable Device for Presbyopia
- Results

In attention to quality-of-life issues among aging baby boomers grows, there is an increased interest in the development of new presbyopia treatments. Intracorneal implants or corneal inlays, which have been on the backburner in ophthalmology for several years, are quickly becoming a safe and reversible treatment.

Since last decade until today, three inlays have been tested for clinical trials. Acufocus, Biovision and Presbylens. The three of them show several differences (Figs 7.1A to C).

- They are made of different materials
- Different mechanism of action to accomplish its refractive purpose

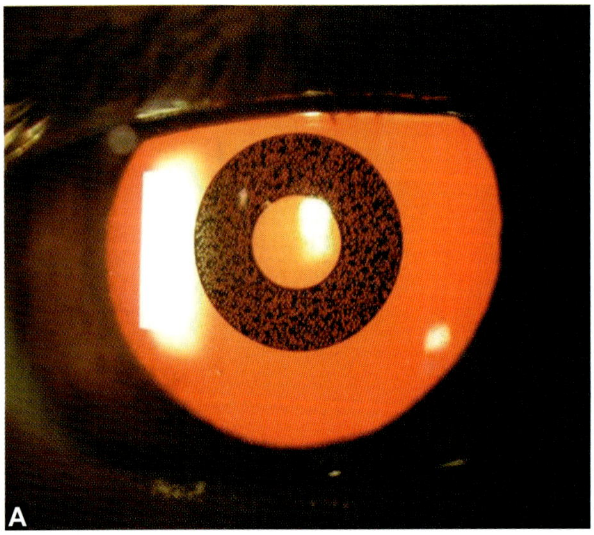

AcuFocus slit-lamp photopharmacologically dilated pupil

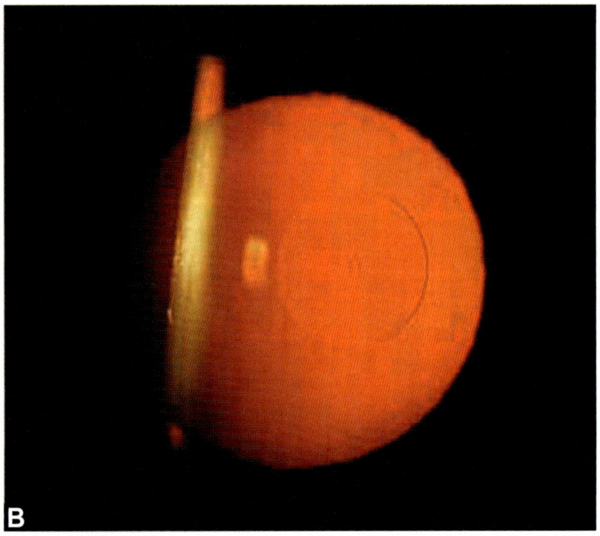

Biovision slit-lamp photopharmacologically dilated pupil

Figs 7.1A and B

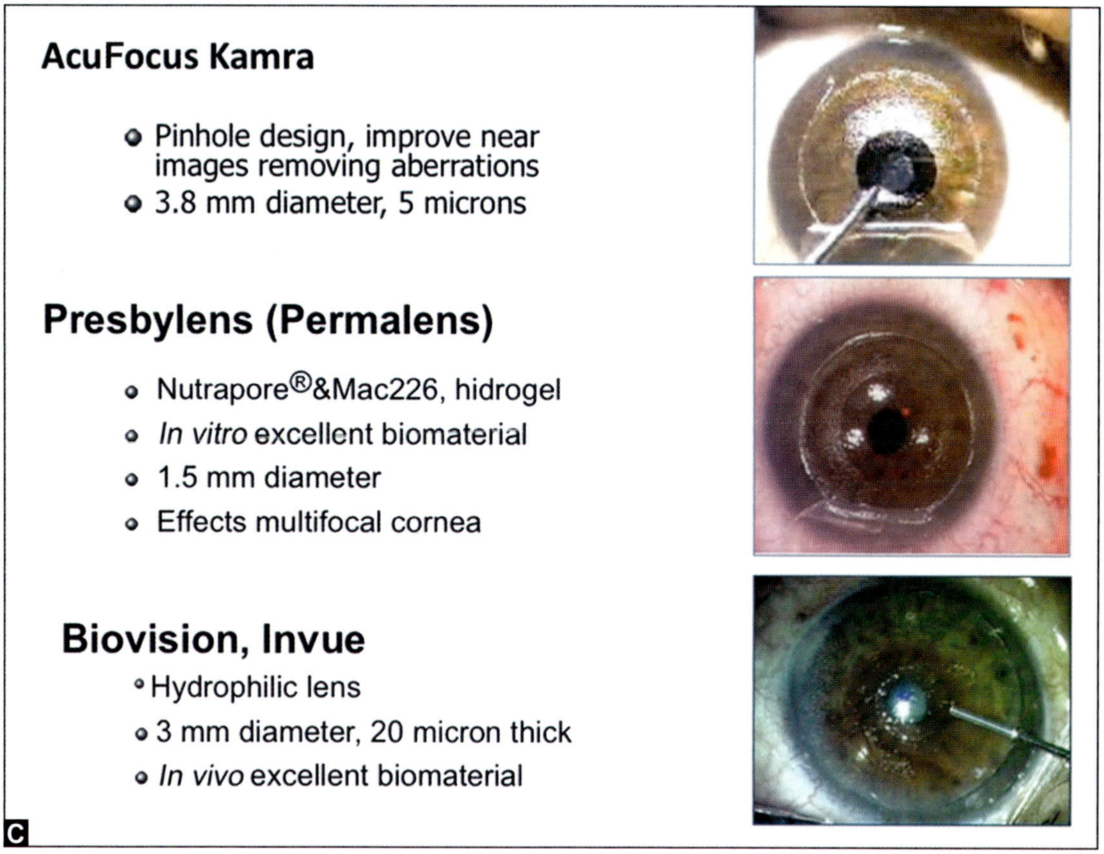

Figs 7.1A to C Differences between three inlays tested during clinical trials to correct presbyopia, in order AcuFocus, Presbylens, Invue

- Thickness and diameter is also different.
- Technique of implantation

SYNTHETIC KERATOPHAKIA HISTORY

- José Ignacio Barraquer (Fig. 7.2) represents the father of modern Keratophakia, he started his attempts in 1949
- At the beginning he tried with Cryolate lenticules to modify the corneal structure
- In 1964 he used glass and plexiglass as a corneal inlay material
- However, there were some failures related to biocompatibility issues and final necrosis.
- Richard Lindström (Fig. 7.3) implanted polisulphone inlays, 10 years ago
- Early on, few of them became opaque
- At the end he found hidrogel inlays demonstrated biocompatibility
- No histological evidence of inflammatory reaction.

Fig. 7.2 José Ignacio Barraquer is the father of modern Keratophakia and many other refractive procedures

Fig. 7.3 Richard Lindström, Michael Blumenthal, Vladimir Feingold myself in Acapulco during a training session for inlay implantation, August 2005

FEATURES AND SURGICAL TECHNIQUE

At early stage the AcuFocus and Presbylens inlays were designed to be implanted under a corneal flap, either performed with microkeratome or Femtosecond laser. The Biovision inlay was the only inlay originally designed to be implanted trough a corneal pocket, again dissected either with pocket maker, a mechanical microkeratome capable of creating a pocket, or trough a pocket carved by IntraLase. It is important to make clear that at present time, all three corneal inlays are implanted only in the non dominant eye.

From my personal experience, I believe that corneal flap (Fig. 7.4A) is not the right way to place a corneal inlay because the implantation will be very shallow, it should compromise anterior nutrition of the cornea, there will be dry eye issues, and the lens has decentration risk because, as we know early postoperative time the flap adhesion is not tight enough to hold the lens in place.

In contrast if we are able to dissect a corneal pocket with IntraLase Femtoseconds laser we have several advantages to inject a corneal inlay: it can be placed deeper. The right place is the middle stroma in order to avoid anterior nutrition problems and neutrophic corneal disease, as well. Besides the corneal pocket will hold the inlay steady, because there is no space for movement (Figs 7.4B and 7.5).

ACUFOCUS INLAY FEATURES

The AcuFocus Corneal Inlay (ACI) is an investigational device in the United States, and is under IDE status (Investigational Device Exemption). In the United States, this investigational product can only be used in pre-marketing clinical trials approved by the FDA.

Designed to improve near vision in patients with presbyopia, and it has several advantages: easily implanted, minimal impact on distance vision and represents one of few presbyiopia removable procedures.

ACI work using the same principle used in camera lenses to increase range of vision. Small aperture increases depth of focus. The small aperture created by the ACI reduces the blur when viewing near objects (Figs 7.6 and 7.7). Light rays pass through the small aperture over a small angle, which increases the depth of focus. It is important to mention that distance vision is minimally affected.

The AcuFocus lens is 5 μm thin, has a 3.8 mm outer diameter and has a 1.6 mm pinhole in the center (Figs 7.8 and 7.9). It blocks 90 percent of the light. Therefore, in bright light, you get a pinhole effect to enhance acuity, in dim light the pupil's bigger than 3.8 mm so the eye gets more light from around the side, but the pinhole still gives good quality of vision. It's implanted in one eye, either in a corneal pocket or below a LASIK flap.

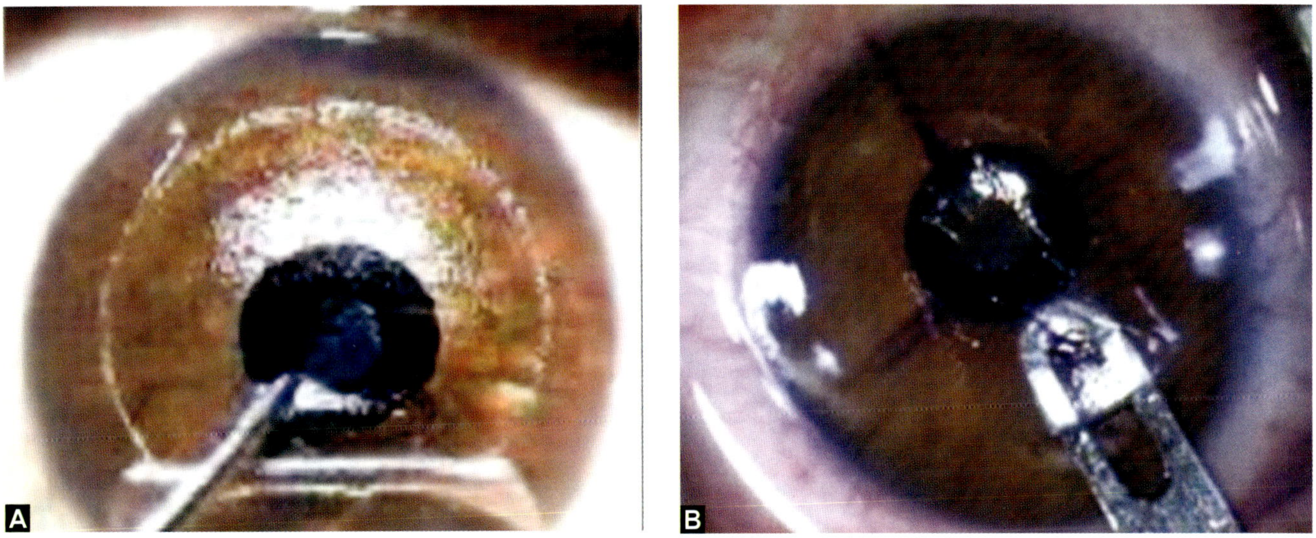

Figs 7.4A and B (A) AcuFocus implanted under a corneal flap; (B) AcuFocus inlay through a IntraLase pocket

IntraLase pocket benefits

- Safe dissection to implant a corneal inlay
- Computerized parameters programming
- Avoid dry eye related to corneal flap
- Prevents neutrophic keratopathy
- When we place an inlay at the middle stroma we evade unexpected topographic changes
- Stay clear of inlay movement, there is no space for movement.
- No astigmatism induction.

Fig. 7.5 Corneal inlay IntraLase pocket implantation technique benefits

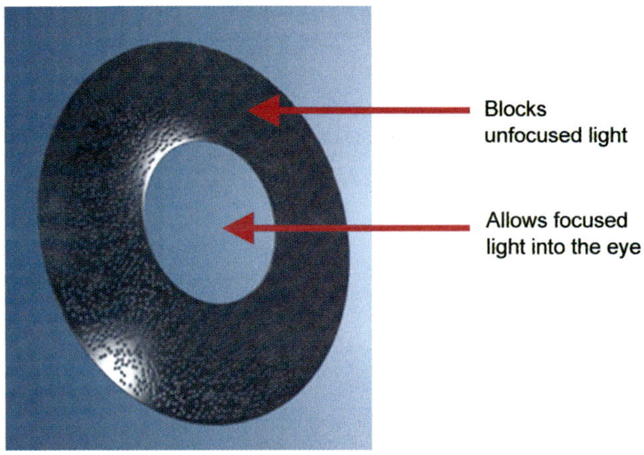

Fig. 7.7 The small aperture created by the ACI reduces the blur when viewing near objects

Fig. 7.6 ACI work using the same principle used in camera lenses to increase range of vision

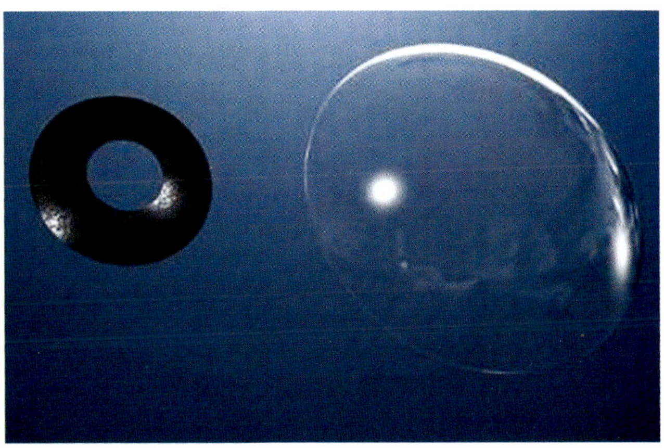

Fig. 7.8 3.8 mm diameter–smaller than a contact lens

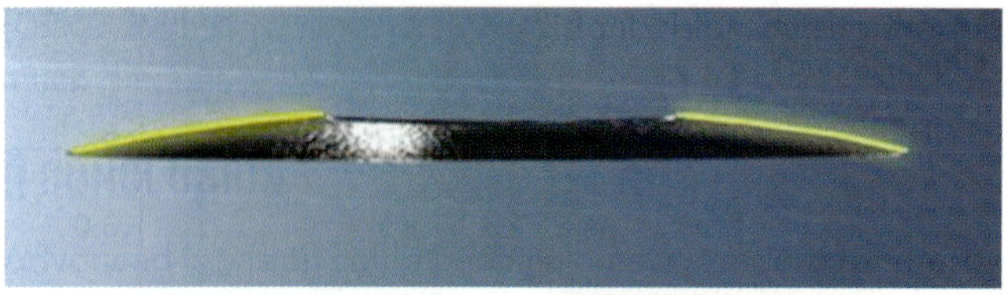

Fig. 7.9 ACI corneal inlay in profile, 5 microns thin

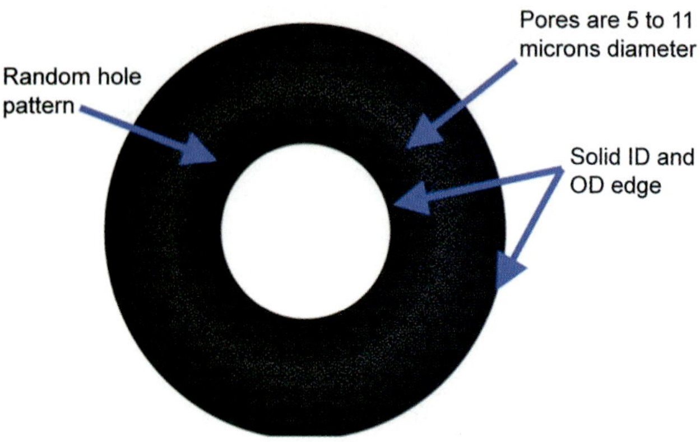

Fig. 7.10 AcuFocus corneal inlay perforations, facilitate corneal nutrition

Protocol for AcuFocus implantation in emmetropes presbyopic patients.

- Include presbyopic emmetropic patients between +0.50 to -0.50 of SE.
- Astigmatism less than 0.75 Dt.
- Endothelial cell count a least 2000 cel/mm^2
- Normal anterior segment and retina exam
- Rule out dry eye (normal BUT and Shirmer test)
- IOP and visual fields within normal limits
- Normal corneal topography, Orbscan or Pentacam normal also
- Select non-dominant eye using blurring vision tolerance test
- Discuss risk and benefits.
- Avoid cases with unrealistic expectations.

SURGICAL INTRALASE SURGICAL TECHNIQUE STEP-BY-STEP

Step 1: OZ marking at visual axis using 1 ° Purkinje image.

Step 2: Horizontal alignment line as a guide for the IntraLase cone mask.

Step 3: Temporal displacement of IntraLase suction ring, notice 2 mm nasal conjunctiva exposure.

Step 4: InraLase cone mask aligned to corneal OZ mark and horizontal line guide, and IntraLase pocket dissection.

Step 5: Sinsky initial opening, spatula stromal pocket dissection.

Step 6: AcuFocus inlay in the injector, inlay injection though a pocket.

Step 7: AcuFocus inlay release and final centration checking.

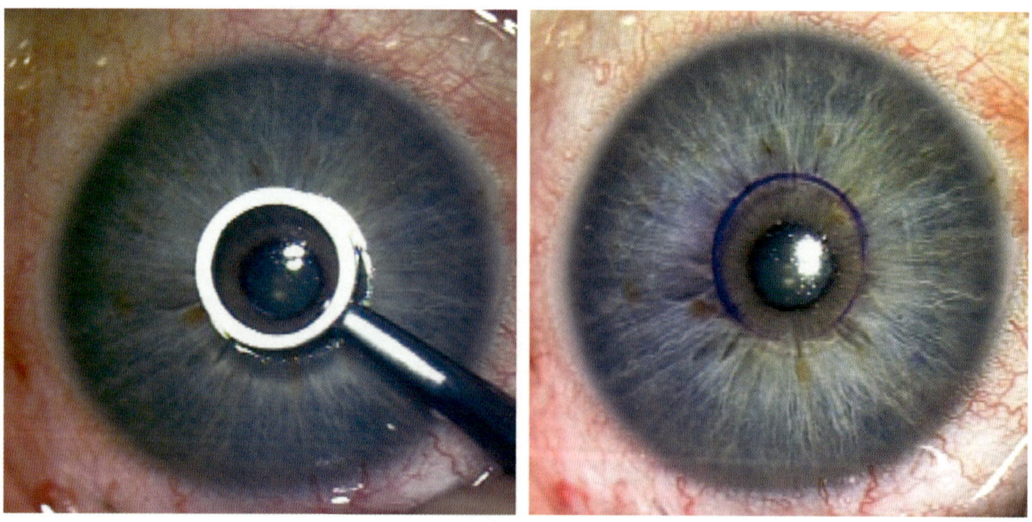

Step 1

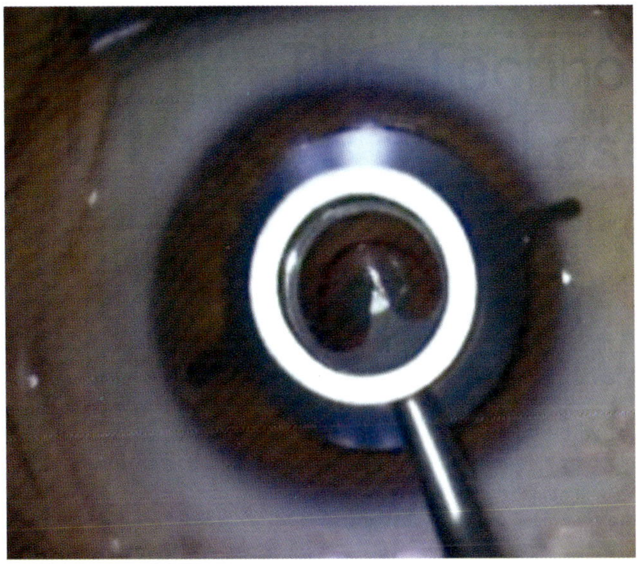

Step 2

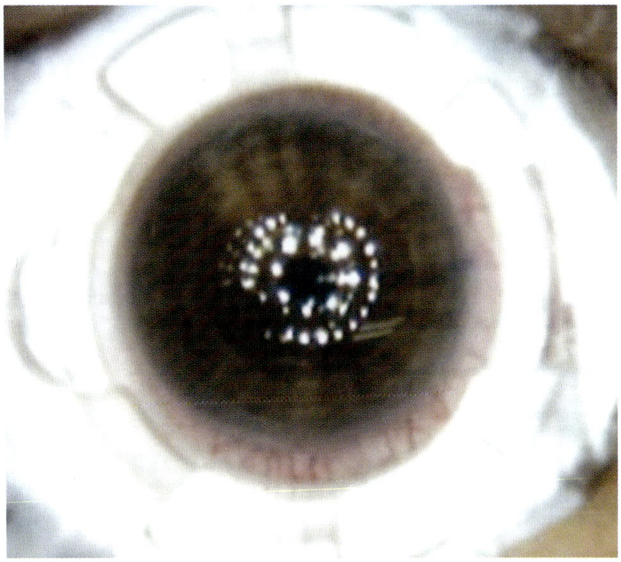

Step 3

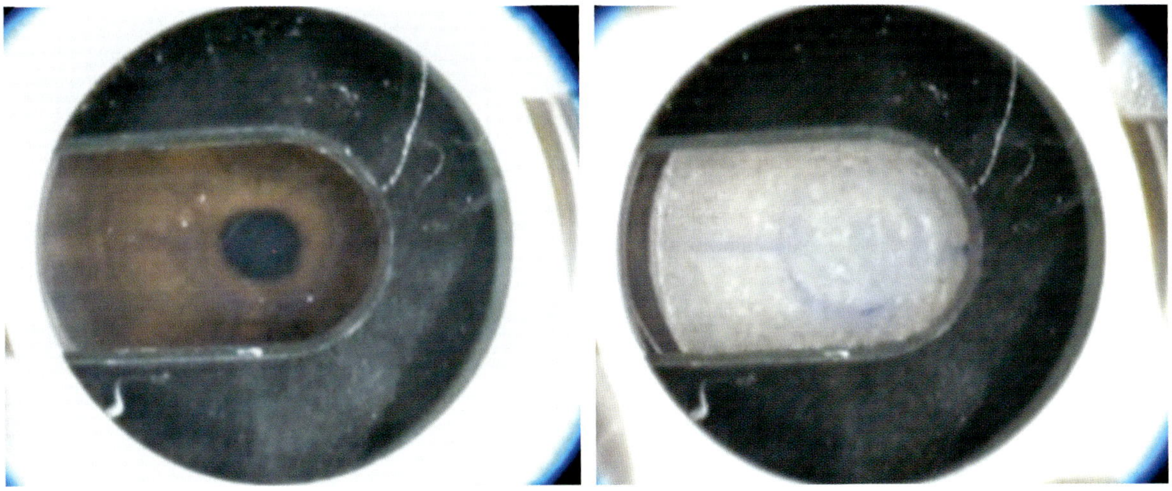

Step 4

Kamra AcuFocus has incorporate the Acutarget, technology. This machine allows to center the Kamra inlay aperture in accordance to the visual axis of the patient for better results. The Acutarget permits:

1. Identify the precise situation of the visual axis in relation to the pupil.
2. Kappa angle determination
3. Precise measurements in micros
4. Centration control
5. Fine tuning adjustments

Today we have used the Kamra technology for presbyopia in four different scenarios: (videos of each and presented)

1. Emmetropic presbyopic patient
2. Post-LASIK emmetropic patient
3. Sim-LASIK (simultaneous LASIK and Kamra)
4. Simultaneous LASIK enhancement and Kamra implantation

INVEU INLAY FEATURES AND SURGICAL TECHNIQUE

The microlens is made of water-permeable, biocompatible hydrogel that is about 10 μm thick at the edge, with a center

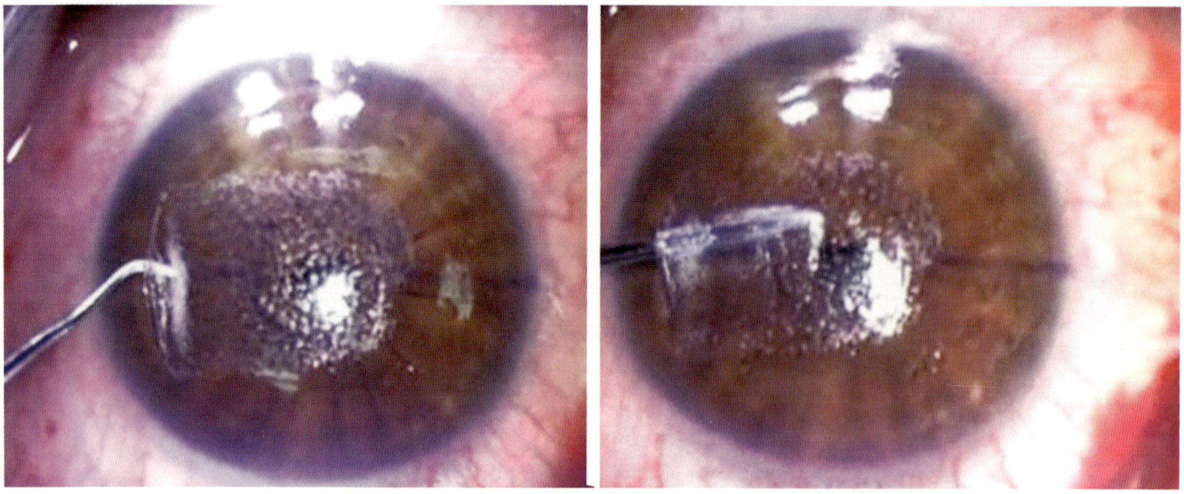

Step 5

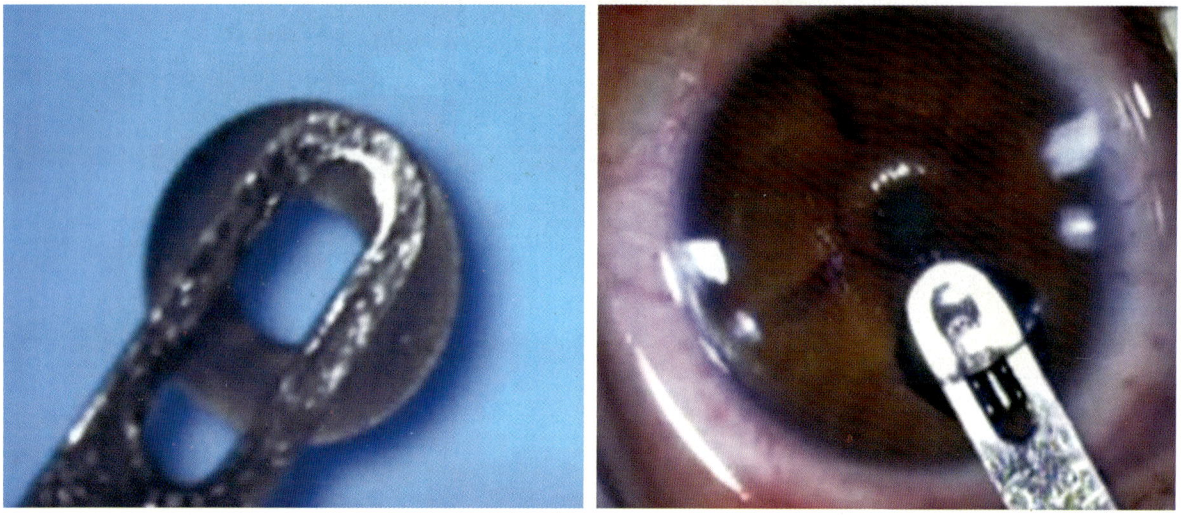

Step 6

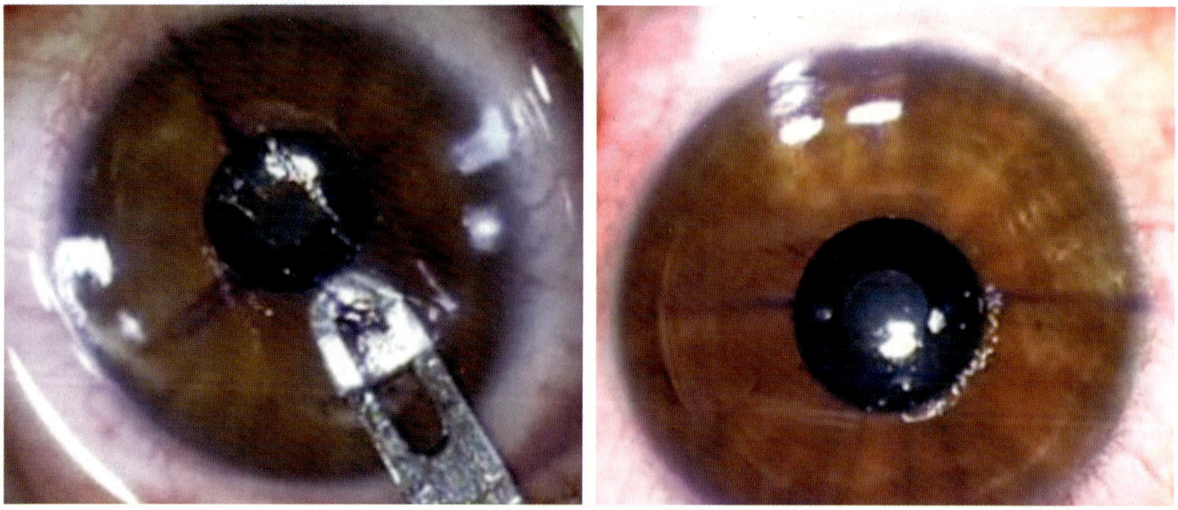

Step 7

thickness that varies depending on its add power, which is in an annular ring around a small central clear zone. The overall diameter is in the 3.6 to 3.8 mm range.

This microlens creates a multifocal cornea, out of an emmetropic cornea by adding plus power to the cornea. So it is designed for the emmetropic presbyope, whose cornea is healthy and has no astigmatism. This includes patients who have been made emmetropic through refractive or cataract surgery. With the lens in place in the cornea, the center zone is for reading and or the peripheral zone is for dis-

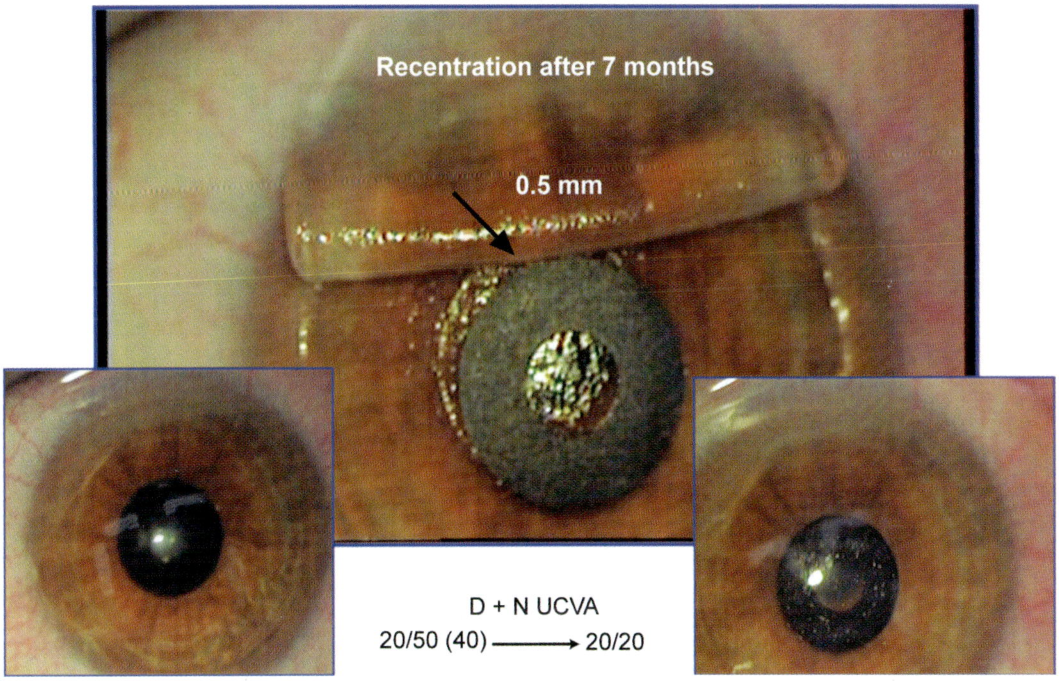

Fig. 7.11 Importance of centration

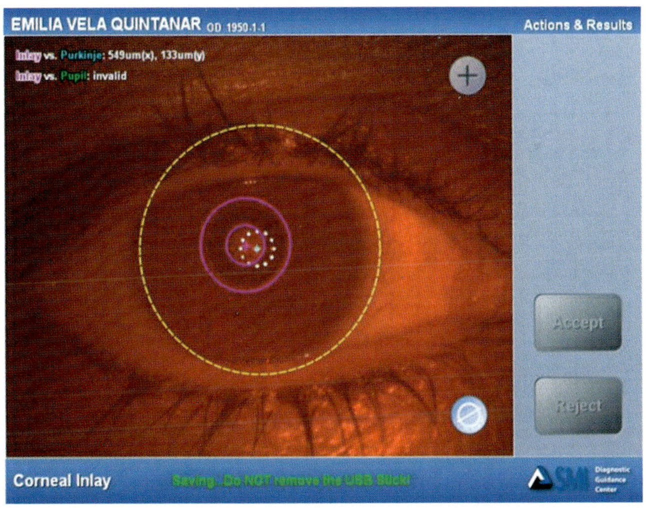

Fig. 7.12 AcuTarget™ postoperative evaluation of decentered case

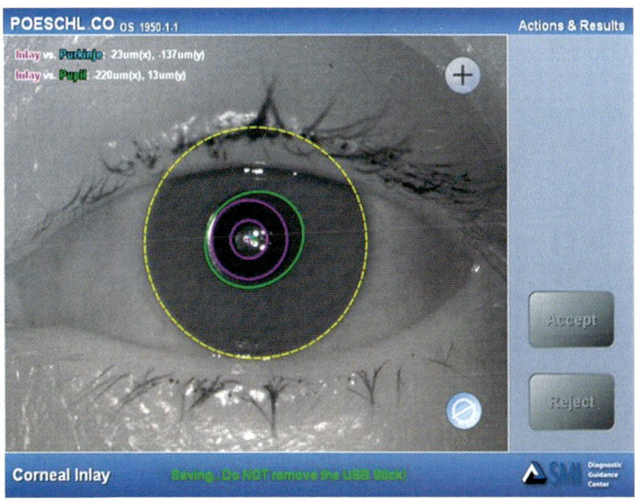

Fig. 7.13 AcuTarget™ postevaluation of operative centered case

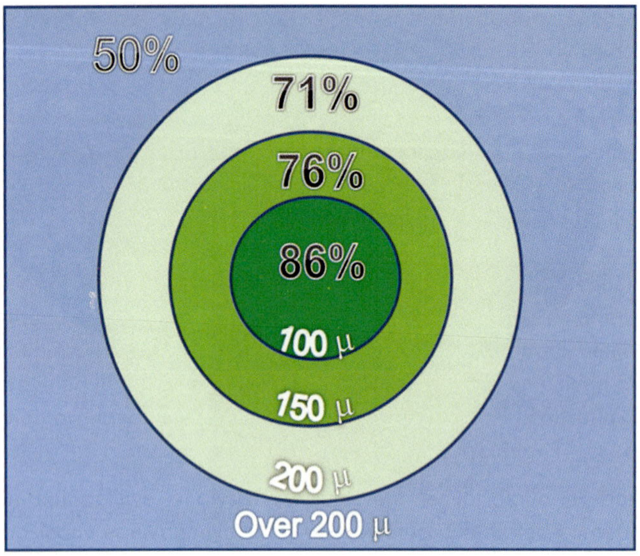

Fig. 7.14 Percent chance of achieving UCNVA of 20/25 (J2) or better, 3 months/49 patients

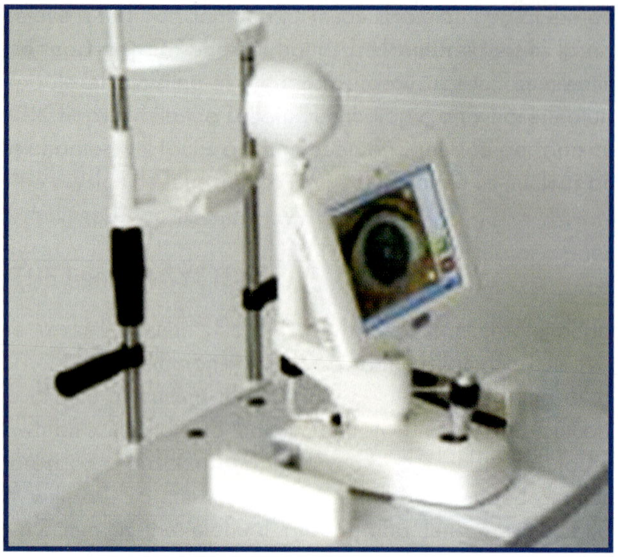

Fig. 7.15 The KAMRA™ AcuTarget™ system

tance, and the mid-peripheral of the lens is for near tasks also.

In general, the benefit of having a multifocal cornea is that the patient has a significant improvement in his ability to read. The drawback is that there's a small loss of contrast sensitivity, or quality of distance vision. There can also be an induction of mild night-vision symptoms, primarily halo, as occurs with other multifocal optics.

Since the emmetropic presbyope has good vision to begin with, other than his near-vision deficiency, it will be easy to demonstrate what to expect with a corneal implant. The nice thing about this device is that, though it is hard to demonstrate multifocal optics to a patient with a cataract, because he can not see, when a presbyopic patient walks in with 20/20 or 20/15 vision, you can demonstrate the visual effect of the corneal implant using a multifocal contact lens.

- *The effect of pupil size:* It is pupil-size dependent on the extremes, it turns out that every multifocal intraocular lens has some pupil dependence. He says that, when a patient's pupil is very small, his visual system with the Inveu in place is distance-dominant. If the pupil is in the middle range, about 3.5 or 3.6 mm, the typical size for reading, optics favor near work. If the pupil dilates to 5.5 mm, such as in dim light conditions, including driving, the eye is distance-dominant. Dr. Ionnis Pallikaris has called that Inveu microlens system produces an intelligent reading vision depending to the pupil physiology.
- *Decentration:* Once the lenses are put in the pocket, created either with Visitome 20/10 microkeratome or Femtosecond laser (Figs 7.10 and 7.11), they tend not to decenter. But the trick is putting them in the right place, so there are evolving techniques and instruments aimed at putting the implants perfectly in the visual axis (Figs 7.12A and B). So, decentration definitely would be problematic, since if it is decentered by 0.3 mm, the optics will degrade somewhat.
- *Ocular tolerance:* This microlens appears to be well-tolerated. You can get some interface reaction with any material you put in the cornea. But there has not been enough to cause forward light scatter.

In summary, with the Biovision Inveu microlens the patients retain quite good distance vision, and have good intermediate and excellent near vision. There is a mild loss of contrast and mild night-vision symptoms, but it is only implanted in one eye. It is the extremely rare patient who is bothered by the optical aberration problems.

Inveu forceps is capable to inject the microlens by pushing a condoled trigger which facilitates the Inveu inlay release; at end proper centration at the visual axis is mandatory.

Sequence of Biovision Inveu Inlay implantation in a pocket carved with IntraLase Femtosecond laser.

Step 1: Programming IntraLase Femtosecond laser parameters. 250 m depth, 9.5 diameter, 30° side cu, using Sánchez León's mask to create only a corneal pocket. First image nasal to temporal laser raster, and second image side cut 30° creation.

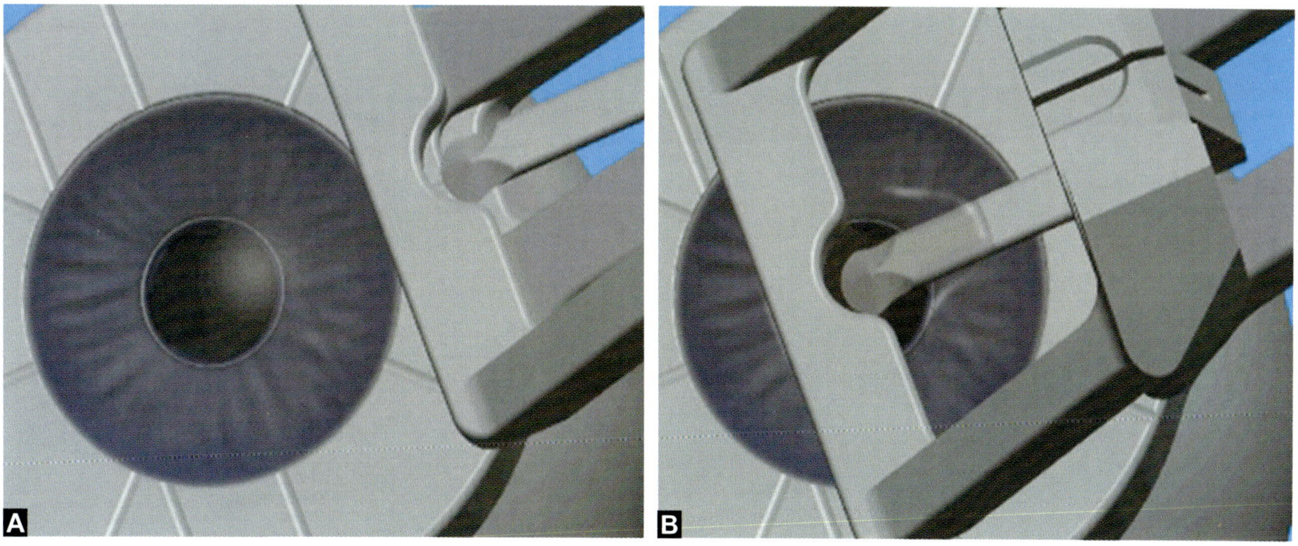

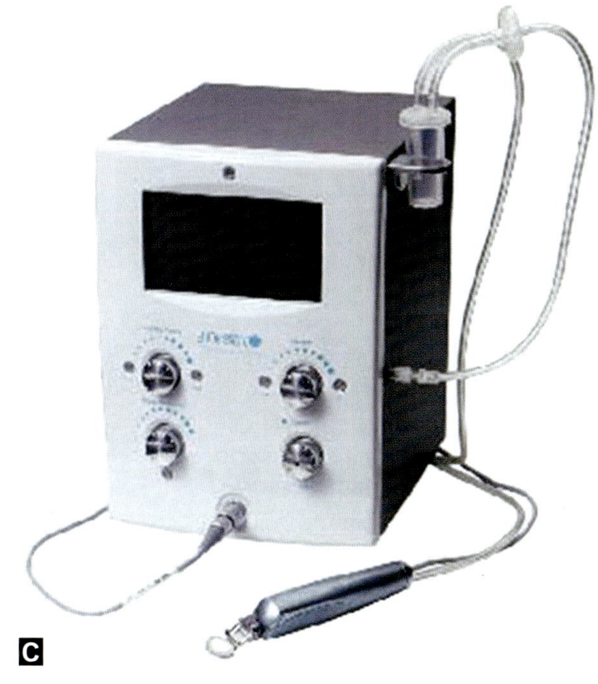

Figs 7.16A to C (A and B) To create a mechanical pocket, the surgeon uses Inveu's Visitome 20-10 microkeratome with a unique blade cartridge assembly design. An automating the pocket-making procedure simplifies the operation for the surgeon; (C) Invue's Visitome 20-10 microkeratome

Step 2: Sinsky hook is used to start dissection of the pocket entrance, followed by IntraLase Rhein spatula corneal pocket dissection.

Step 3: Lens guide grasp the intracorneal lens. It comes with a inserter which pushes the microlens inside and at the end of the pocket.

Step 4 and 5: Retraction of inserter to release the inveu microlens, cenrtration at the visual axis, using first Punkinje image is critical.

THE PRESBYLENS™ CORNEAL INLAY: AN ADJUSTABLE DEVICE FOR PRESBYOPIA

The Presbylens™ Corneal Inlay (CI) is produce by ReVision Optics, Inc. (Lake Forest, California)) is designed to correct presbyopia by changing the shape of the eye's corneal surface, in other word changes the topography and reproduces a multifocal system. The lens is made of a piece of proprietary polymer that is 1.5 mm in diameter and designed to fit beneath a corneal flap in the non-dominant eye of a plano-

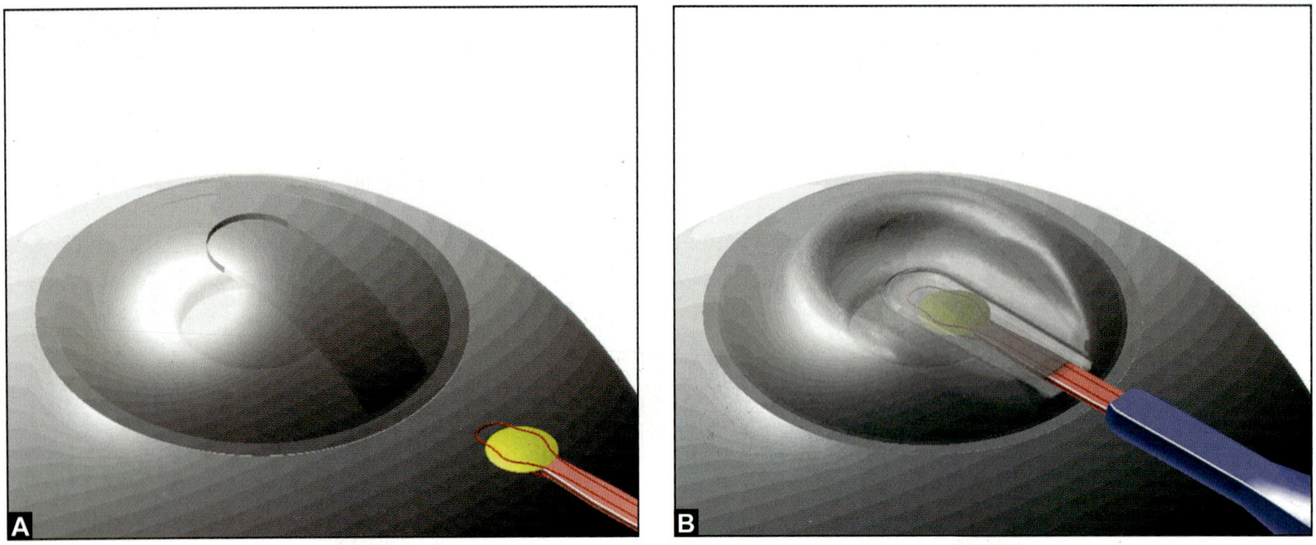

Figs 7.17A and B Inveu forceps holding the invue inlay ready for injection

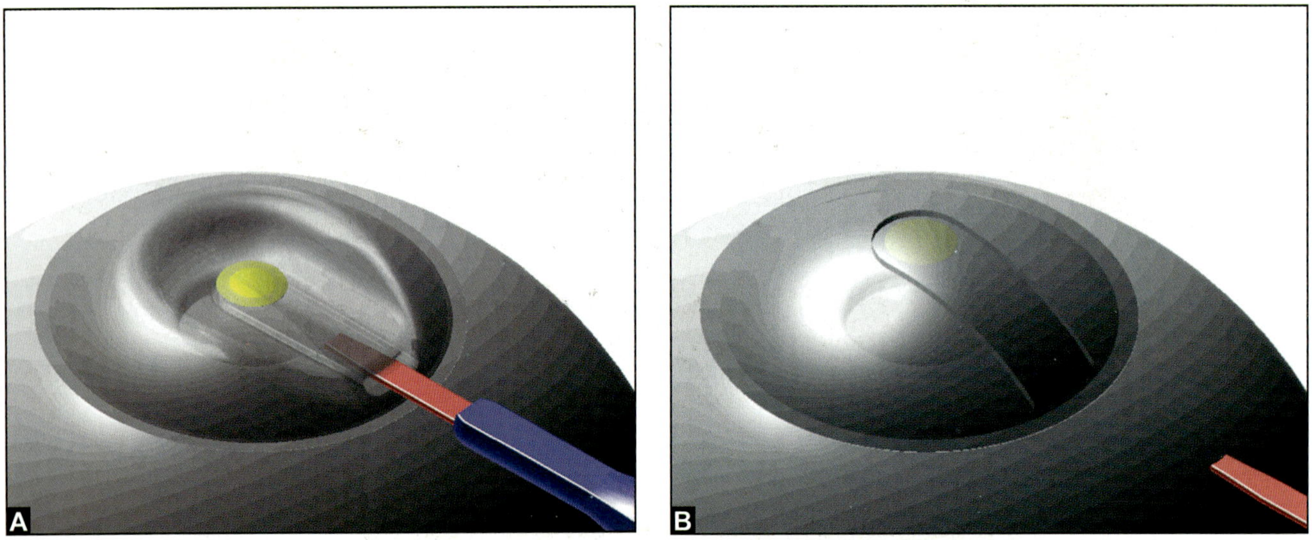

Figs 7.18A and B

presbyope. There, it changes the curvature of the cornea enough to enhance near vision, minimizing any decrease in the distance vision. The Presbylens is similar in nature to a soft contact lens; however the presbylens is inserted just under the eye's outer surface.

The presbylens is a mere 1.5 mm in diameter and it is ultra thin (approximately half the thickness of a sheet of paper); these dimensional factors as well as the proprietary bioengineering of the material are intended to make the presbylens biocompatible within the cornea.

Three-fourths (3/4) of your eyes focusing power (refractive power) comes from the cornea; the outermost tissue of the eye. The unique properties of the cornea and the optical difference between the cornea and air (corneal/air interface) allow for very small changes in the surface of the cornea to have large effects on the eye's focusing power.

These optical characteristics of the cornea are essentially why LASIK works; LASIK removes a small portion of the corneal tissue to change the surface of the cornea and improve your vision.

Similarly, the Presbylens Corneal Inlay (CI) is also designed to take advantage of the cornea's optical properties, but by adding a small inlay just under the surface of the cornea. Once the 1.5 mm corneal inlay is inserted, it is designed

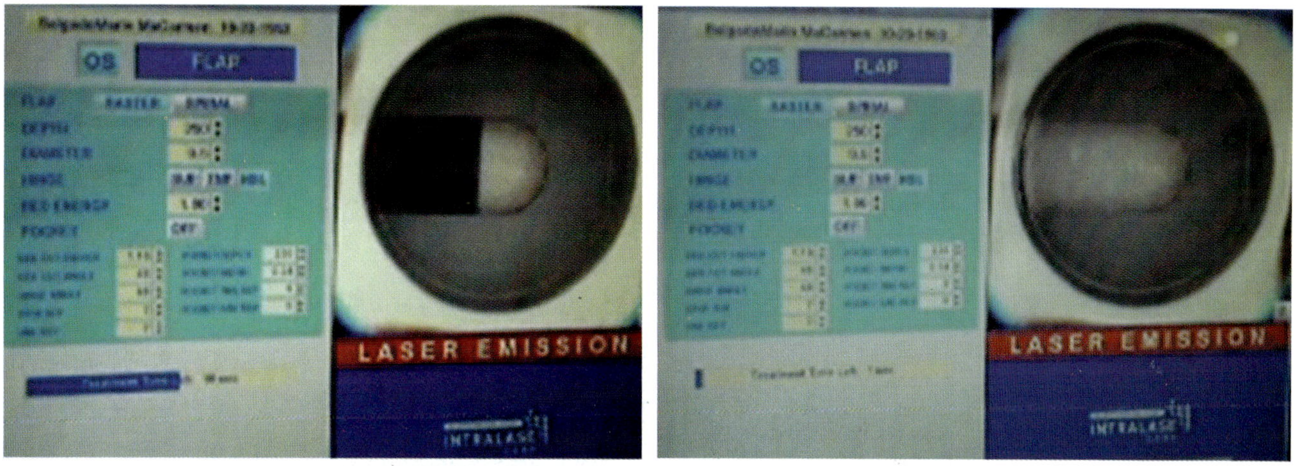

Step 1

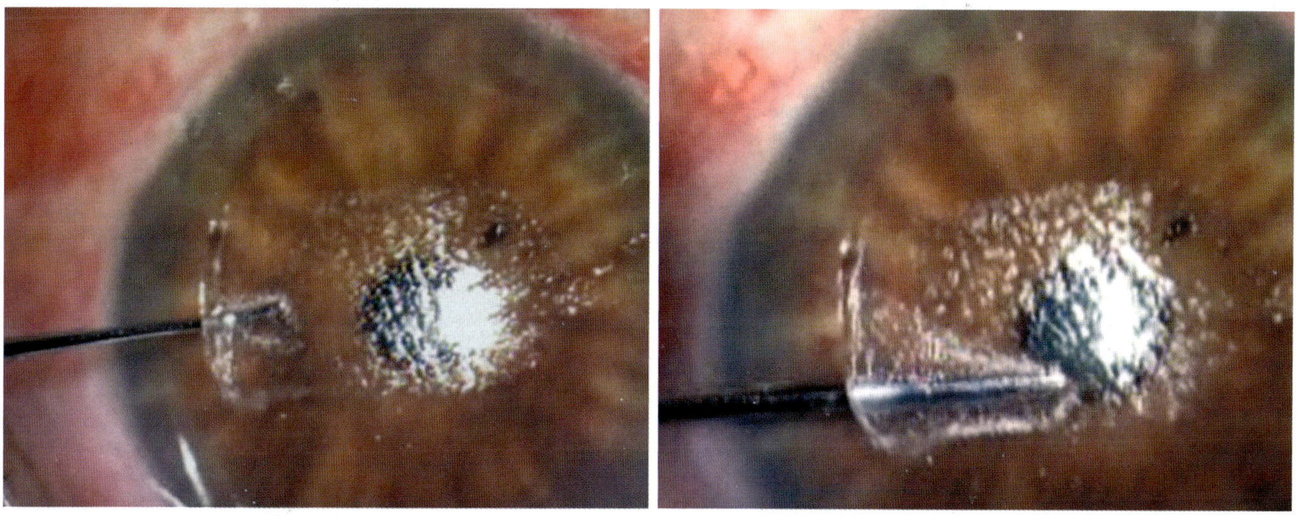

Step 2

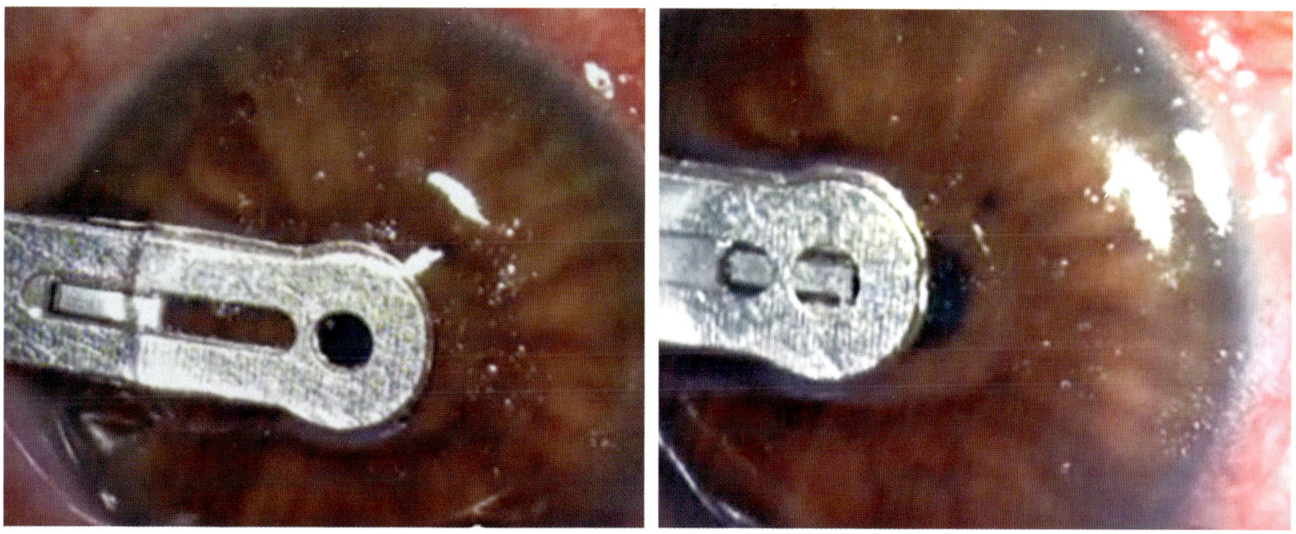

Step 3

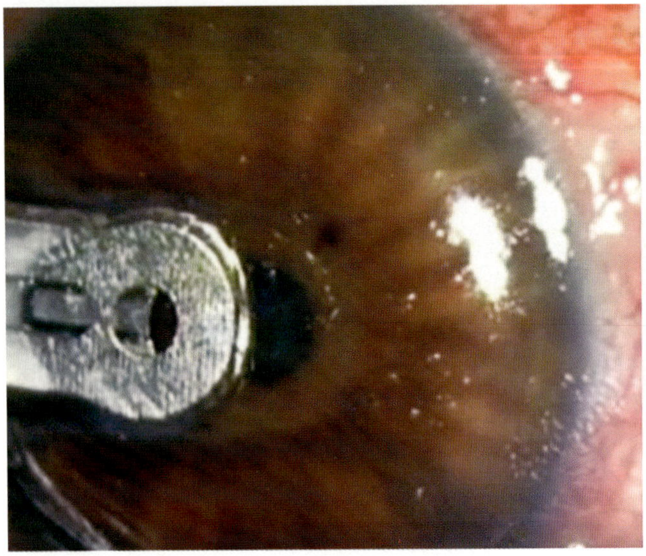

Step 4

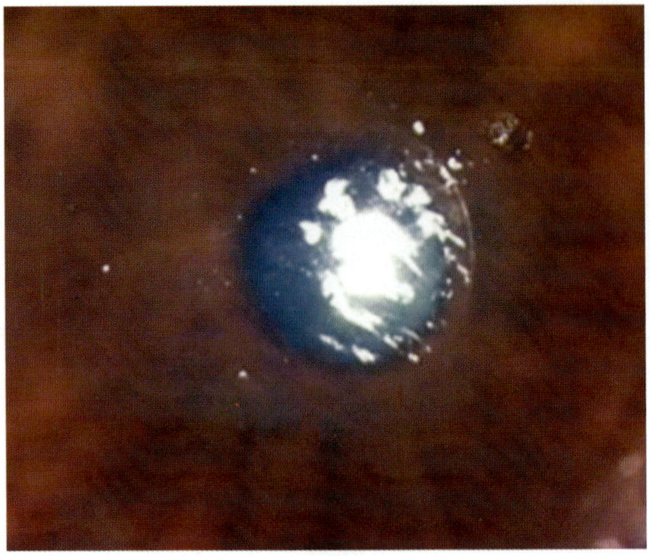

Step 5

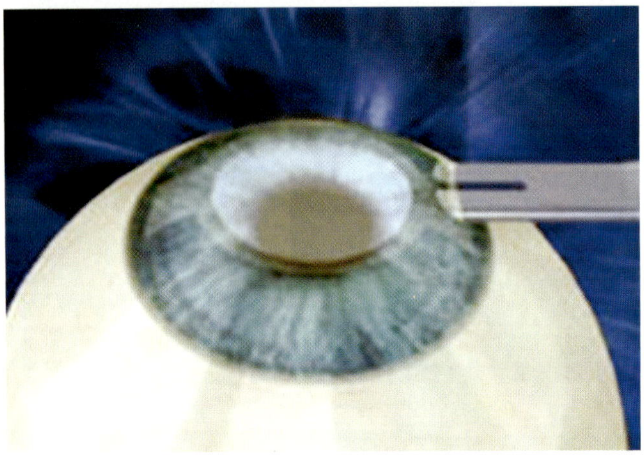

Fig. 7.19 Steps 1 to 6

to change only a small central area of the cornea to increase near focusing power with little decrease in intermediate and distance focusing as the figure above illustrates.

The lens is situated beneath what is essentially a LASIK flap, though plans are in the works to create a lamellar space for it that will allow it to sit even more snugly.

Surgically, it is a fairly straightforward, quick implantation. The flap is usually 120 to 130 μm thick, and you can create it with either a microkeratome or a femtosecond laser. When the lens comes from the manufacturer, it is already loaded into a delivery device. So, it is then a matter of sliding the lens off in the center of the pupil, letting it dry for a few seconds, and then folding the flap back over top of the lens.

RESULTS

Inveu Results

Inveu corneal inlays results were presented at 12th Winter Refractive.

Meeting of the ESCRS with the most recent design of the inlay in a series of 27 patients. Among eyes that had reached a follow-up of 12 months, distance visual acuity was 20/30 or better in 80 percent, and 20/40 or better in all eyes. In addition, uncorrected near visual acuity was J1 or better in 85 percent of eyes and J3 or better in all eyes. Furthermore, all eyes had a near and distance visual acuity of 20/20 or better with best correction for distance.

Wavefront analysis reconstruction at different pupil sizes of eyes implanted with the device show that the multifocal optics of the inlay work in conjunction with the constriction and dilation of the pupil that occurs during accommodative and dis-accommodative effort. That is, when a patient focuses on a distant object, the resulting pupil dilation delivers more light through the distance component of the re-shaped corneal surface, he explained. Conversely, constriction of the pupil, as when focusing on near objects, delivers more light through the re-shaped corneal surface's near component.

Figure wavefront reconstruction at different pupil sizes. At 6.5 mm the wavefront refraction when pupil dilates and light travels trough the hole system in and out of the Inveu microlens o refraction is close to 0 (–0.20 Dt), at 3.5 mm reconstruction when the pupil constricts the light travels only trough the Inveu microlens and the optic system becomes myopic of –1.25 Dt to produce near vision gain, this is what we call we

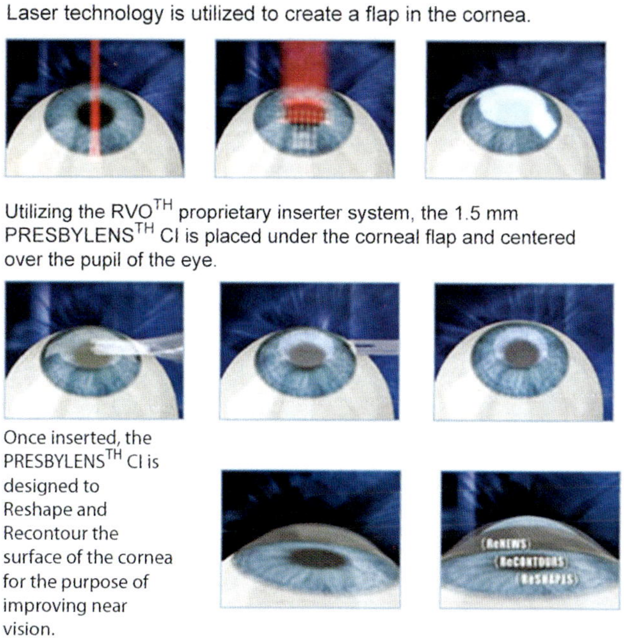

Fig. 7.20

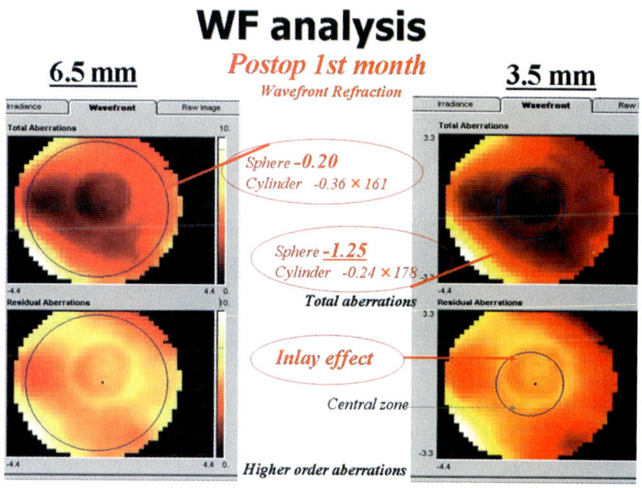

Fig. 7.21

an "intelligent near vision system" which works in relation to pupil physiology.

Unlike conductive keratoplasty, presbyLASIK or accommodating IOLs, the INVEU corneal inlay procedure is reversible. It is also minimally invasive and produces only minimal side effects.

BIBLIOGRAPHY

1. Pallikaris I, et al. Cataract and Refractive Surgery Today Europe, Sep 2007: 53-54.
2. Lindström R. Review of Ophthalmology. Vol. No: 12:09: 9/1/2005.
3. Dishler Review of Ophthalmology J. Vol. No: 16:03Issue: 3/1/2009.
4. Sánchez F. Treating Presbyopia from Inside the Cornea Vol. No. 14:09 Issue 9/1/2007.
5. Sánchez F, Neuhann T. Corneal inlay increases spectacle independence in presbyopes. Eurotimes: 08, June, 2009.
6. Creating Femtosecond Pocket Under Previous LASIK Flap for Correction of Presbyopia with Corneal Inlay, ASCRS, 2012.

8 Surgical Treatment of Keratoconus and LASIK Ectasia with Intrastromal Corneal Ring Segments (Keraring®) and Femtoseconds Laser Implantation

Francisco Sánchez León (México)

Chapter Outline

Keraring intrastromal corneal ring segments, are implantable precision devices for the correction of surface irregularities and reduction of refractive errors associated to keratoconus and other corneal ectasias.[1-4]

Unlike other intracorneal rings, Keraring was specially designed for the treatment of corneal ectatic disorders, providing greater and better corneal regularization and refractive effects.

Keraring implantation accomplish the corneoplastique concept[9] (Fig. 8.1) which attempt to improve the patient's optics, topography, visual acuity and reduce or eliminate the refractive error in order to stabilize the cornea, avoiding or delaying the need for corneal transplantation.

The present chapter will describe advantages, indications, contraindications, mechanism of action, manual and IntraLase technique, nomograms, and postoperative care.

Advantages

Clinically Proven Safety and Efficacy

Keraring holds the largest history of use and longest follow-up in the world: Over 100.000 eyes implanted in 45 countries with follow-up as long as 16 years. Independent clinical studies confirm Keraring's safety and efficacy.[4,5,8]

Reversibility

Keraring may be explanted at any time, allowing the cornea to revert to its preoperative shape. The procedure is reversible.

Adjustability

If necessary, topographic and refractive results may be adjusted by exchanging or repositioning the implant. This usually happens in our experience in less than 5 percent of virgin the cases, and around 10 percent of post-LASIK ectasia patients.

Quick Visual Recovery

Keraring's minimally invasive surgical technique allows the patient to quickly resume regular activities. Topographic and refractive changes start right after the implantation and stable results are achieved within 3 months on average.

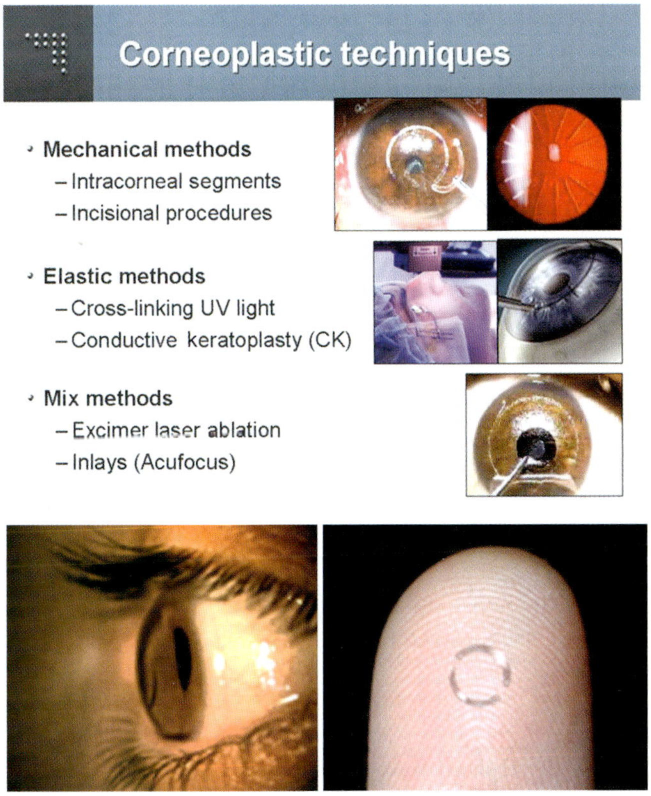

Fig. 8.1

Exclusive Prismatic Design[4] (Fig. 8.2)

Keraring's unique design generates a prismatic effect by which the light refracted in the implant is reflected out of the eye, thus reducing glare and halos.

Greater Refractive Potential

Keraring's design and implantation zone provide greater potential of refraction change, allowing correction of high degrees of myopia and astigmatism.

Does not Compromise Corneal Transplantation

Should lamellar or penetrating keratoplasty may be required, these can be performed as per the usual techniques even in the presence of the implant, which will be removed with the recipient graft. Keraring does not interfere with corneal.

Indications[2,3,5]

- Keratoconus with poor BSCVA and contact lens intolerance
- Progressing keratoconus
- Pellucid marginal degeneration

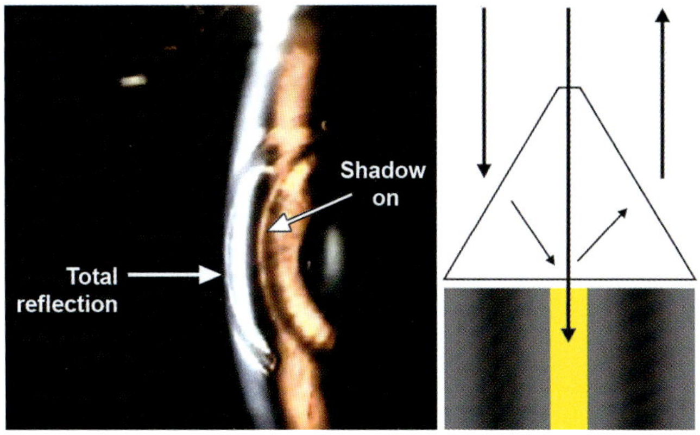

Fig. 8.2 Keraring prismatic effect

- Corneal ectasia following LASIK, PRK, LASEK, EPI-LASIK
- Irregular astigmatism following radial keratotomy
- Irregular astigmatism following penetrating keratoplasty
- Corneal surface irregularities following trauma.

Contraindications

- Acute keratoconus with K reading > 70 D
- Severe central corneal opacities
- Hydrops
- Following penetrating keratoplasty with decentred graft
- Severe atopic disease
- Recurring corneal erosion syndrome
- Patient's high expectation for uncorrected emmetropia.

Mechanism of Action[2,5,6] (Figs 8.3 and 8.6)

- Central corneal flattening through tissue addition in the midperiphery: Preserves corneal integrity and spares the central.

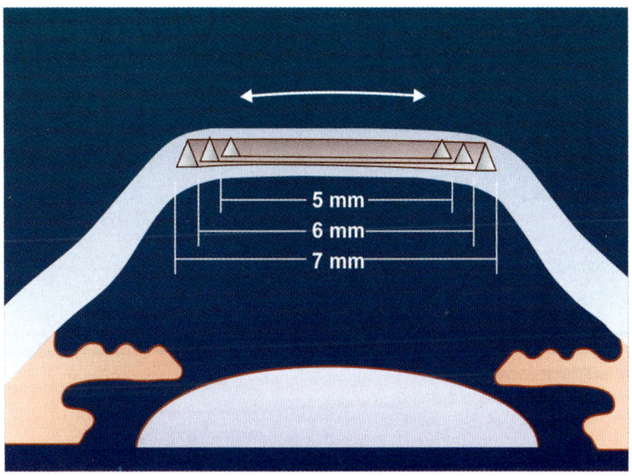

Fig. 8.3 Central corneal flattening through tissue addition in the midperiphery

- Corneal topography regularization, preserving the natural prolate profile: Reduces optical aberrations, improves visual acuity and contact lens tolerance
- Displacement of corneal apex toward the central pupil: Improves corneal optics
- Stabilization of ectasia progression: Avoids or delays corneal transplantation for unspecified time.

SURGICAL TECHNIQUE AND SPECIFICATIONS

Surgical Technique

Keraring implantation is an outpatient procedure performed under topical anesthesia. Special surgical instrumentation has been developed by Mediphacos for manual and femtosecond laser[5,10] assisted techniques.

How to Make Keraring Implant Surgical Plan

1. Evaluate cornea through biomicroscopy.
2. Determine steepest corneal meridian
   - Using manual keratometer or topographer (SIM-K).
3. Study topography map and determine how the ectasia is distributed on the cornea in relation to the steepest meridian.
4. Measure manifest refraction and determine spherical equivalent refraction of the subject eye.

Manual Dissection and Implantation Technique[9]

Step 1: Central marking on microscope light reflex (visual axis), not center of the pupil.

Step 2: Marking of 5/7 mm, and steepest meridian.

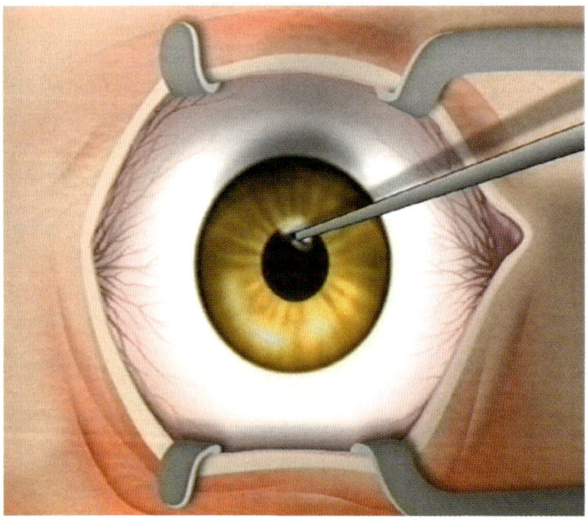

Step 1

Step 3: Incision at steepest meridian:

- 80 percent depth of local pachymetry
- Micrometer diamond knife with square blade (1mm)

Step 4: Creation of intrastromal pockets with Suarez Spreader:

- Depth equal to incision

Step 5: Sequential tunnel creation using right and left spatula:s

- In case there is great tissue resistance to delaminating suspect superficial delaminating
- Repeat incision and pocket
- Keep the cornea stretched by pulling up the spatulas while advancing (delaminating)

Step 6: Segment implantation with Albertazzi forceps:

- Segment implanted with flat side down
- Insert each segment all the way to the end of the tunnel
- Keep the tips far from the incision.

Step 7: Well implanted segments must be:

- Symmetrical position
- Deep stroma
- Tips far from the incision

KERARING INTRALASE TECHNIQUE IMAGE ANIMATION

Keraring IntraLase Technique—Advantages

The femtosecond laser has distinct advantages over mechanical ring dissector microkeratome. I prefer the IntraLase femtosecond laser (Advanced Medical Optics, Santa Ana, California).

Short time: The femtosecond laser creates a ring channel in a shorter time than a mechanical technique. What takes the mechanical ring dissector approximately 8 minutes may be accomplished with the femtosecond laser in approximately 8 seconds (Fig. 8.9).

Reduced risks: Creating channels with the femtosecond laser means less (1) epithelial defect, (2) discomfort, and (3) risk of inflammation/infection. In a mechanical technique, the risk of epithelial defect–in turn causing discomfort, inflammation and/or infection–is much more than when a femtosecond laser is used. Rabinowitz et al[3] studied the epithelial defects in two patient sets (i.e. mechanical ring dissector against femtosecond laser). On the first postoperative day, 50 percent and 15 percent of patients in the mechanical ring dissector and femtosecond groups, respectively, had significant epithelial defects along with moderate to severe discomfort. IntraLase offers a tighter dissection which avoid segment migration, compare to manual dissection when sometimes sutures are required (Fig. 8.10).

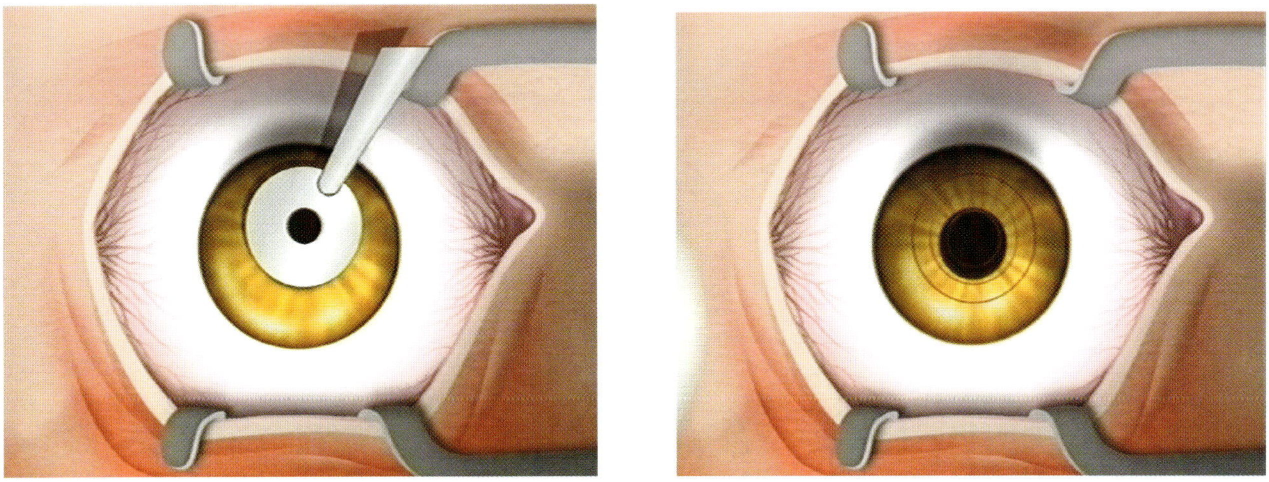

Step 2

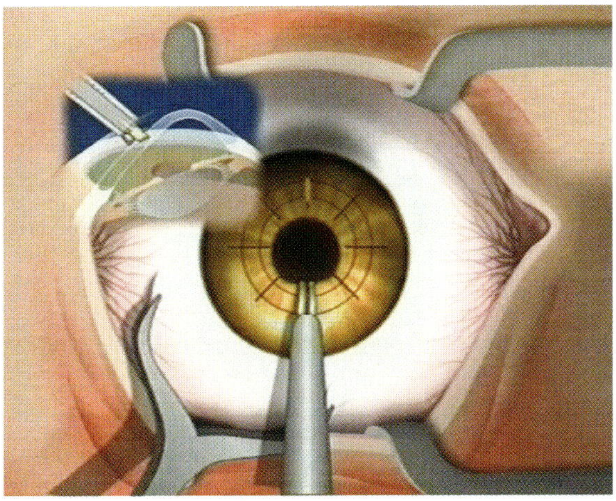

Step 3

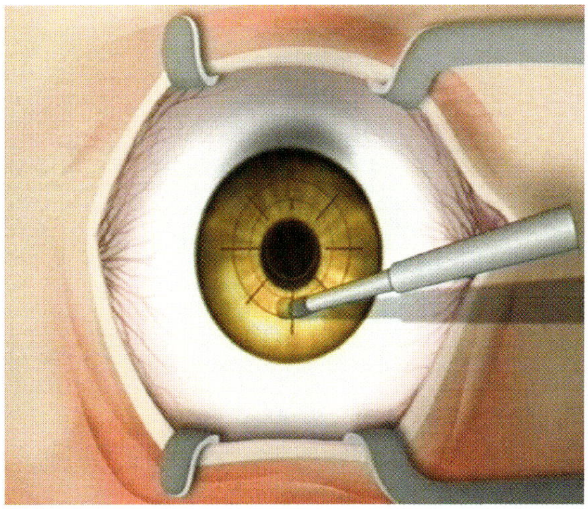

Step 4

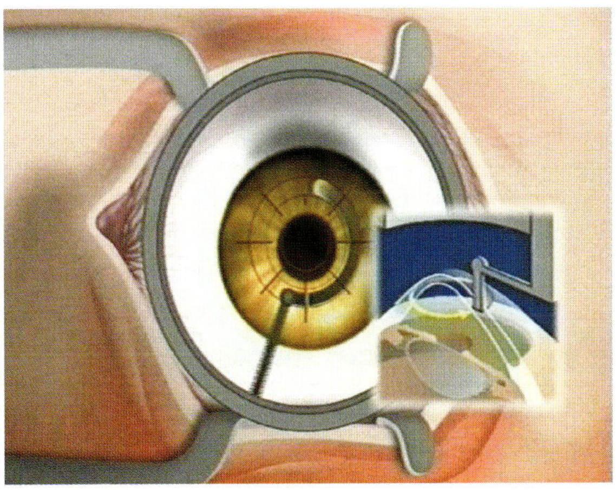

Step 5

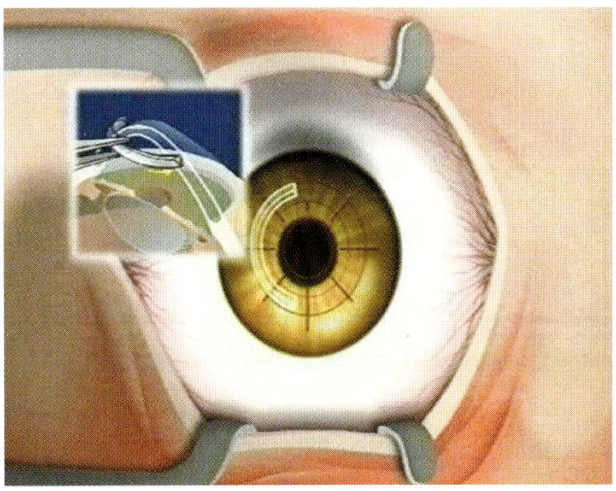

Step 6

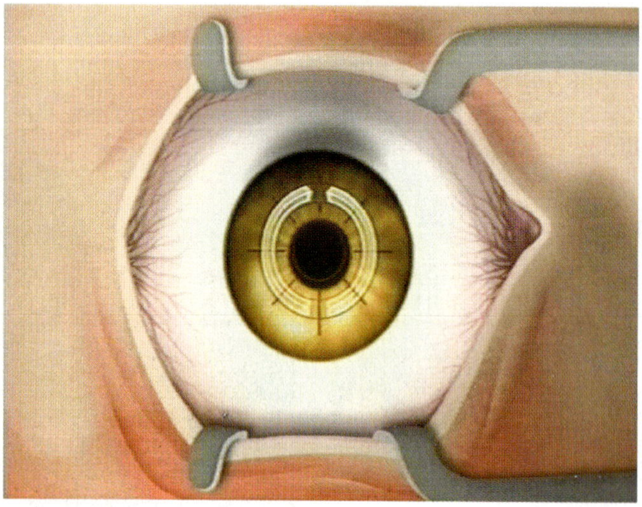

Step 7

Fig. 8.4 IntraLase animation, the laser passes trough the external layer of the cornea until it reaches a precise depth in the cornea

Fig. 8.5 IntraLase creates a O_2, N, CO_2 bubbles of to carve the ring dissection

Fig. 8.6 Keraring remodeling, cornea architecture changes after implantation

Fig. 8.7 Cross-linking only necessary in the scenario of undercorrection or unstable cornea

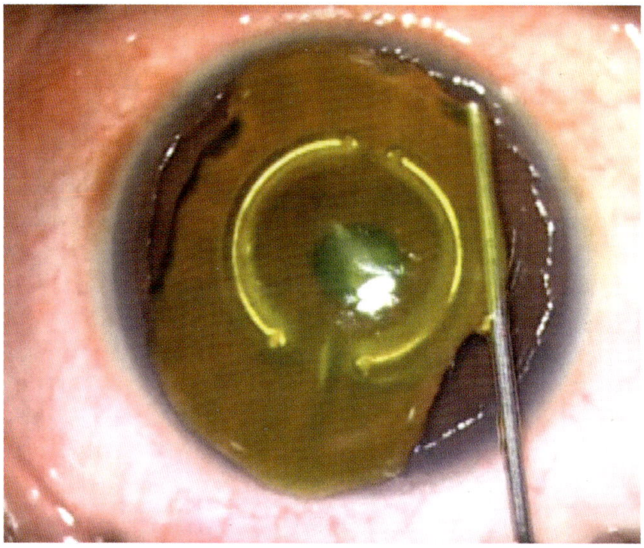

Fig. 8.8 Combined corneoplastique technique: Corneal scrapping followed by riboflavin instillation for cross-linking, 4 months after Keraring implantation[9]

Uniform dissection up to 400 μm: Digitally changing the depth and the inner and outer diameters of the channel is easily viewed on the femtosecond laser screen. This may avoid the risk of perforation compared with mechanical ring dissector. Using a mechanical technique, Kanellopoulos et al[4] presented one case of anterior chamber perforation in 20 patients.

Good centration: Another big advantage of the femtosecond laser for intracorneal ring segment implantation is its predictable centration, which is important especially for rings with

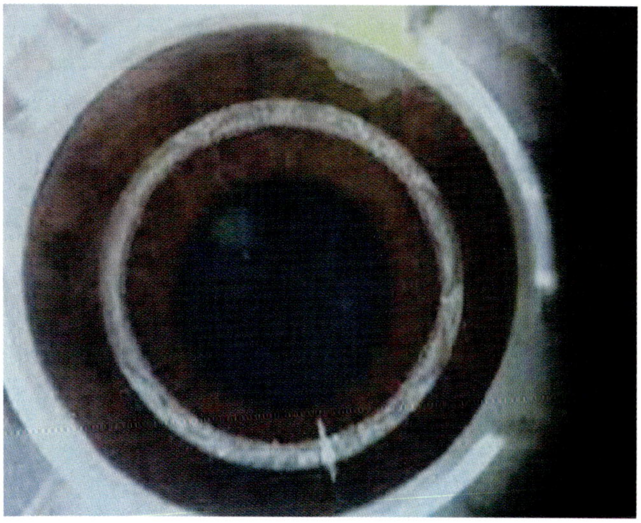

Fig. 8.9 IntraLase channel creation takes approximately 8 seconds

Fig. 8.10 The U-suture technique is an effective method to stop segment migration

5 mm optical zones (i.e, Keraring). Centralization may be more easily achieved using a marker before applying the laser. After placing the vacuum ring and applanation to the cornea, we choose the geometric axis of the cornea to center the procedure. Some ring companies advise that the ink mark need to be placed at the geometric axis of the cornea; other companies advise the Purkinje reflex as the center. Myself, however, prefer somewhere in between the Purkinje reflex and pupil center.

Better symmetry: Better symmetry provides a better outcome. In a mechanical Intacs implantation method, if the vacuum is lost after the first channel creation, the second channel may be created at a different depth and location. This maneuver is even more difficult for manual Keraring implantation. Alternately, femtosecond laser channels are created simultaneously in 6 seconds, and the chance of loosing vacuum is low.

Aseptic technique: The femtosecond laser gives a chance for aseptic surgery, because we have sterile disposable suction rings and cones for each eye.

Customizing tunnel size: Customization of the tunnel is possible with a femtosecond laser. As previously mentioned, the depth of the channel as well as the inner and outer channel diameters is digitally changed with ease in femtosecond laser techniques. If we decrease the inner diameter of the channel after ring implantation, the ring pulls the channel distally. This effect may increase with narrower channels. Rabiniovitz et al[3] showed that narrower ring segment channels produced a greater change in visual acuity and refraction.

Coskunseven conducted a study with three groups according to inner and outer diameters.[5] In group two, they made a narrower channel, and in group three, the channel width was same as group two, but the inner diameter was smaller. They found that a smaller inner diameter and narrower channel resulted in more effective results versus the other groups.

More effective: The femtosecond laser is more effective for intracorneal ring segment implantation than a mechanical technique. Several studies compared mechanical versus femtosecond laser techniques. Rabinowitz et al showed that with the exception of the change in surface regularity index, the laser group performed better in all parameters. When we compared our IntraLase results with mechanical spreader results at 1 year[6,7] we concluded that results with the IntraLase were better.

KERARING INTRALASE PROCEDURE STEP-BY-STEP

Step 1: Centration of the procedure using a inked Sinsky hook at the first Punkinje reflex.

Step 2: IntraLase Ring Programming:

- Depth 80 percent
- Inner diameter 4.8 mm
- Outer diameter 5.3 mm
- Entry cut length 0.8 to 1.0 mm
- Entry cut thickness 1 micron
- Axis according to steepest topographic meridian.

Steps 3 and 4:

- Suction ring applying
- Followed by ducking the cone attached to the robotic IntraLase arm into the suction ring.

Steps 5 and 6:

- Overlay centering and adjusting using IntraLase overlay software
- Ring dissection, takes 6 seconds.

Step 7: Keraring segments insertion using the Albertazzi forceps each side of the incision entry.

Step 8: Using the Sinsky hook, final touch segment position is achieved.

Step 9: Keraring symmetry checking first image 160° similar segment implantation, and second image nonsimilar length segment implantation.

NOMOGRAMS AND CONSULTANCY

Based on statistical analysis of thousands of results and valuable feedback from ophthalmic surgeons, nomograms are being constantly updated. Mediphacos provides complimentary technical consultancy services to support Keraring users worldwide. We rather to use Keraring than other intracorneal rings because we have different segments to fulfill each particular type of refraction and different topographic patterns (Figs 8.12 to 8.16).

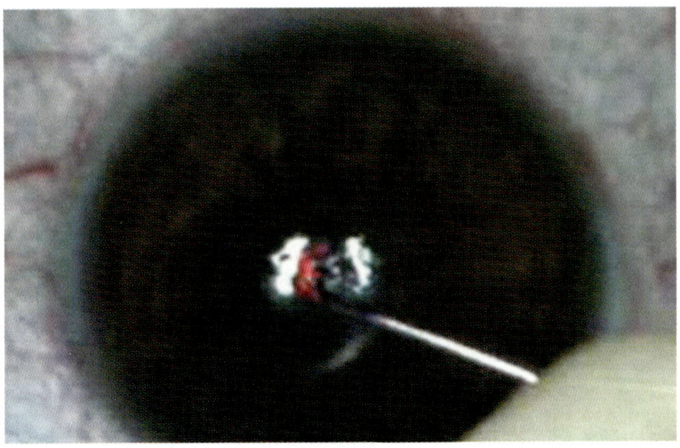

Step 1

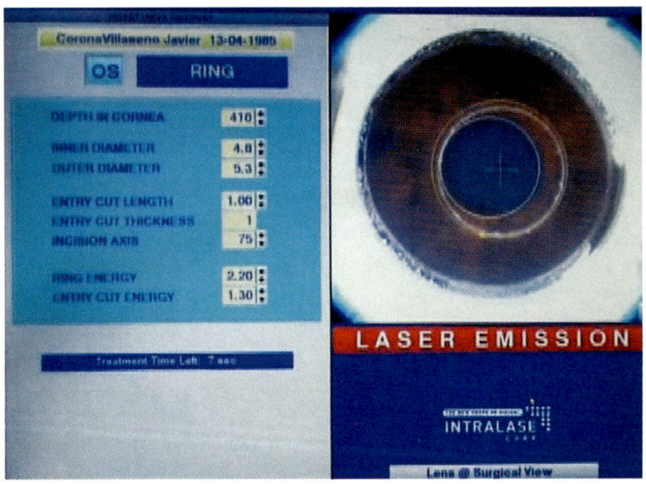

Step 2

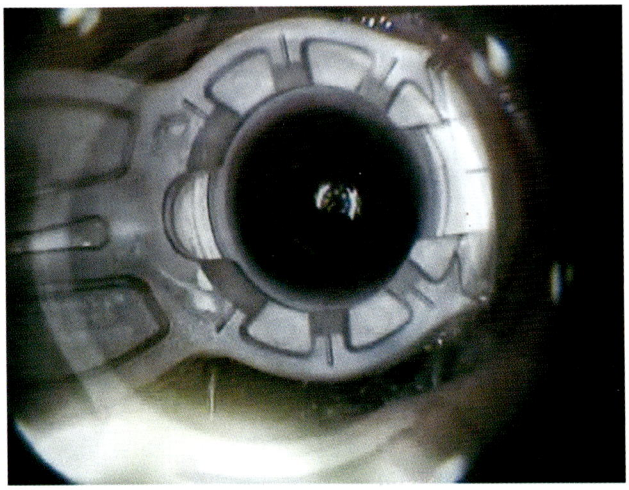

Step 3

Step 4

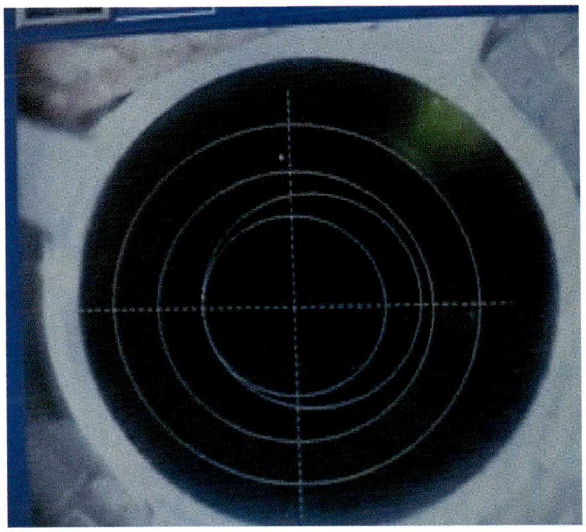

Step 5

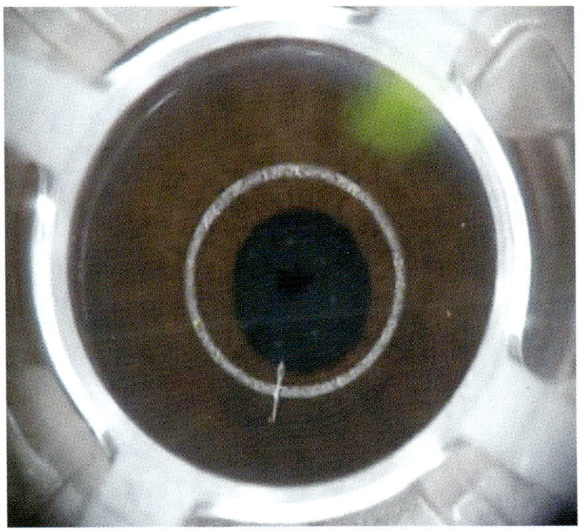

Step 6

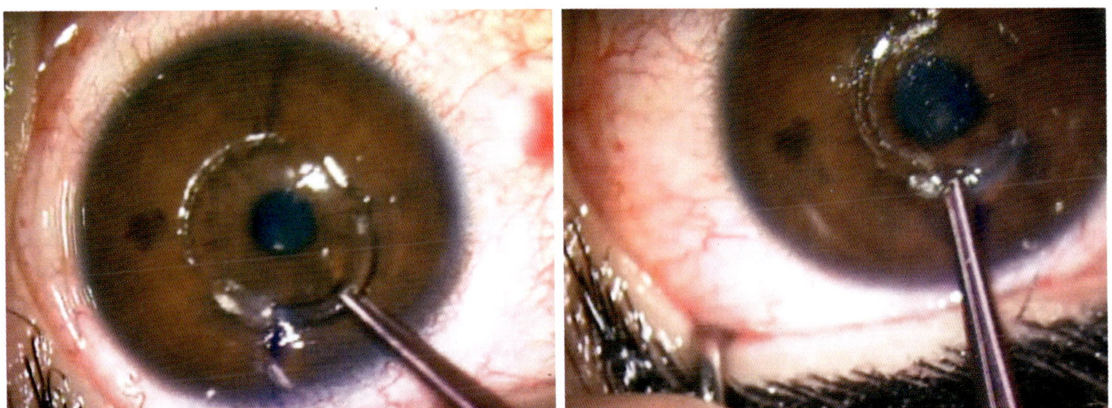

Step 7

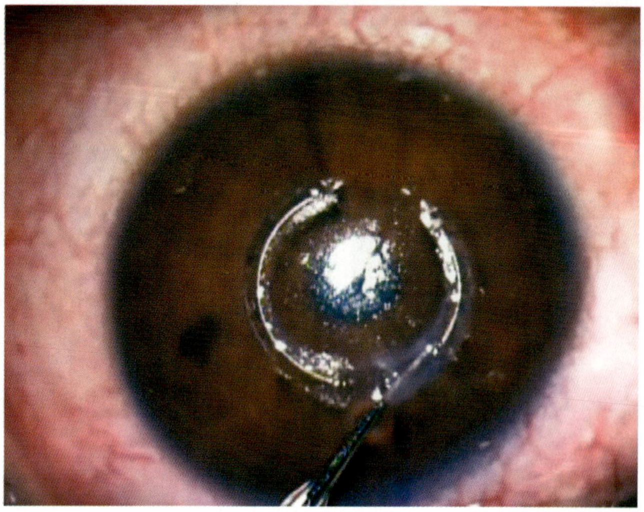

Step 8

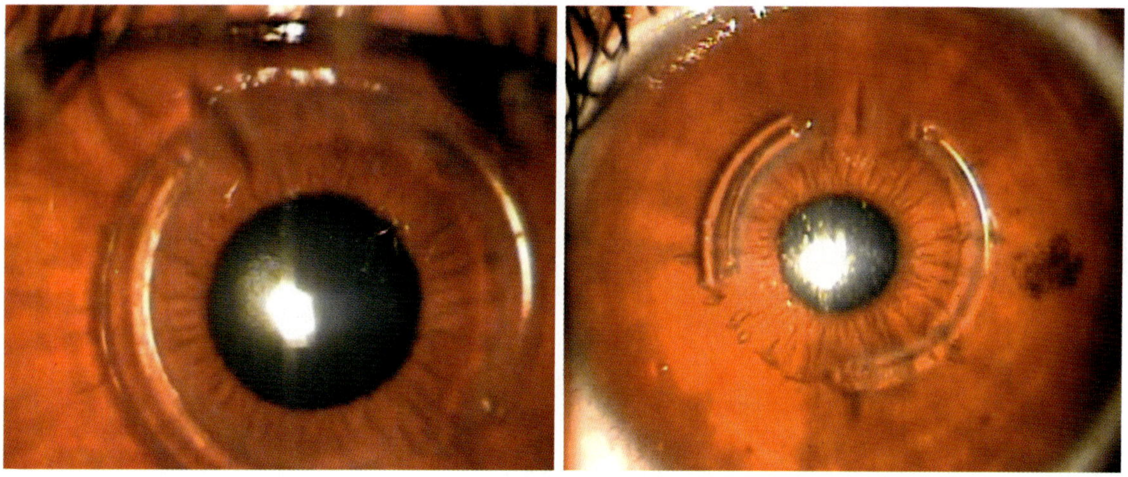

Step 9

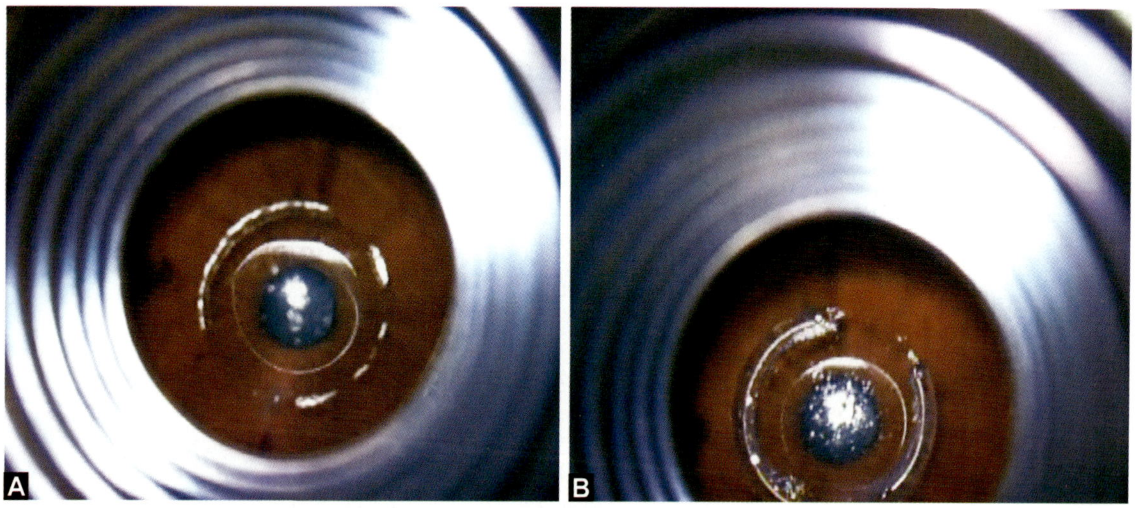

Figs 8.11A and B Maloney ring reflex irregular oval preoperative image, round reflex at the postoperative image

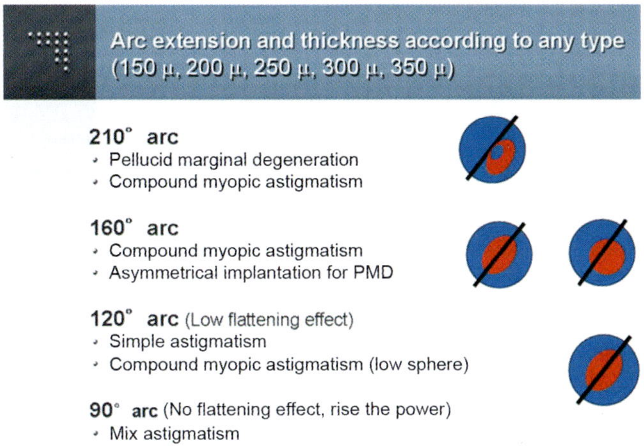

Fig. 8.12 Different arc of length for different refractions an topographic patterns

KERARING SPECIFICATIONS (FIG. 8.13)

- Material: Medical-grade PMMA
- Fabrication: Manufacturing processes audited and certified as per ISO-9001, ISO13485, CE Mark and GMP.
- Variable diameters: 5.0 mm (standard), 5.5 mm, 6.0 mm
- Variable thicknesses: From 150 μm to 350 μm in 50 μm steps.
- Variable arc lengths: 90°, 120°, 160° (standard), 210°, 240°.

SET OF INSTRUMENTS (FIG. 8.17)

- Produced and distributed by Mediphacos. (www.mediphacos.com.br)

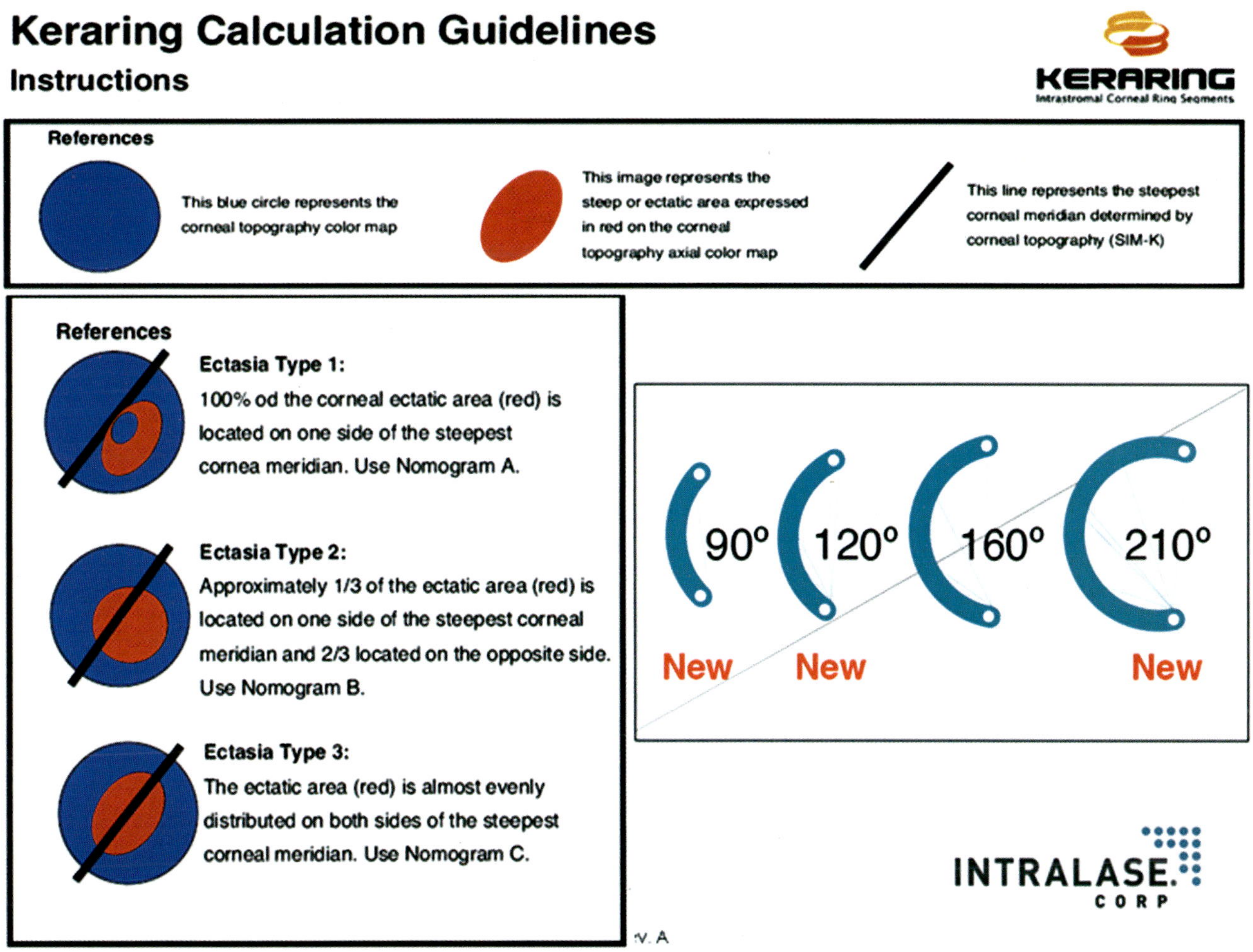

Fig. 8.13 Nomogram explanation, and different arc of length segments

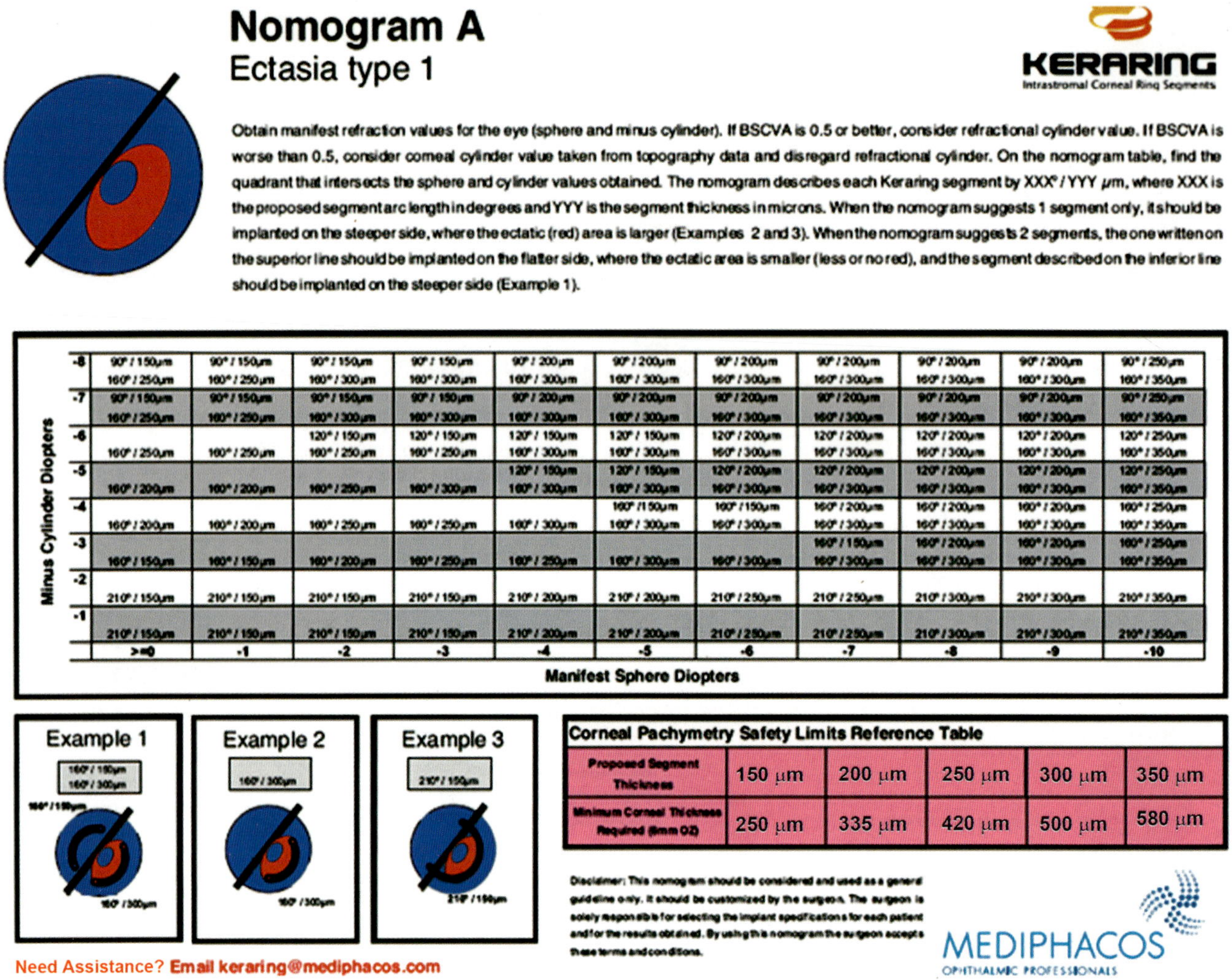

Nomogram A
Ectasia type 1

KERARING
Intrastromal Corneal Ring Segments

Obtain manifest refraction values for the eye (sphere and minus cylinder). If BSCVA is 0.5 or better, consider refractional cylinder value. If BSCVA is worse than 0.5, consider corneal cylinder value taken from topography data and disregard refractional cylinder. On the nomogram table, find the quadrant that intersects the sphere and cylinder values obtained. The nomogram describes each Keraring segment by XXX° / YYY µm, where XXX is the proposed segment arc length in degrees and YYY is the segment thickness in microns. When the nomogram suggests 1 segment only, it should be implanted on the steeper side, where the ectatic (red) area is larger (Examples 2 and 3). When the nomogram suggests 2 segments, the one written on the superior line should be implanted on the flatter side, where the ectatic area is smaller (less or no red), and the segment described on the inferior line should be implanted on the steeper side (Example 1).

Minus Cylinder Diopters \ Manifest Sphere Diopters	>=0	-1	-2	-3	-4	-5	-6	-7	-8	-9	-10
-8	90°/150µm 160°/250µm	90°/150µm 160°/250µm	90°/150µm 160°/300µm	90°/150µm 160°/300µm	90°/200µm 160°/300µm	90°/200µm 160°/300µm	90°/200µm 160°/300µm	90°/200µm 160°/300µm	90°/200µm 160°/300µm	90°/200µm 160°/300µm	90°/250µm 160°/350µm
-7	90°/150µm 160°/250µm	90°/150µm 160°/250µm	90°/150µm 160°/300µm	90°/150µm 160°/300µm	90°/200µm 160°/300µm	90°/200µm 160°/300µm	90°/200µm 160°/300µm	90°/200µm 160°/300µm	90°/200µm 160°/300µm	90°/200µm 160°/300µm	90°/250µm 160°/350µm
-6	160°/250µm	160°/250µm	120°/150µm 160°/250µm	120°/150µm 160°/250µm	120°/150µm 160°/300µm	120°/150µm 160°/300µm	120°/200µm 160°/300µm	120°/200µm 160°/300µm	120°/200µm 160°/300µm	120°/200µm 160°/300µm	120°/250µm 160°/350µm
-5	160°/200µm	160°/200µm	160°/250µm	160°/300µm	120°/150µm 160°/300µm	120°/150µm 160°/300µm	120°/200µm 160°/300µm	120°/200µm 160°/300µm	120°/200µm 160°/300µm	120°/200µm 160°/300µm	120°/250µm 160°/350µm
-4	160°/200µm	160°/200µm	160°/250µm	160°/250µm	160°/300µm	160°/150µm 160°/300µm	160°/150µm 160°/300µm	160°/200µm 160°/300µm	160°/200µm 160°/300µm	160°/200µm 160°/300µm	160°/250µm 160°/350µm
-3	160°/150µm	160°/150µm	160°/200µm	160°/250µm	160°/250µm	160°/300µm	160°/300µm	160°/150µm 160°/300µm	160°/200µm 160°/300µm	160°/200µm 160°/300µm	160°/250µm 160°/350µm
-2	210°/150µm	210°/150µm	210°/150µm	210°/150µm	210°/200µm	210°/200µm	210°/250µm	210°/250µm	210°/300µm	210°/300µm	210°/350µm
-1	210°/150µm	210°/150µm	210°/150µm	210°/150µm	210°/200µm	210°/200µm	210°/250µm	210°/250µm	210°/300µm	210°/300µm	210°/350µm

Example 1: 160°/150µm, 160°/300µm
Example 2: 160°/300µm
Example 3: 210°/150µm

Corneal Pachymetry Safety Limits Reference Table

Proposed Segment Thickness	150 µm	200 µm	250 µm	300 µm	350 µm
Minimum Corneal Thickness Required (6mm OZ)	250 µm	335 µm	420 µm	500 µm	580 µm

Disclaimer: This nomogram should be considered and used as a general guideline only. It should be customized by the surgeon. The surgeon is solely responsible for selecting the implant specifications for each patient and for the results obtained. By using this nomogram the surgeon accepts these terms and conditions.

Need Assistance? Email keraring@mediphacos.com

MEDIPHACOS
OPHTHALMIC PROFESSIONALS

Fig. 8.14 Nomogram A, ectasia type 1

POSTOPERATIVE CARE

- Therapeutic soft contact lens for 24 hours
- Paracetamol 750 mg every 8 hours or Ketorolac 10 mg every 8 hours if required
- Sodium Diclofenac drops: 1 drop every 15′ for one hour. 1 drop every 8 hours for 7 days.
- Moxifloxacin drops: 1 drop every 6 hours for 14 days.
- Prednisolone acetate 1 drop every 8 hours for 14 days
- Lubricating drops: According to necessity for unlimited time
- Do not suspend antibiotics until incision is completely healed

CUSTOM KERARING AND INTRALASE TOPOGRAPHIC AND REFRACTIVE EXAMPLES

As we observed in all these cases the four corneoplastique concepts were achieved.[9]

- Improve optics
- Improve the topography
- Reduce the refraction
- Improve UCVA and BCVA

COMPLICATIONS

Intraoperative complications are rare:

- Gas bubble posterior breakthrough, skill surgeon can insert the segments

Nomogram B

Ectasia type 2

KERARING
Intrastromal Corneal Ring Segments

Obtain manifest refraction values for the eye (sphere and minus cylinder). If BSCVA is 0.5 or better, consider refractional cylinder value. If BSCVA is worse than 0.5, consider corneal cylinder value taken from topography data and disregard refractional cylinder. On the nomogram table, find the quadrant that intersects the sphere and cylinder values obtained. The nomogram describes each Keraring segment by XXX° / YYY µm, where XXX is the proposed segment arc length in degrees and YYY is the segment thickness in microns. When the nomogram suggests 1 segment only, it should be implanted on the steeper side, where the ectatic (red) area is larger (Examples 2 and 3). When the nomogram suggests 2 segments, the one written on the superior line should be implanted on the flatter side, where the ectatic area is smaller (less red), and the segment described on the inferior line should be implanted on the steeper side (Example 1).

Minus Cylinder Diopters	>=0	-1	-2	-3	-4	-5	-6	-7	-8	-9	-10
-8	90°/150µm 160°/250µm	90°/150µm 160°/250µm	90°/150µm 160°/300µm	90°/150µm 160°/300µm	90°/200µm 160°/300µm	90°/200µm 160°/300µm	90°/200µm 160°/300µm	90°/200µm 160°/300µm	90°/200µm 160°/300µm	90°/200µm 160°/300µm	90°/250µm 160°/350µm
-7	90°/150µm 160°/250µm	90°/150µm 160°/250µm	90°/150µm 160°/300µm	90°/150µm 160°/300µm	90°/200µm 160°/300µm	90°/200µm 160°/300µm	90°/200µm 160°/300µm	90°/200µm 160°/300µm	90°/200µm 160°/300µm	90°/200µm 160°/300µm	90°/250µm 160°/350µm
-6	160°/250µm	160°/250µm	120°/150µm 160°/250µm	120°/150µm 160°/250µm	120°/150µm 160°/300µm	120°/150µm 160°/300µm	120°/200µm 160°/300µm	120°/200µm 160°/300µm	120°/200µm 160°/300µm	120°/200µm 160°/300µm	120°/250µm 160°/350µm
-5	160°/200µm	160°/200µm	160°/250µm	160°/150µm 160°/250µm	120°/150µm 160°/300µm	120°/150µm 160°/300µm	120°/200µm 160°/300µm	120°/200µm 160°/300µm	120°/200µm 160°/300µm	120°/200µm 160°/300µm	120°/250µm 160°/350µm
-4	160°/200µm	160°/200µm	160°/250µm	160°/150µm 160°/250µm	160°/150µm 160°/300µm	160°/150µm 160°/300µm	160°/200µm 160°/300µm	160°/200µm 160°/300µm	160°/200µm 160°/300µm	160°/200µm 160°/300µm	160°/250µm 160°/350µm
-3	160°/150µm	160°/150µm	160°/200µm	160°/250µm	160°/150µm 160°/250µm	160°/150µm 160°/300µm	160°/200µm 160°/300µm	160°/150µm 160°/300µm	160°/200µm 160°/300µm	160°/200µm 160°/300µm	160°/250µm 160°/350µm
-2	210°/150µm	210°/150µm	210°/150µm	210°/150µm	210°/200µm	210°/200µm	210°/250µm	210°/250µm	210°/300µm	210°/300µm	210°/350µm
-1	210°/150µm	210°/150µm	210°/150µm	210°/150µm	210°/200µm	210°/200µm	210°/250µm	210°/250µm	210°/300µm	210°/300µm	210°/350µm

Manifest Sphere Diopters

Example 1

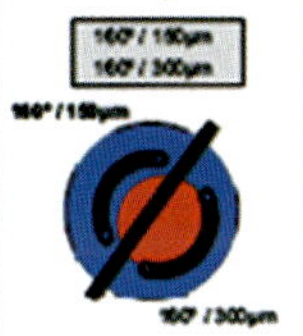

Example 2

Example 3

Corneal Pachymetry Safety Limits Reference Table

Proposed Segment Thickness	150 µm	200 µm	250 µm	300 µm	350 µm
Minimum Corneal Thickness Required (6mm OZ)	250 µm	335 µm	420 µm	500 µm	580 µm

Disclaimer: This nomogram should be considered and used as a general guideline only. It should be customized by the surgeon. The surgeon is solely responsible for selecting the implant specifications for each patient and for the results obtained. By using this nomogram the surgeon accepts these terms and conditions.

Need Assistance? **Email keraring@mediphacos.com**

MEDIPHACOS
OPHTHALMIC PROFESSIONALS

Fig. 8.15 Nomogram B, ectasia type 2

Postoperative Complications

- Chronic inflammation or infection inside the channel (1%)
- Migration of the intracorneal ring segments is rare (2.6%) (Fig. 8.24), vascularization of the wound (0.5%)
- Corneal melting/exposed segments (1%)

KERARING AND INTRALASE CONCLUSION: ORTHOPEDIC OR REFRACTIVE PROCEDURE

- Nomogram defined by sphere and cylinder
- Exchangeable procedure to enhance the refractive residual error
- Adjustuble procedure
- Different arc extension and thickness to fullfill different refractions and topography patterns
- Patients expectations achievement
- Improvement of postoperative UCVA vs preoperative BCVA
- IntraLase FS 60 Hz improve the refractive result of this corneal procedure[3,5,8-12]

Nomogram C

Ectasia type 3

KERARING
Intrastromal Corneal Ring Segments

Obtain manifest refraction values for the eye (sphere and minus cylinder). If BSCVA is 0.5 or better, consider refractional cylinder value. If BSCVA is worse than 0.5, consider corneal cylinder value taken from topography data and disregard refractional cylinder. On the nomogram table, find the quadrant that intersects the sphere and cylinder values obtained. The nomogram describes each Keraring segment by XXX°/ YYY µm, where XXX is the proposed segment arc length in degrees and YYY is the segment thickness in microns.

Minus Cylinder Diopters											
-8	90° / 250µm 90° / 250µm	90° / 250µm 90° / 250µm	120° / 300µm 120° / 300µm	120° / 300µm 120° / 300µm	160° / 300µm 160° / 300µm	160° / 300µm 160° / 300µm	160° / 300µm 160° / 300µm	160° / 300µm 160° / 300µm	160° / 300µm 160° / 300µm	160° / 300µm 160° / 300µm	160° / 300µm 160° / 300µm
-7	90° / 250µm 90° / 250µm	90° / 250µm 90° / 250µm	120° / 250µm 120° / 250µm	120° / 250µm 120° / 250µm	160° / 300µm 160° / 300µm	160° / 300µm 160° / 300µm	160° / 300µm 160° / 300µm	160° / 300µm 160° / 300µm	160° / 300µm 160° / 300µm	160° / 300µm 160° / 300µm	160° / 300µm 160° / 300µm
-6	90° / 200µm 90° / 200µm	90° / 200µm 90° / 200µm	120° / 250µm 120° / 250µm	120° / 250µm 120° / 250µm	160° / 300µm 160° / 300µm	160° / 300µm 160° / 300µm	160° / 300µm 160° / 300µm	160° / 300µm 160° / 300µm	160° / 300µm 160° / 300µm	160° / 300µm 160° / 300µm	160° / 300µm 160° / 300µm
-5	90° / 200µm 90° / 200µm	90° / 200µm 90° / 200µm	120° / 200µm 120° / 200µm	120° / 250µm 120° / 250µm	160° / 250µm 160° / 250µm	160° / 250µm 160° / 250µm	160° / 300µm 160° / 300µm	160° / 300µm 160° / 300µm	160° / 300µm 160° / 300µm	160° / 300µm 160° / 300µm	160° / 300µm 160° / 300µm
-4	120° / 200µm 120° / 200µm	120° / 200µm 120° / 200µm	160° / 200µm 160° / 200µm	160° / 200µm 160° / 200µm	160° / 250µm 160° / 250µm	160° / 250µm 160° / 250µm	160° / 300µm 160° / 300µm	160° / 300µm 160° / 300µm	160° / 300µm 160° / 300µm	160° / 300µm 160° / 300µm	160° / 300µm 160° / 300µm
-3	120° / 150µm 120° / 150µm	120° / 150µm 120° / 150µm	160° / 150µm 160° / 150µm	160° / 200µm 160° / 200µm	160° / 200µm 160° / 200µm	160° / 200µm 160° / 200µm	160° / 250µm 160° / 250µm	160° / 250µm 160° / 250µm	160° / 300µm 160° / 300µm	160° / 300µm 160° / 300µm	160° / 300µm 160° / 300µm
-2	120° / 150µm 120° / 150µm	120° / 150µm 120° / 150µm	160° / 150µm 160° / 150µm	160° / 150µm 160° / 150µm	160° / 200µm 160° / 200µm	160° / 200µm 160° / 200µm	160° / 250µm 160° / 250µm	160° / 250µm 160° / 250µm	160° / 300µm 160° / 300µm	160° / 300µm 160° / 300µm	160° / 300µm 160° / 300µm
-1	120° / 150µm 120° / 150µm	120° / 150µm 120° / 150µm	160° / 150µm 160° / 150µm	160° / 150µm 160° / 150µm	160° / 200µm 160° / 200µm	160° / 200µm 160° / 200µm	160° / 250µm 160° / 250µm	160° / 250µm 160° / 250µm	160° / 300µm 160° / 300µm	160° / 300µm 160° / 300µm	160° / 300µm 160° / 300µm
	>=0	-1	-2	-3	-4	-5	-6	-7	-8	-9	-10

Manifest Sphere Diopters

Corneal Pachymetry Safety Limits Reference Table					
Proposed Segment Thickness	150 µm	200 µm	250 µm	300 µm	350 µm
Minimum Corneal Thickness Required (6mm OZ)	250 µm	335 µm	420 µm	500 µm	580 µm

Disclaimer: This nomogram should be considered and used as a general guideline only. It should be customized by the surgeon. The surgeon is solely responsible for selecting the implant specifications for each patient and for the results obtained. By using this nomogram the surgeon accepts these terms and conditions.

Need Assistance? Email kerarimg@mediphacos.com

MEDIPHACOS
OPHTHALMIC PROFESSIONALS

Fig. 8.16 Nomogram C,ectasia type 3

Fig. 8.17 Keraring instruments implantation set

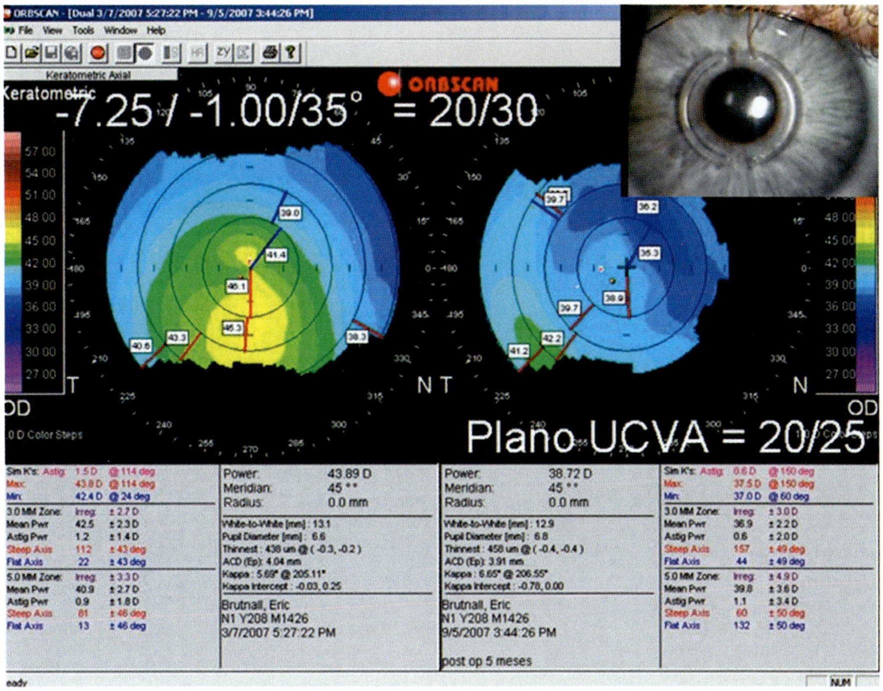

Fig. 8.18 160° arc of length/300 μ thickness symmetric segments implantation

- Nomogram C, ectasia type 3
- Preoperative -7.25/-1.00 X 35°= BCVA 20/30
- Postoperative Plano = UCVA 20/25

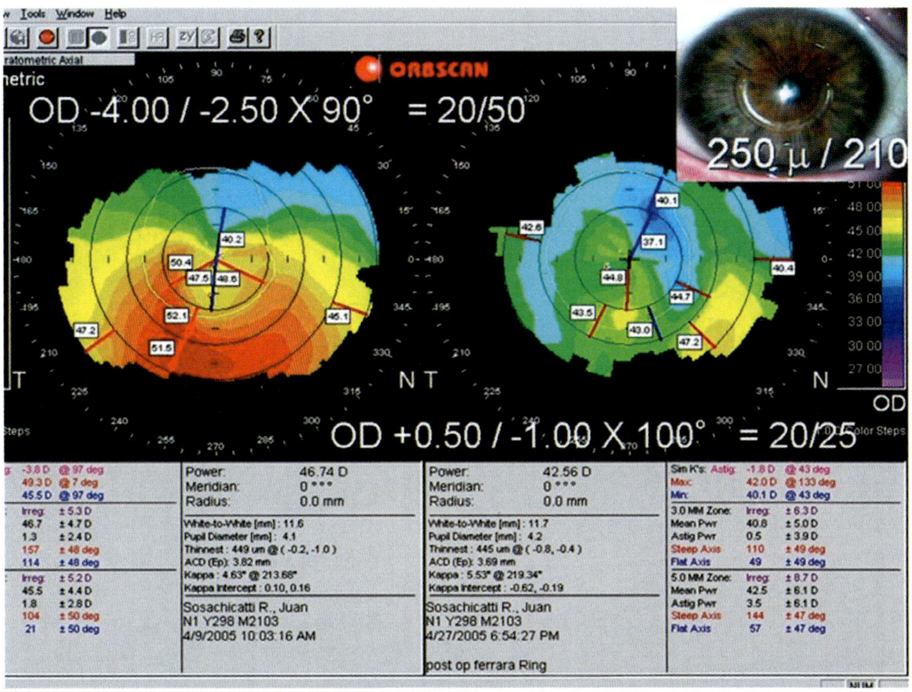

Fig. 8.19 210° arc of length, 250 μ thickness, only one inferior segment implantation

- Nomogram A, ectasia type 1
- Preoperative -4.00/-2.50 X 90° = BCVA 20/50
- Postoperative +0.50/-1.00 X 100° = BCVA 20/25

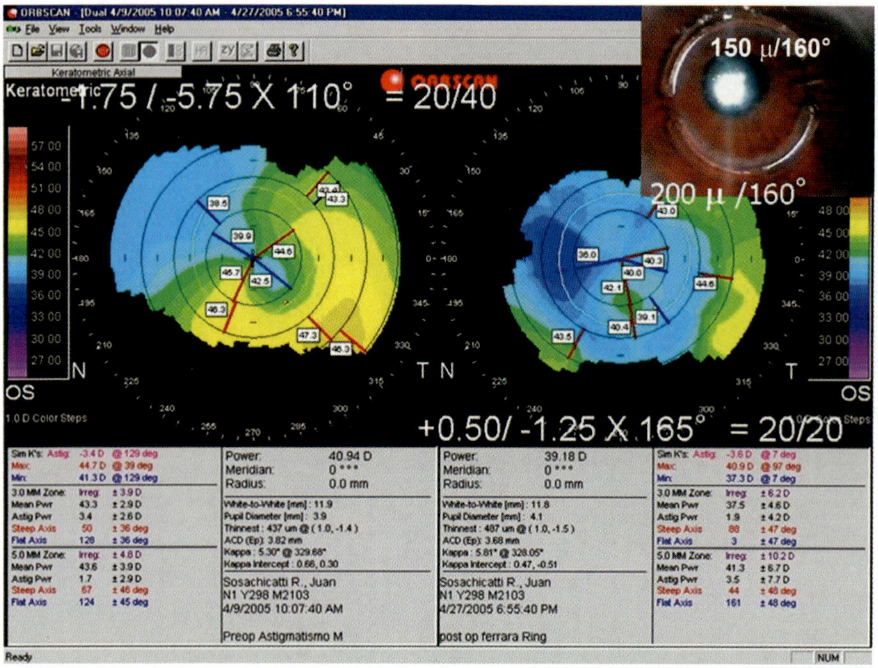

Fig. 8.20 160° arc of length asymmetric implantation 150 µ superior, 200 µ inferior

- Nomogram B, ectasia type 2
- Preoperative -1.75/-5.75 X 110° = BCVA 20/40
- Postoperative +0.50/-1.25 X 165° = BCVA 20/20

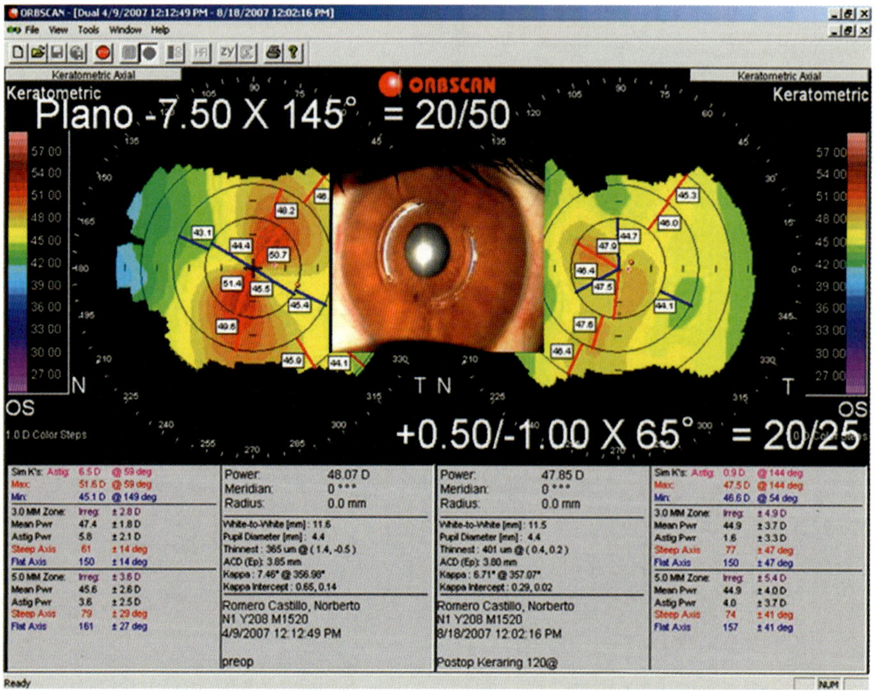

Fig. 8.21 120° arc of length, 200 µ segment symmetric implantation

- Nomogram C, type 3, notice the minimal corneal flattening effect
- Preoperative Plano/-7.50 X 145° = BCVA 20/50
- Postoperative +0.50 /-1.00 X 65°

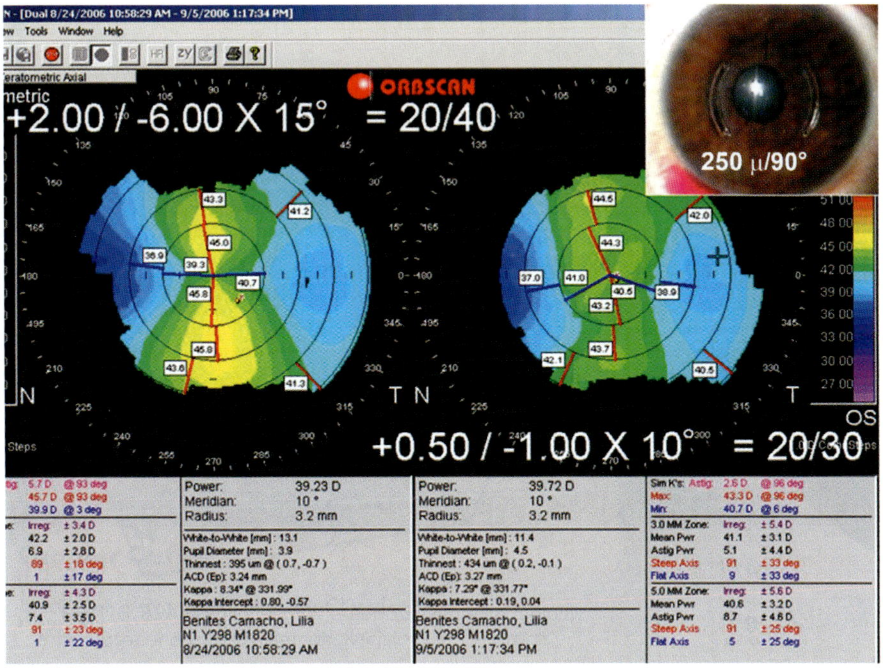

Fig. 8.22 90° arc of length, 250 μ thickness symmetrical segments implantation

- Nomogram C, type 3
- Preoperative +2.00/-6.00 X 15° = BCVA 20/40
- Postoperative +0.50/-1.00 X 10° = BCVA 20/30

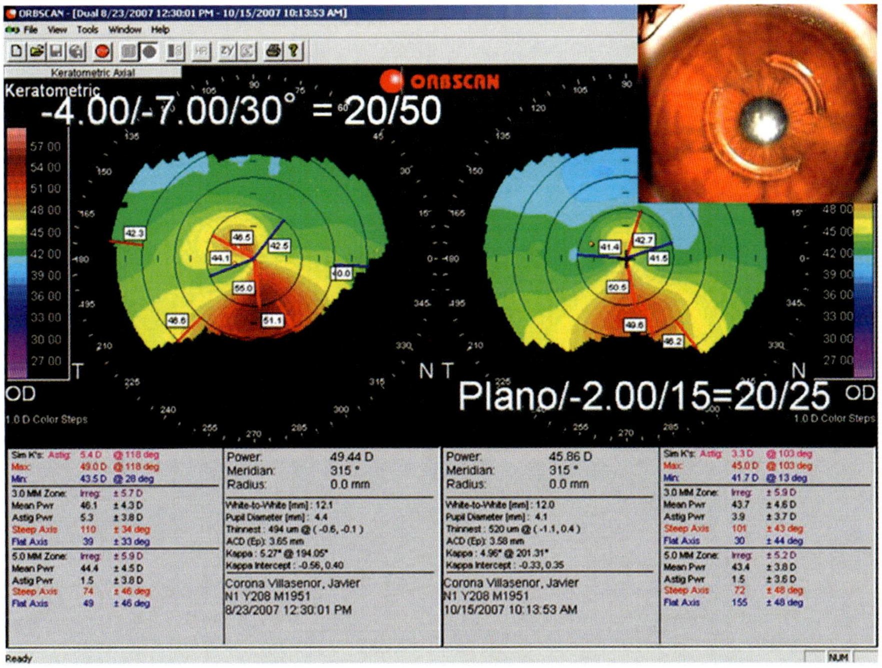

Fig. 8.23 Asymmetric arc of length implantation inferior 160°, 90° superior

- Nomogram B, ectasia type 2
- Preoperative -4.00/-7.00 X 30° = BCVA 20/50
- Postoperative Plano/-2.00/15° = BCVA 20/25

Fig. 8.24 The U-suture technique is an effective method to stop segment migration

REFERENCES

1. Flemming JF, Reynolds AE, Kilmer L, et al. The intrastromal corneal ring: two cases in rabbits. J Refract Surg. 1987;3:227-32.
2. Colin J, Cochener B, Savary G, Malet F. Correcting keratoconus with intracorneal rings. J Cataract Refract Surg. 2000;26:1117-22.
3. Rabinowitz YS, Li X, Ignacio TS, Maguen E. Intacs Inserts using the femtosecond laser compared to the mechanical spreader in the treatment of keratoconus. J Refract Surg. 2006;22:764-71.
4. Kanellopoulos AJ, Pe LH, Perry HD, Donnenfeld ED. Modified intracorneal ring segment implantations (Intacs) for the management of moderate to advanced keratoconus: efficacy and complications. Cornea. 2006;25:29-33.
5. Coskunseven E. Modification of the parameters in intralase to improve effect in keratoconus patients with Intacs. Paper presented at the XXIV Meeting of the European Society of Cataract and Refractive Surgeons. London; September 11, 2006.
6. Colin J, Cochener B, Savary G, et al. INTACS insert for treating keratoconus: one-year results. Ophthalmology. 2001;108:1409-14.
7. Siganos CS, Kymionis GD, Kartakis N, et al. Management of keratoconus with Intacs. Am J Ophthalmol. 2003;135:64-70.
8. Coskunseven E. Results of IntraLase Intacs in 300 Keratoconus Eyes. Paper presented at the 11th Winter Refractive Surgery Meeting of the European Society of Cataract and Refractive Surgery. Athens, Greece; February 3, 2007.
9. Gark Ashok. Mastering Corneal Collagen Cross Linking Techinques: Guliani 56-63 , Lovisolo 98-109; 2009 Jaypee
10. Efekan Kuskunseven. Cataract and Refractive Surgery Today Europe, 32-34, October 2007
11. Levinger S, Pokrov R. Keratoconus managed with intacs: one-year results. Arch Ophthalmol. 2005;123:1308-14.
12. Colin J. European clinical evalution: use of intacs for the treatment of keratoconus. J Cataract Refract Surg. 2006;32:747-55.

9

LASIK Flap Creation with FEMTEC

C Banu Cosar (Turkey)

Chapter Outline

- Corneal flap creation with femtosecond laser systems
- FS laser flap complications
- Femtec

INTRODUCTION

Currently, four femtosecond (FS) lasers are FDA-licensed for LASIK: Intralase (Abbott Medical Optics), VisuMax (Carl Zeiss Meditec AG), Femtec (Technolas Perfect Vision), and Femto LDV (Ziemer Ophthalmic Systems AG).[1]

All four FS laser systems use ultra-short laser pulses, and the interaction process for cutting the corneal tissue is based on photodisruption. However, significantly different technical details and clinical experience are have to be considered.[1]

In this chapter, femtosecond laser-assisted LASIK is discussed, technical features of Femtec laser system are summarized, and a video showing the author's personal technique with Femtec is presented.

CORNEAL FLAP CREATION WITH FEMTOSECOND LASER SYSTEMS

The femtosecond ophthalmic lasers are photodisruptive lasers with a pulse duration in the femtosecond (10^{-15}) range. Photodisruption begins with laser-induced optical breakdown, when a strongly focused and short-duration laser pulse generates a high-intensity electrical field, thereby causing the formation of a mixture of free electrons and ions that constitutes the plasma state. The optically generated hot plasma expands with supersonic velocity and displaces surrounding material. As the plasma's expansion slows, the initial supersonic displacement propagates through the surrounding tissue as a shock wave. The rapid expansion of the generated plasma quickly decreases its temperature, and the vaporized tissue forms a cavitation gas bubble in the focal volume of the laser beam. The cavitation bubble consists mainly of CO_2, N_2, and H_2O, which can diffuse out from the tissue through normal mechanisms. The ablation was found to be nonthermal due to the extremely short interaction time.[2]

FS laser pulses create a lamellar stromal cut in a "raster" or a "spiral" patern (Figs 9.1A to C). Then the side cut is performed with pulses going in the "posteroanterior" direction.[3]

Advantages of FS Laser Flap Creation

1. Reduced incidence of complications like button-hole, epithelial abrasion, short flap, free cap, blade marks, and irregular cut
2. Greater options in flap diameter, flap thickneess, side cut angle, hinge position and hinge length
3. Absence of moving parts as in a microkeratome
4. Increased flap safety and flap thickness predictability
5. Capability to cut thinner flaps.[4,5]

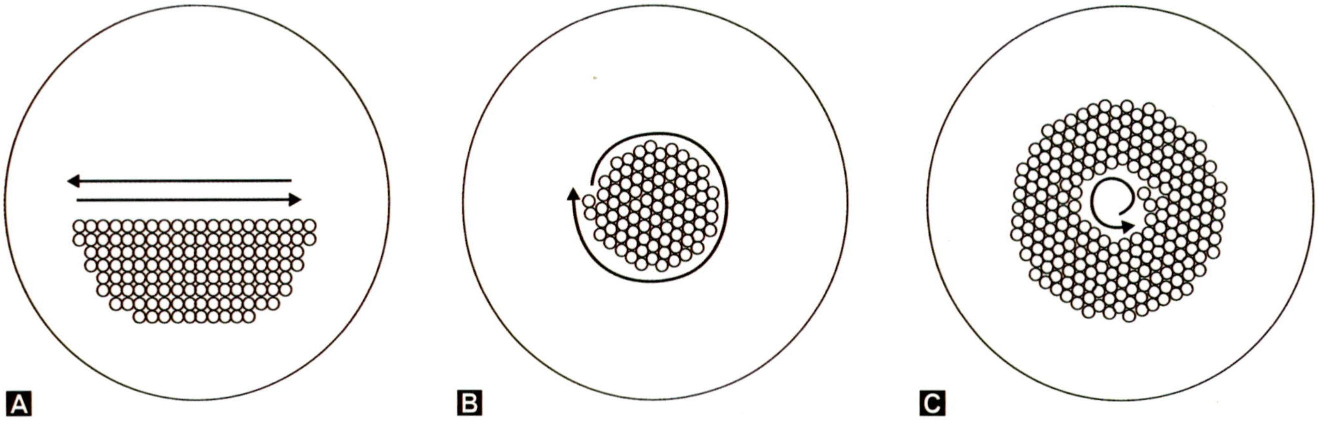

Figs 9.1A to C Laser spot firing patterns with FS lasers. (A) Raster pattern; (B) Centrifugal spiral pattern; (C) Centripetal spiral pattern

Disadvantages of FS Laser Flap Creation

1. Increased cost
2. Larger physical size
3. Lack of portability.[3]

FS LASER FLAP COMPLICATIONS

1. Cavitation gas bubbles confounds the ability of the surgeon and the excimer laser eye tracker to locate the pupil by cavitation gas bubbles. These bubbles can be moved away from the center by pushing with a cannula (Fig. 9.2). Today, the bubble layer is decreased by the use of raster or centripetal spiral patterns, peripheral gutters evacuating the bubbles away from the center, and less energy level with faster firing rates.[3]
2. *Transient light sensitivity syndrome (TLSS):* TLSS is seen days/weeks after LASIK procedure. The visual acuity is good but there's increased ligt sensitivity. There's no positive findings in the clinical examination. It resolves in weeks with topical corticosteroids. The exact mechanism is unknown. However, it is believed to be due to biochemical response of the keratocytes to near-infrared laser energy or inflammatory response of the adjacent tissues to gas bubbles.[6]
3. *Diffuse lamellar keratitis (DLK):* DLK was relatively common after initial FS laser surgeries. However, with the advent of faster firing systems, DLK incidence decreased. Today, DLK limited to the peripheral cornea, because of increased energy use for side cuts, is still encountered. This lamellar inflammation which is different from microkeratome related-DLK is believed to be the microscopic tissue damage aggrevated by ocular surface inflammatory mediators.[7]

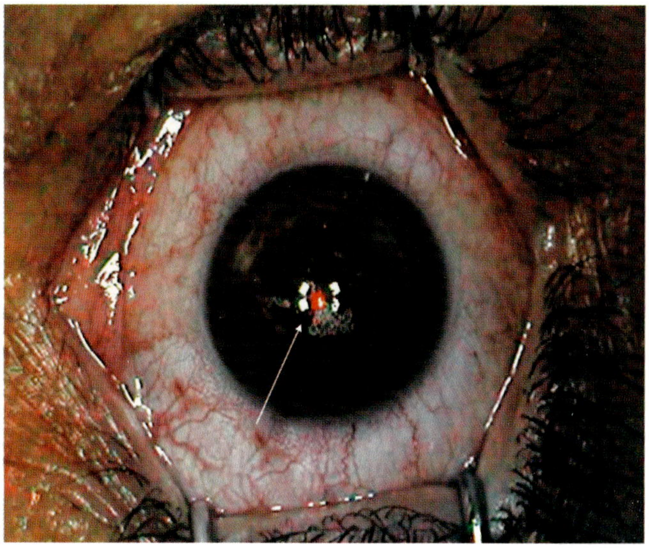

Fig. 9.2 Opaque bubble layer after FS laser flap cut

4. *Rainbow glare:* This optical side effect due to the light scattering from posterior surface of the interface is observed in 19 percent of Femto-LASIK patients. Visual acuity is not effected in most patients. The incidence of this complication has been decreased by the improvements in the focusing optics.[8]
5. *Loss of vacuum in the suction ring:* Loss of vacuum by femtosecond laser is not as serious a problem as with the mechanical microkeratome. The vacuum ring can be reinserted and the flap could be recut. If loss of vacuum occurs during side cut, a smaller new cut is performed. Premature vacuum loss causes less change in the corneal shape with curved applanation systems.[3]

Table 9.1 Technical features of Femtec

Wavelength (nm)	1040
Pulse width (fs)	>500
Spot size (μm)	>1
Repetition rate (kHz)	40, 80
Pulse energy (μJ)	>1

FEMTEC

With Femtec, contact between the laser and the eye is established by a concave shaped disposable contact glass called "patient interface". By lifting up the bed/patient, the interface is inserted into the suction ring and then locked with the ring. The correct position of the interface is indicated by a light. The technical features of Femtec laser system is shown in Table 9.1.[1] Femtec-assisted LASIK is safe and efficient with successful refractive results (Cosar CB, unpublished data).

REFERENCES

1. Lubatschowski H. Overview of commercially available femtosecond lasers in refractive surgery. J Refract Surg. 2008;24:102-7.
2. Cosar CB, Sener AB. The FEMTEC laser in LASIK flap creation. CRS Today. 2004;2:27-9.
3. Soong HK, Malta JB. Femtosecond lasers in ophthalmology. Am J Ophthalmol. 2009;147(2):189-97.
4. Kezirian GM, Stonecipher KG. Comparison of the IntraLase femtosecond laser and mechanical microkeratomes for laser in situ keratomileusis. J Cataract Refract Surg. 2004;30(4):804-11.
5. Durrie S, Kezirian GM. Femtosecond laser versus mechanical microkeratome flaps in wavefront-guided laser in situ keratomileusis: prospective contralateral eye study. J Cataract Refract Surg. 2005;31(1):120-6.
6. Stonecipher KG, Dishler JG, Ignacio TS, Binder PS. Transient light sensitivity after femtosecond laser flap creation: Clinical findings and management. J Cataract Refract Surg. 2006;32(1):91-4.
7. Gil-Cazorla R, Teus MA, de Benito-Llopis L, et al. Incidence of diffuse lamellar keratitis after laser in situ keratomileusis associated with the IntraLase 15 kHz femtosecond laser and Moria M2 microkeratome. J Cataract Refract Surg. 2008;34(1):28-31.
8. Krueger RR, Thornton IL, Xu M, et al. Rainbow glare as an optical side effect of IntraLASIK. Ophthalmology. 2008;115(7):1187-95.

10

Advanced Isovision: Predictable PresbyLASIK

Frederic Hehn (France)

Chapter Outline

- Purpose
- Methods
- Materials
- Results
- Discussion

INTRODUCTION

Commonly presbyLASIK use monovision to correct presbyopia. Some patients can get satisfaction, but some others become unhappy with their monovision. Monovision is a compromise, because it takes count about the dominant eye for distant vision and dominated eye for near vision. Therefore, this optical system gives not a natural vision: loss of contrast sensitivity, visual fatigability, bad fusion and poor binocularity. At least monovision is limited by the time, because the maximum tolerated ametropia is about 1.5 to 2.0 diopters, then monovision becomes unavailable for patients more than 55 years old. Monovision is a refractive correction then it concern only the LOAs (low order aberrations of Zernike polynomas). These refractive treatments are based on the modification of C3, and C5 polynomas to correct astigmatism, and the modification of the C4 polynoma (spherical equivalent SE) , called the first defocus to correct the sphere.

Others refractive ways use multizones treatments with the first very central part of the cornea for near vision, closed to a second one for intermediate vision until the third zone for distant vision. Or the contrary strategy distant vision on the central zone, then intermediate zone and at least near vision. This multizones strategy is a big compromise because the quality and visual performance depend on the pupil size, then of the lightness conditions.

From centuries ophthalmologists (Helmholtz, Decartes-Fresnel law $n_1 sin_1 = n_2 sin2$) knew that positive spherical aberrations (SA) were bad for the distant visual acuity at night with a lot of halos and glare, due to a big PSF (Point Spread function). That's the reason why the optical zones (OZ) and transition zones (TZ) have been enlarged in cases of myopic patients. Recently new ways have been explored to use negative SA in order to get an intermediate vision. Negatives SA appear when the cornea becomes hyperprolate, then the Q value of asphericity becomes negative so. The first one has been the advanced monovision of Theo Seiler in 2007, the second one the blended vision of Dan Reinstein[2] in 2009. In the both techniques in one eye (the dominated) for Theo Seiler,[3] in the both eyes for Dan Reinstein it is created a sufficient amount of negative SA due to an hyperprolate shape of the cornea, to increase the depth of field. This increasing of the depth of focus is used to upgrade the quality of binocularity, then the patient's confort. The both techniques use micromonovision. Spherical aberrations belong to HOAs (High Order Aberrations; they are called the second defocus, C12 Zernike polynoma. We propose advanced isovision, this new technique uses neither monovision nor eye dominance.

PURPOSE

The purpose of this article is double. The first one is to demonstrate that it is possible to get not only a good intermediate vision, but the both so good intermediate and near vision only in using negative SA. Effectively the use of negative spherical aberration is a natural way, because it looks like the crystalline lens does during accommodation. The second one is to demonstrate the possibility of estimate the real final Q value of the cornea and match it with the high customized parameters of the patient's eyes. In accordance with the pupil dynamic size, anterior chamber depth, biomechanics and so on... In fact the real challenge is to get a predictable final corneal asphericity Q value (QF), because if Q is too much negative for instance -1.80 probably the quality of vision will decrease for the both distant and near due to the too much amount of HOA; at the contrary if Q is not enough negative, for instance -0.40 probably the depth of focus will be not sufficient and the near vision impossible without glasses.

METHODS

At first we observed what happens during accommodation (Figs 10.1 and 10.2). Based on the Cheng[1] and Co results in 2004, we can say that during accommodation the anterior shape of the crystalline lens turns to prolate into an hyperprolate shape, with Q < 0. These authors studied accommodation in young optometrist students in using a Hartmann-schack aberrometer. They remarked that the only significant change in HOAs concern the C12, or Z(4,0) called spherical aberrations (SA). Accommodation produces always a decreasing of the SA, with a very strong relationship (R=1) -0.043µ rms of SA per diopter of accommodation, then -0.130µ for 3D. Then we guess if a hyperprolate corneal shape would be able to compensate totally the presbyopia?

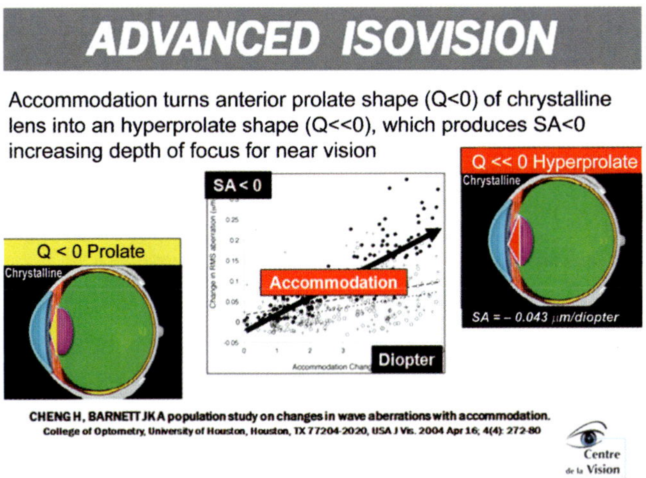

Fig. 10.1 SA during accommodation

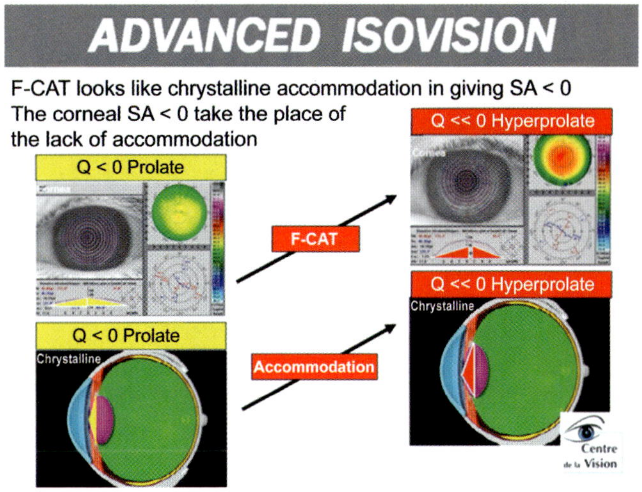

Fig. 10.2 Hyperprolate cornea looks like crystalline

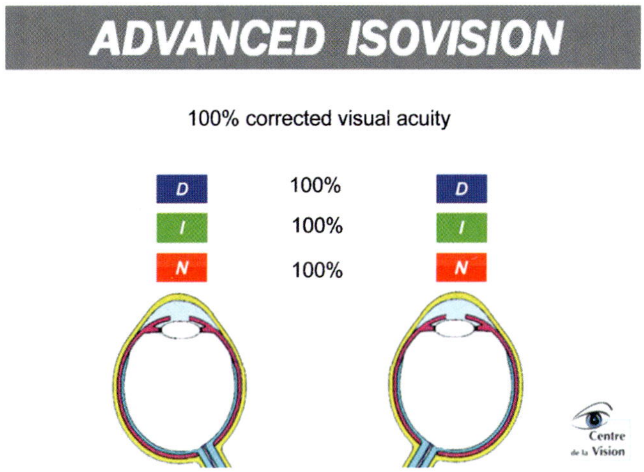

Fig. 10.3 Isovision 100% correction at any distance

We decided to treat the two eyes in the same manner called 'isovision (Fig. 10.3). We choose to treat totally each eye for distance vision as a refractive treatment, and gave the patient a near and intermediate vision in using a Q asphericity modification. Then, we called our technique 'Advanced Isovision ' (Figs 10.4 and 10.5).

MATERIALS

We are using the Wavelight Allegretto 400 Hz Eye Q since 2004. This plateform offers several CAT (Customized Ablation Treatment), the Topographic T-CAT, the Aberrometric A-CAT. The F-CAT (F = Q factor) permit to modify the Q target in order to increase the negative Q asphericity of the cornea.

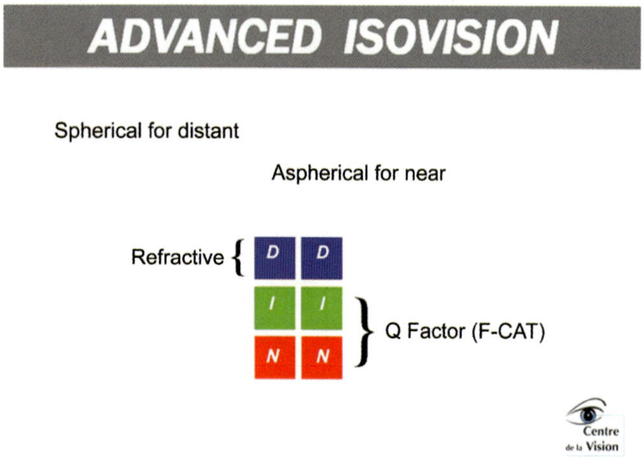

Fig. 10.4 Distant is refractive and Q factor is for near

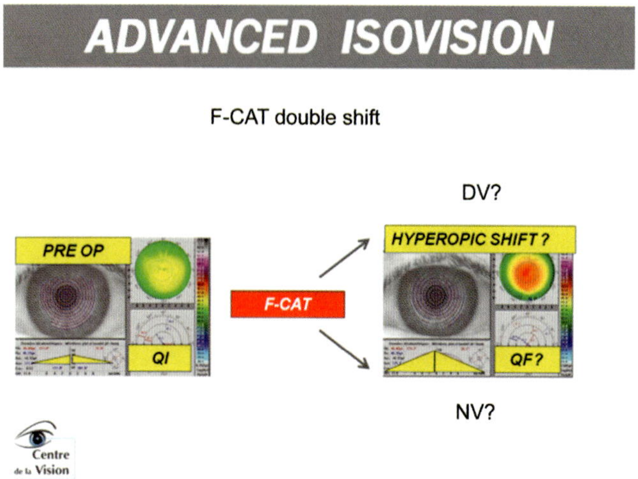

Fig. 10.5 F-CAT gives a double shift

This F-CAT takes place especially for myopic patient in order to limit the oblate profile of the cornea, then limit positives SA , then upgrade the vision quality after a myopic treatment, especially for driving at night. With F-CAT we can choice a negative asphericity from +1.00 to –1.00 , then the maximum delta Q available using the F-CAT is –2.00.

When the Q target (QT) is push to a negative value, F-CAT gives a double shift: a refractive shift and an aspheric shift. With the hyperopic shift, there is no predictability for the both distant and near vision; with the aspheric shift, the final real asphericity value QF is not controlled, then there is no predictability about the created amount of depth of focus, then no predictability for near and intermediate vision.

Hyperopic shift (Fig. 10.5): For instance a preoperative plano refraction with a QT = –1.00 produces approximately an hyperopic shift of +2.00 Diopters. The clinical nomogram of Luiz Assis (2008) Gives a target sphere TS to compensate the hyperopic shift. For this author TS depends on the spherical equivalent (SE) for myopic and hyperopic patient. $TS_{(Assis)}$ = f(SE). This nomogram becomes not enough predictable if K readings are more than 45 D or less than 41 D. In 2008 we present our nomogram called F-CAT_N, which takes count about SE and K reading for a fixed QT value of -1.00. $TS_{(Hehn)}$ = f(SE,K, Q-1). To build this nomogram, we used our technique called the 'C4 CONSTANT ' method. For that, we studied the Zernike polynomas in using the 'Zernike-Rms' button available in the treatment window of the F-CAT software. When we modified the Q value, we remark that the C4 change so, then we search a TS to erase this change. We found a TS for each diopter from –8.00 to +4.00 , and for the K reading for 36 D to 50 D, we obtain a tabular, then a final equation: TS = f (x,k, Q-1) with x= SE ; k = (K1+K2) / 2). With this nomogram called F-CAT_N, we obtained the emmetropia, but at that time we did not yet got the predictability of the QF.

Aspheric shift (Fig. 10.5): Then we worked on, and considered the preoperative and postoperative status of the cornea. A preoperative 'Initial' cornea get a KCI (Initial Central Keratometry), a KPI (Initial Peripheral Keratometry) and a QI (Initial Q value), a postoperative 'Final' cornea get a KCF, KPF, QF. KPI = f(QI;KCI), KCF = f(KCI;SE) , and KPF = f(KPI;TS), Then we can calculate the QF. Then the QF became predictable we called the new treatment F-CAT_NP. "N" for nomogram to get a predictable emmetropia, and "P" for predictable QF value (Fig. 10.6).

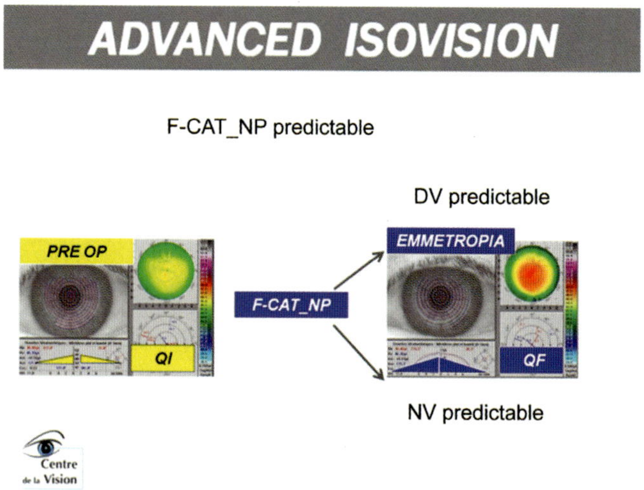

Fig. 10.6 F-CAT_NP gives emmetropia and near

Advanced isovision software: We build a software to calculate all the data and obtain a perfect F-CAT_NP treatment with predictable distant and near vision from -8.00 to +4.00 SE.

RESULTS

The results are very impressive, because the rate of predictability is higher than ever obtain before with previous PresbyLASIK techniques. We present our results about 60 eyes of 30 patients with a follow-up of 3 months(Figs 10.7 and 10.8). For distant vision the both eyes work very well together. Because there is no monovision at all, each eye can achieve 20/20 monocular and the patient can get more than 20/20 binocular. For distant vision there is no compromise (Fig. 10.9). Most of the patient can drive themselves their car to go to the office for the post surgery control at day one. Ninty percent of the patients can read a book without glasses only few minutes after the laser procedure. The depth of focus is simply better than any multifocal IOL (refractive mutizones or diffractive) as we can see on the defocus curve below (Fig. 10.11). Especially there is no gap at all for intermediate vision. Each eye can achieve J2 or j1. For near vision there is no compromise (Fig. 10.10). Therefore it is possible to treat also a pseudophakic patient, implanted previously with a monofocal IOL. Then it is possible to perform a cataract surgery very easily before or after 'advanced isovision' and maintain good results of the both distant and near vision. The quality

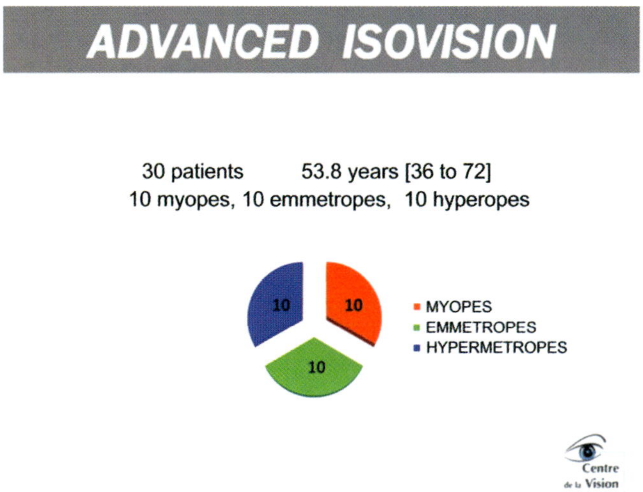

Fig. 10.7 30 patients , 3 months follow-up

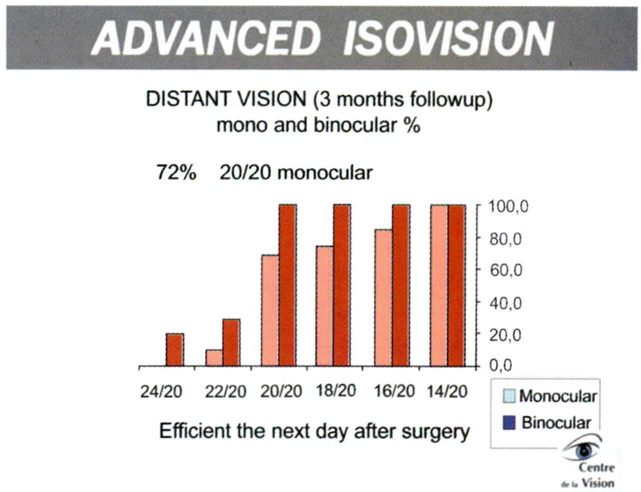

Fig. 10.9 No compromise for distant vision

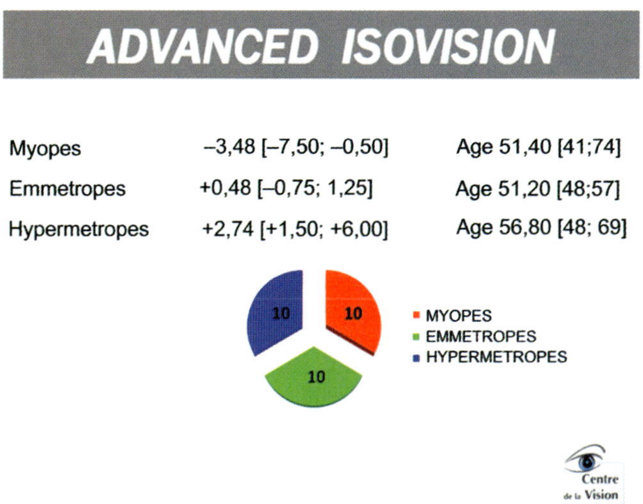

Fig. 10.8 10 myopic, 10 emmetropic, 10 hyperopic

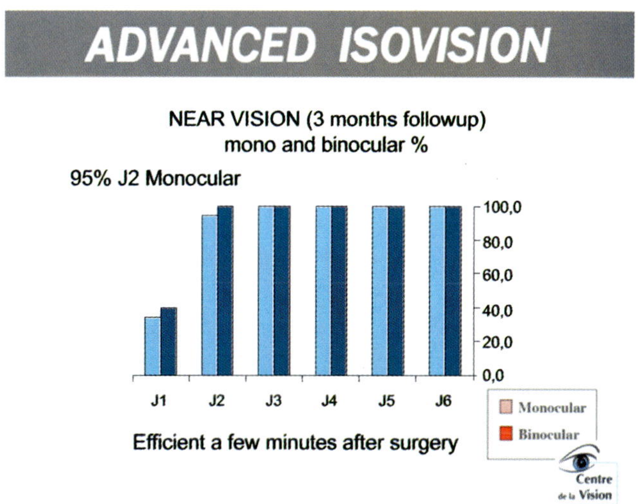

Fig. 10.10 No compromise for near vision

of vision at night is excellent because the PSF of the treated cornea is very small, and there is no loss of contrast due to this technique (Fig. 10.12). The patients said: " I can see naturally, as I was twenty years old."

Results about myopic, emmetropic, hyperopic patients (Figs 10.7 and 10.8)

Visual acuity distant and near, mono, binoc (Figs 10.9 and 10.10)

Defocus curve and Contrast sensitivity (Figs 10.11 and 10.12).

DISCUSSION

How does it work? The effect of F-CAT_NP does not modify the refraction, let it plano if it is the case. F-CAT_NP does not change the central power of the cornea, creates not at all myopia in any part of the cornea, there is no central island. Then, F-CAT_NP just decreases the corneal power in is middle-peripheral part at 6.5 mm width from the very center. This decreasing power equals the same value than the TS, on a 6.5 mm OZ. For instance if TS = -2.00 and peripheral preoperative KPI reading is 42.0 D, the new peripheral KPF value becomes 40.0 D.

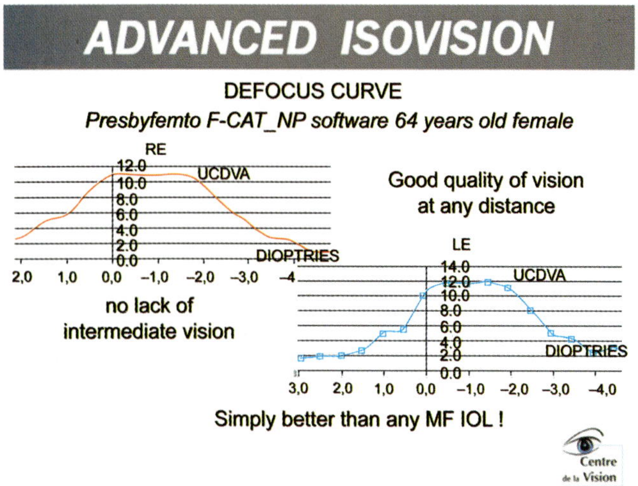

Fig. 10.11 Defocus curve better than any MF-IOL

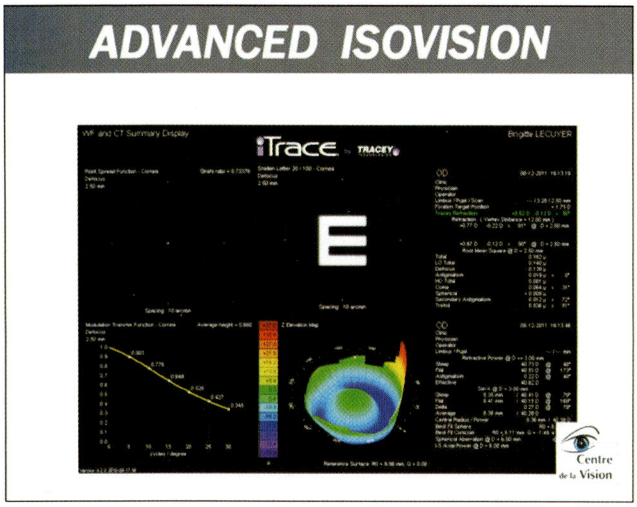

Fig. 10.12 F-CAT_NP gives good PSF and MTF

Why there is no compromise ? (Fig. 10.14) Each eye can achieve 20/20 because the distant vision is due to the final curvature KCF of the central part of the cornea, the mean final K reading KCF gives emmetropia. Then distant vision is based on the first defocus. The C4 coefficient polynoma, the Z(2;0) of Zernike from the level two. The quality of distant vision is good because it is builded on a large 6.5 mm OZ. Each eye can achieve J1 because the near vision depends on the QF (Final Q value). Q value is build by the variation of the K reading from the center to the periphery, generally 30° from part to part of the visual axis, or 8.00 mm OZ, we considered Q at 30°. Q becomes < 0 , the cornea hyperprolate, that is produce SA, which increase the depth of field. SA–are the second defocus, the C12 coefficient polynoma, the Z(4;0) of Zernike from the level four. The quality of near vision is good because it is builded on a large 6.5 mm OZ. For the same KCF we can obtain different QF in using advanced monovision, the Q negative and SA, seem to be a additional optical property. C4 and C12 belong not to the same level of the Zernike polynoma, then they could be not antagonist but synergic, then it appears possible to get 20/20 and J1 monocular in each eye without compromise!

Myopic patient: It is well known. In this case, that the laser turn the cornea from prolate into an oblate shape and produce some amount of SA +. After the distant treatment the Q asphericity will positive. In this case, the topolyzer Q value measurement will be not in accordance with the estimated QF calculated by the software. Nevertheless, in myopic eyes F-CAT_NP creates a real Q negative asphericity in the center part of the cornea, then a sufficient amount of SA to get a good near vision. Globally the cornea becomes oblate, but prolate in the center. The topolyzer cannot measure the central asphericity and find a Q > 0. Of course it is not possible to turn after the distant myopic treatment an oblate cornea (for instance +1.00), into a hyperprolate cornea (for instance -1.00), because it would be consume too much tissue. In this case the software consider after any myopic treatment the Q value = 0 before the F-CAT_NP.

Hyperopic patient: It is well known In this case, that the laser turn the cornea from prolate into an hyperprolate shape and produce some amount of SA. After the distant treatment the Q asphericity will be more negative. Thus sometimes, the distant vision will be sufficient to achieve the Q ideal, and it will not be necessary to add an F-CAT treatment. The software calculates automatically the Q value after the distant treatment and proposes if the F-CAT has to be considered or not. In this case the topolyzer Q value measurement will be in accordance with the estimated QF calculate by the software.

Latent hyperopia: In case of difference between the refraction with and without cycloplegia, the software take count about this. This situation is very frequent for patient between 40 and 55 years old, with such amount of efficient residual accommodation. It is a big problem, in one hand, if we treat the cycloplegic refraction, the patient probably becomes myope, and get poor distant vision; on the other hand, if we treat the subjective refraction without cycloplegia, perhaps the near vision will be not sufficient. Then we choice to treat the refraction without, but we increase the negative Q value in order to enhance the depth of focus. Never mind the patient has to be advertised that he probably needs an enhancement to correct the residual hyperopia revealed after the laser surgery.

Emmetropic patient: That is a big double challenge. The first one for distant vision, because this kind of patient does not support to get any loss of line after the laser. In this study we consider that emmetropic patient range of SE is −1.00 to +1.50. At first it will be necessary to get the rightest data as possible of the patient. Surgeon has to research the maximum preoperative visual acuity, with and without correction, and not stop the refraction at 20/20 if patient can get more, sometimes 22/20, or 24/20. If we want to obtain patient's satisfaction, we are obliged to restore an UCDVA comparable to their previous BCDVA or more. That the reason why, the software take count about that and fixe a target sphere of +0.25 instead of plano for distant vision.

Pseudophakic patient: In case of pseudophakic patient with a monofocal IOL, it will be possible to give the patient a hyperprolate shape of is cornea to get Q-1.00 in case of an ideal pupil dynamic sizes 2.0 mm, 4 mm and 6 mm. This presbyLASIK will restore this eye a sufficient depth of focus to get a good near and intermediate vision (Fig. 10.13).

Importance of pupil size: We are studying the dynamic pupil with the pupillometer device of the TOPOLYZER (Wavelight topograph from OCULUS). The device measure under different flash of brightness the pupil response : min, mean and max pupil size for photopic, mesopic and scotopic conditions. The quality of the near vision depends on min and mean pupil, and the quality of the distant vision depends on mean and max. If the pupil max is very large, more than 6.5 mm, the patient has to be advise about risk of halos and glare at night. If mean pupil is always very small, less than

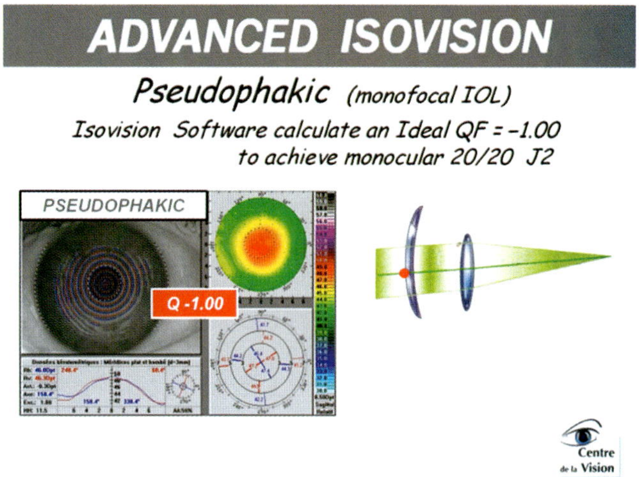

Fig. 10.13 F-CAT_NP and pseudophakic monofocal

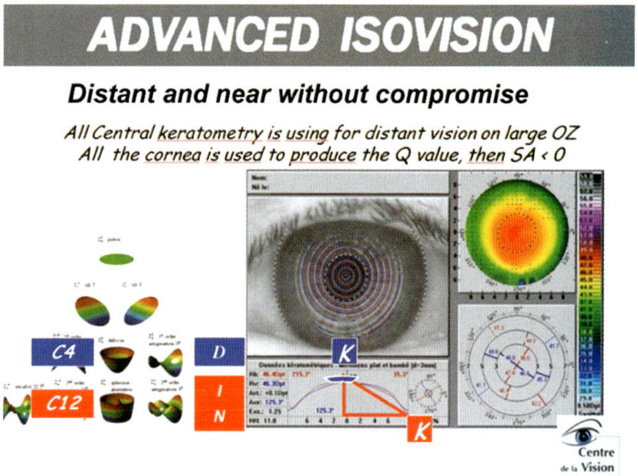

Fig. 10.14 Level 2 and level 4 of Zernike polynomas

3 mm, patient has to be advise that Q asphericity could build not enough depth of focus to get a satisfied near vision. If min pupil is too large more than 3 mm probably the quality of near vision decreases. Near reflex : myosis, accommodation, convergence. At the contrary of SA+, the SA− are compatible with the near reflex, especially the convergence and the myosis.

Biomechanics considerations: We take count about the ORA ocular response analyzer to include or not our patient, and as an help to identify the patient at risk of keratoconus. Probably, elasticity CRF and viscosity CH can explain some cornea that do not respond to the laser treatment and remains flat (0.3%), with no possibility to get a satisfied near vision and some others which becomes too much hyperprolate

(0.5%) with myopia and poor distant vision due to the too much amount of HOA.

CONCLUSION

Advanced isovision could become a new standard for presby LASIK, because with it, patients can achieve 20/20 J1 monocular without compromise, without using monovision, without using eye dominance. Furthers studies have to confirm these preliminary results.

REFERENCES

1. Cheng H. A population study on changes in wave aberrations with accommodation. J Vis. 2004;4(4):272-80.
2. Reinstein D. LASIK for hyperopic astigmatism and presbyopia using micro-monovision with MEL80 platform. J Refract Surg. 2009;25(1):37-58.
3. Seiler T. Q-factor customized ablation profile for the correction of myopic astigmatism. J Cataract Refract Surg. Vol 32, 2006.

11 The LensAR Laser System™; The Technology for Femtosecond Laser Assisted Refractive Cataract Surgery and Presbyopia Reversal

Carlo Francesco Lovisolo (Italy), Eugenio Lipari

Chapter Outline

- The LensAR System

INTRODUCTION

Since the mid-Sixties, when photodisruption with the Nd:YAG laser was first introduced to dissect intraocular membranes and synechiae, to treat posterior capsule opacities and to perforate the iris in angle closure glaucoma,[1-3] the use of plasma-forming lasers to perform certain phases of cataract extraction has been advocated as potentially safer and more effective than conventional surgical procedures. Later on, it became clear that the mechanisms leading to evaporation of the material within the focal region after the impact with the laser beam (plasma formation or optical breakdown) were driving a shock wave with temperature and pressure rise and expansion of a cavitation bubble that could cause unwanted side effects and complications to adjacent tissues, whose damage range depended on the pulse energy. To reduce pulse energy by maintaining the plasma generation turned out to be possible only by changing the pulse duration from the nanosecond (10^{-9} seconds) to the pico (10^{-12}),[4] and then to the femtosecond (10^{-15} seconds) range. The small amplitude and quick decay of the shock wave produced with these low level energies strongly reduced the cavitation and the amount of tissue evaporated per unit time (i.e. the cutting efficiency) remained stable with minimum collateral damage. Rather than photodisrupting tissue, the laser became suitable to act as an intraocular scalpel for precise anatomic dissection.

Despite authoritative, extensive meta-analysis studies[5] have shown only potential advantages in predictability, in the past decade femtosecond lasers have gained a central role in corneal surgery, worldwide. They have brought to the keratorefractive procedures distinct plus over mechanical microkeratomes, like the creation of more consistently thin, planar LASIK flaps with more adherent and smoother stromal bed surfaces, which result in less iatrogenic aberrations postoperatively: Moreover, the "all laser laser" intrastromal ablation may correct ammetropias by internally reshaping the surfaces, holding promises in correcting presbyopia with a minimally invasive approach. Tunnels and cuts with accurately customized shapes create a better fit for intrastromal ring segments to correct ectatic disorders like keratoconus[6-8] and pellucid marginal degeneration and between donor and recipient corneas in anterior or posterior lamellar or penetrating keratoplasty (with zig-zag, top-hat, mushroom configurations...), allowing the transplant surgeon to place stitches with less tension, thereby reducing the amount of suture-induced aberrations.[9-14]

On the follow of this hype, some companies - *LenSx* (Aliso Viejo, Ca, USA, now Alcon Laboratories, Fort Worth, Texas, USA, the only firm whose clinical indications have been cleared by the FDA), *Catalys Precision* (OptiMedica, Santa Clara, Ca, USA), LensAR (LensAR, Winter Park, Fla, USA) and *Victus CustomLens* (Technolas Perfect Vision TPV München,

Germany) - are currently bringing their technologies to the cataract surgery field in Europe with the declared aim of dominating the marketplace shortly.

Although commercial issues have invariably arisen such as high cost of installation and per operation fees, value proposition and return on investment, several potential benefits, yet to be proven, are appealing to the surgical facilities. With the increasing ageing of the global population, combined with the rising demand of a high level visual outcome (patients badly ask for seeing, without glasses, better than they ever have, worldwide), the inevitable fate of the most manually performed surgical procedure, cataract extraction, is to become a completely refractively oriented type of procedure. The same requirements for safe, consistent, predictable outcomes of a LASIK practice shall be applied to the near future "opaque or clear lens exchange" surgery. Actually, I do not know a single patient who is not sensitive enough to perceive the benefits of procedure performed in a relaxed (bladeless), comfortable setting, with a computer-controlled accuracy, real-time 3D visualization, which can be tailored to his or her individual specifications. And how can we contest the collective imagination, which assumes that replacing the surgeon's hands with an automated laser-robot obviously translates in less risk of iatrogenic complications like corneal burns, endothelial cell loss, iris damage, posterior capsule interruption with vitreous loss? Greater safety and increased efficacy are therefore better warranted, in the hope of having patients seeing 20/20 uncorrected and J1 a few hours postoperatively; but also efficiency should be positively affected, because the femtosecond is more accurate, reproducible and consistent than any human skill.[15-24] The coming true prediction of specific difficulties and complications associated to the technique by grey-haired ophthalmologists, like the feared capsular block syndrome with preoperative posterior capsular rupture and posterior segment crystalline lens dislocation,[25] is little power to fight against the trend towards surgical automation. At least theoretically, the femtosecond laser-assisted cataract surgery can therefore better guarantee:

- The exact diameter and position of the curvilinear, continuous anterior capsulorhexis (Figs 11.1A to D), as the centration of a premium IOL (aspheric, toric or multifocal implant) is a mandatory requirement for its success.[26,27] Though it is true that a good surgeon may easily create a perfectly round and centered capsulorhexis, the reality is that the average surgeon often makes decentered, asymmetrically oblong rhexis (the patient may inadvertently move...) and realizes that it is anything but perfect only when the implanted IOL looks decentered or tilted or until the patient is complaining of quality of vision symptoms. There is no doubt that to get consistent and reproducible anterior capsule opening within 0.1 mm of intended diameters with reproducible 0.5 mm overhang on the edge of the IOL makes the difference on large numbers.
  Use of femtosecond laser for cataract surgery is also well supported by the literature which carries many reports of changes associated with phacoemulsification which may include corneal edema with a resulting loss of corneal clarity and a consequent suboptimal visual outcome at day 1. Compromise to the blood-aqueous barrier can lead to evidence of an inflammatory response including anterior chamber cells and flare and, in the longer term, a reduction in endothelial cell density is observed. Theories for these changes vary and include mechanical interactions with the surgical tools, fluid volume passing through the anterior chamber and production of free radicals during the phacoemulsification process.
- Perfectly structured, located and customized, corneal incisions—primary, secondary, square, T-shaped or arcuate – ensure self-sealing of wounds and allow great control and management of astigmatism (Figs 11.2A and B).
- The nucleus prechopping and softening phases ease and quicken its removal (ideally, only with aspiration), with a significant reduction of the manipulation and overall ultrasound energy and temperature and pressure rise needed, thus reducing the risk of putting excessive stress on zonules (critical in pseudoexfoliation or phacodonesis eyes, for instance) and of postoperative infection and inflammation (Fig. 11.3).
- A very rapid and smooth, if any, learning curve.

The patient is usually placed and moved on a motorized gurney, first to undergo the femto-assisted preconditioning of the eye (incision creation, capsulotomy, and nucleus breakup), which lasts approximately 30 to 60 seconds; then he or she is wheeled into the operating room, prepped and draped, to carry out the rest of the procedure (conventional phacoemulsification of the prechopped nucleus, automated irrigation/aspiration of the cortex, capsule polishing and IOL implantation). It is up to the surgeon to perform two or more laser procedures before moving to the surgical theater. In some facilities, the laser pretreatment is performed in the same intraocular surgery environment, right aside the operating microscope and the phaco-machine, to limit unwanted effects like the pupil becoming smaller and the IOP higher. Only absurd, typically American considerations, can suggest performing the femto-treatment in a far off-site refractive surgery facility, with a clean though not sterile environment, and then getting about the patient to an intraocular surgery center.

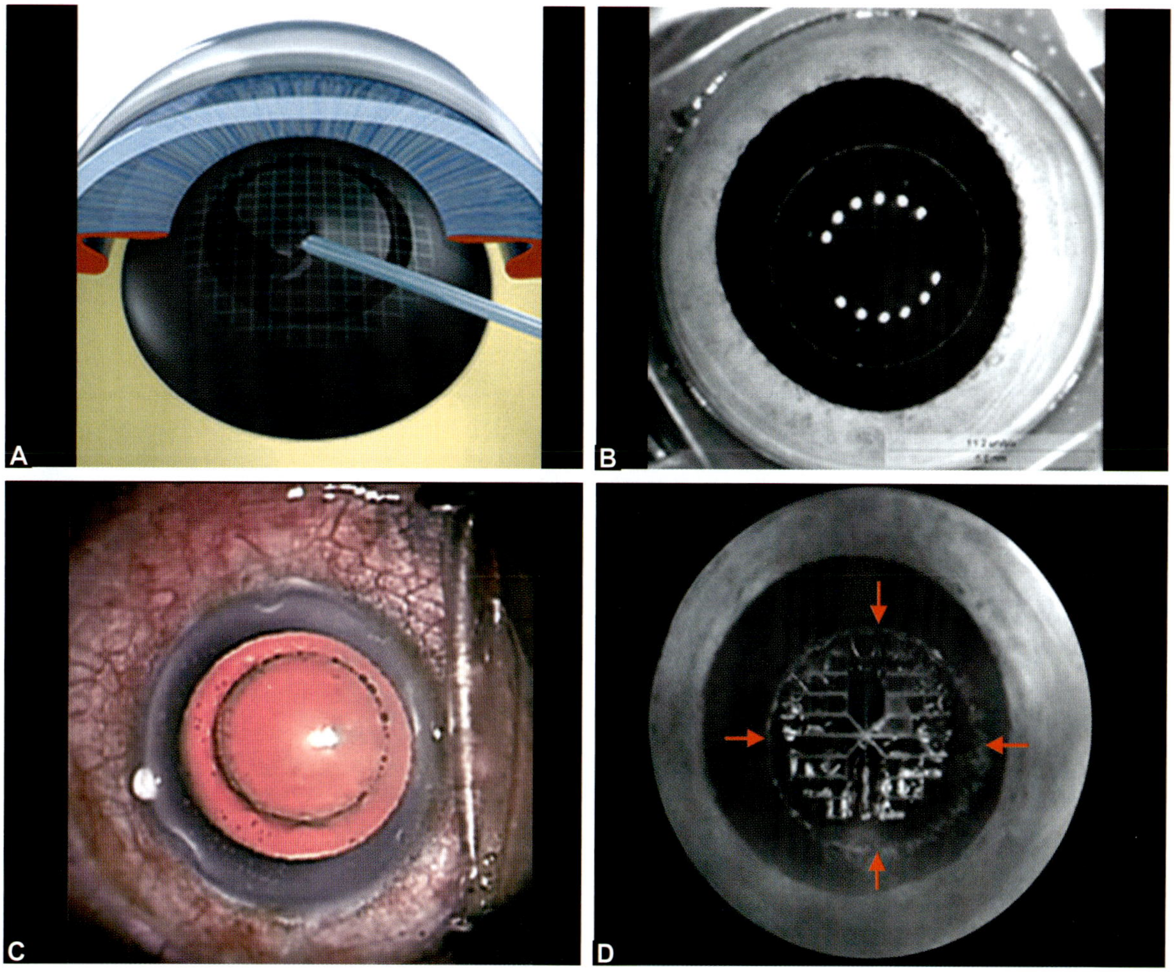

Figs 11.1A to D Femtosecond-laser assisted anterior CCC (continuous curvilinear capsulotomy). (A) Artist's drawing; (B) *In vivo* with infrared; (C) Operating microscope; (D) Cadaver eye, with prechopping algorithm applied

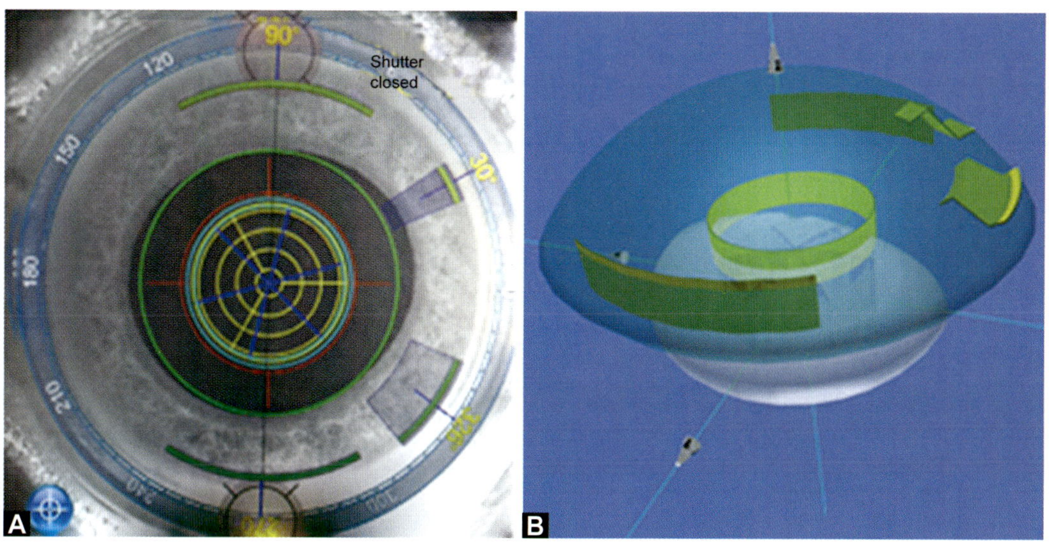

Figs 11.2A and B Femtosecond laser assisted limbar relaxing incisions (LRIs), main wound and side port openings

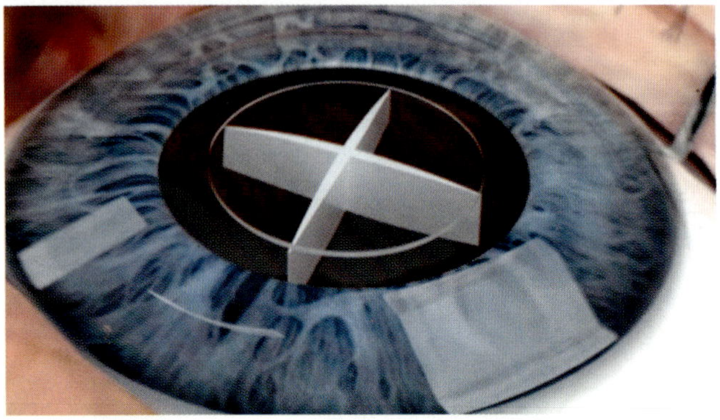

Fig. 11.3 Femtosecond laser-assisted capsulorhexis and cross-cutting of the nucleus

In competitive analysis, the different femto platforms are basically equivalent in terms of providing automated corneal cuts and rhexis. They mainly differ in the way they visualize in three dimensions the ocular structures (anterior chamber depth, thickness of the cornea and the lens) to precisely guide the focalization of the laser beam. This is crucially important to establish the amplitude of the safety zone to prevent damage to the posterior portions of the capsular bag. While *LenSx* and *TPV* are using real-time, high contrast intraoperative optical coherence tomography (OCT) (Figs 11.4A and B),[28] for instance, OptiMedica boasts a liquid interface that does not require corneal applanation and suction, producing comfort to the patient and avoiding distortion of the images of anterior segment tissues. *LensAR* promotes the quality of its proprietary system which integrates Scheimpflug camera imaging and confocal microscopy structured illumination to achieve a tridimensional, high-resolution biometry of the posterior capsule. On another hand, *TPV* remarks that its laser is capable of assisting in cataract surgery as well as creating LASIK flaps, while the *LensAR* technology, on whose features this chapter will focus on, has been designed to perform cuts in the crystalline lens (lentotomy)[29-32] to soften its nucleus and correct presbyopia.

THE LensAR SYSTEM

The LensAR system (Figs 11.5A and B) can create perfect corneal main incisions, side ports and astigmatic cuts (with the choice of cutting through the epithelium or using a sub-Bowman's incision, to better control efficacy and outcome regression characteristics), followed by an ideal capsulorhexis[36] with a 0.16 mm mean deviation from intended diameter (0.42 mm in the manual group) and lens fragmentation-assisted prechopping, to break the nucleus down into small pieces. Emulsification and aspiration of lens fragments from the capsular bag are easier, with a mean reduction in phaco energy documented in clinical studies of about 2/3 (in soft nuclei, even no ultrasounds are used, while in +4 brunescent cataracts around 1/3 of overall energy is saved).[37]

Under competitive analysis with the other systems available on the market, LensAR emphasizes:

- The unique opportunity to treat presbyopia. As early clinical data showed the potential to restore the crystalline lens natural flexibility,[29-30] the LensAR laser system was initially conceived and designed to correct presbyopia. Based on the classic Helmholtz theory for accommodation, in addition to some intraocular anatomic changes of ciliary

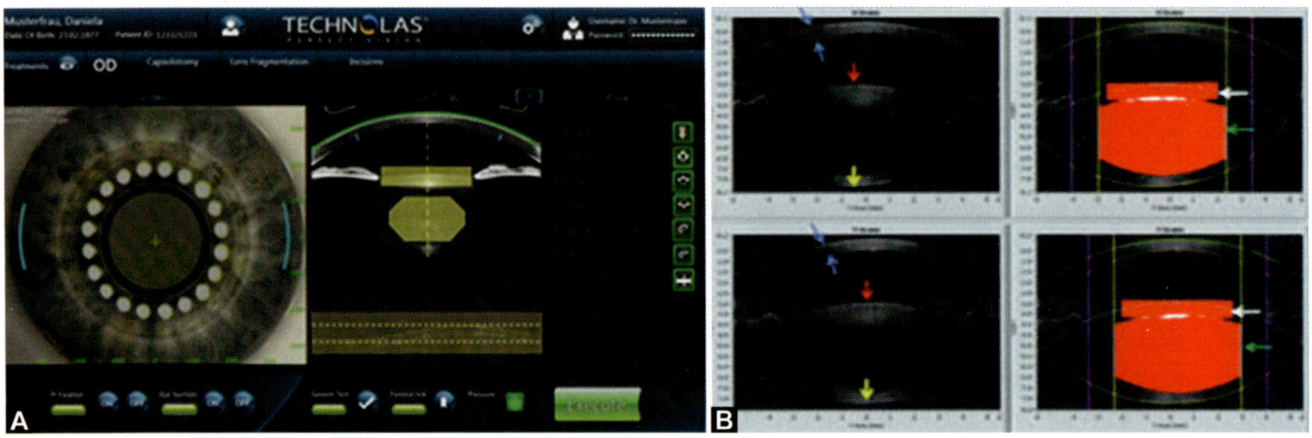

Figs 11.4A and B Intraoperative OCT determines the relevant dimensions of the anterior chamber and provides a no-touch zone of protection to the posterior capsule during the laser-driven phases of softening and segmentation of the nucleus into smaller fragments (*Source:* TPV and LenSx websites)

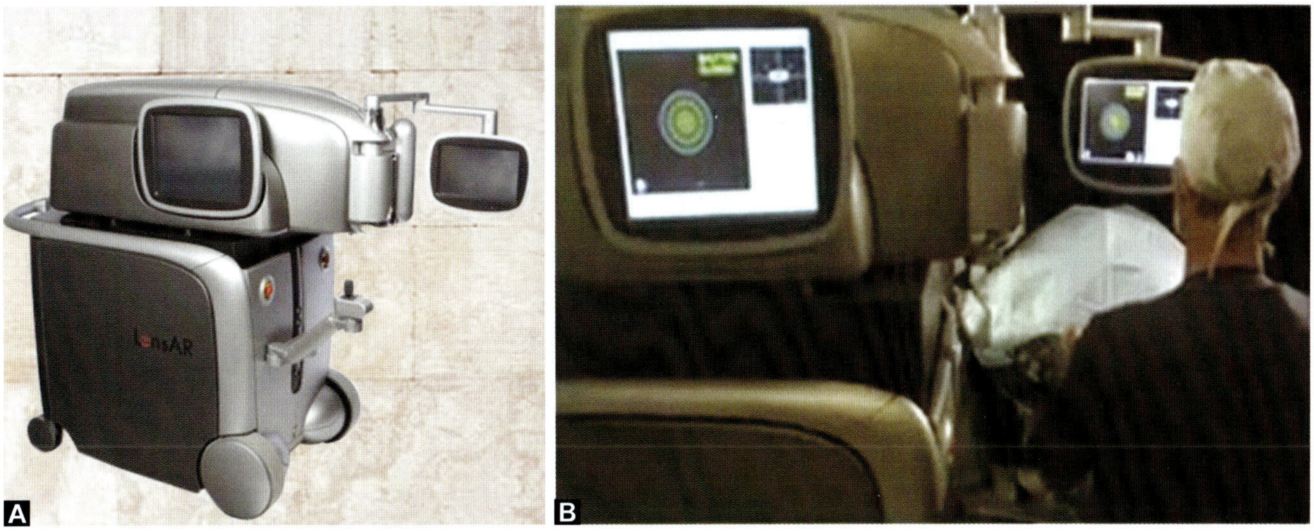

Figs 11.5A and B Brochure (A) Operative; (B) Images of the LensAR system

muscles, increasing sclerosis with correlated loss of elasticity of the lens nucleus and cortex is considered the main cause of the progressive diminishing of the ability to focus on near objects occurring during ageing. Without opening the eye globe (first big advantage), the femtosecond laser delivery of ultrafast pulses creates micro-incisions applied with different algorithms which act as gliding planes inside the crystalline lens cortex- a procedure called lentotomy (Figs 11.6A to C), obtaining an increase in its anterior-posterior thickness accompanied by a decrease of its equatorial horizontal diameter. These geometric changes are thereby increasing the lens flexibility, restoring a true dynamic accommodative capability (second major advantage). In experimental settings (porcine, rabbits and pigs), the visibility of the incisions remained over time, though slightly declining under Scheimpflug camera imaging and using histological techniques, without inducing significant opacities (cataract).[31-35] When market competition in the cataract field became the priority of the company, the research towards presbyopia reversal - currently in the phase of clinical feasibility assessment was slowed down to save resources. All the world customers, however, are strongly asking for this option, that is expected to be launched soon.

- The quality of its proprietary infrared-based high contrast, low noise imaging system, integrating Scheimpflug camera visualization, confocal structured illumination (CSI™) and scanning super luminescent diode technology, which keeps in focus the entire anatomy from the anterior cornea to the posterior capsule and does not rely on image "stitching" to capture the various anatomical structures. The key surfaces are detected automatically by the software and the images are used to construct a 3D model of the eye through ray-tracing. As the capsular opening has been implicated as a major factor in determining the correct effective lens position (ELP) and hence the precise refractive outcome for the patient, LensAR has the unique capability to accurately determine the orientation of the lens with précise tilting evaluation and compensation. Having mapped the ocular structures, the selected capsulotomy and lens fragmentation patterns can then be automatically placed to targeted areas with the desired clearances from the capsule to prevent capsule breakage (Figs 11.7A to C). Instead of putting together, somehow, multiple OCT slices, the 3D high-resolution biometry captures the posterior capsule all on one quick pass and is less sensitive to artifacts and deformations,[36,37] thus helping to keep the location of the anatomy in the safest perspective by creating a thin perinuclear plate just anterior to the cortex, which provides an additional protective layer for the posterior capsule and can easily be aspirated out after nucleus removal.
- The logistic atout of a comfortable docking system and a less bulky, though very robust machine, which can be used in a non-sterile room environment or easily fit in a in a sterile standard operating room, close the phaco machine and operating microscope and can be moved thanks to the locking wheels. Instead, all the competitor

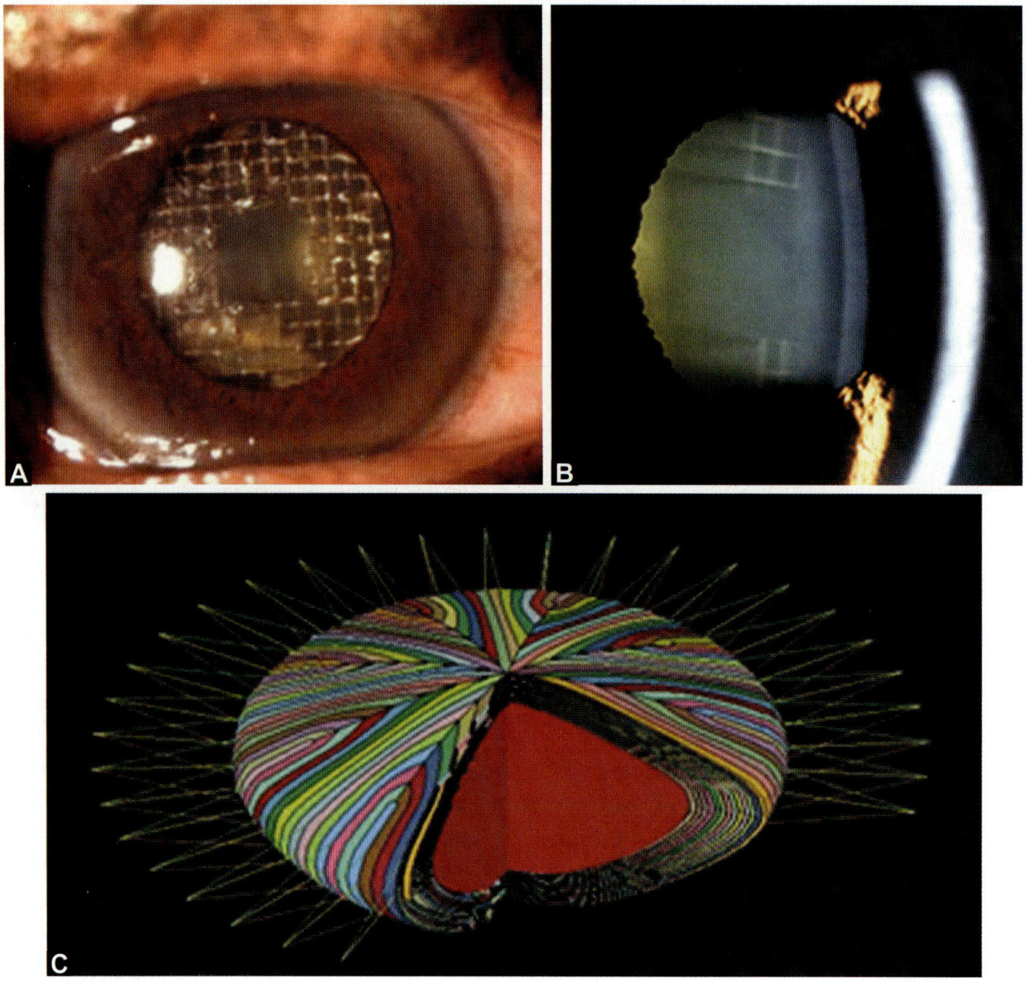

Figs 11.6A to C Femtosecond laser crystalline lens micro-incisions (lentotomy) performed with the LensAR System to reverse presbyopia. Slit lamp: (A) Image at day 1; (B) Day 7; (C) Artists rendering of the treated cortex

systems require a dedicated room and the patient must be moved after the laser procedure in another surgical suite, to perform the removal of the fragmented lens, automated irrigation and aspiration of the cortex, capsule polishing, and IOL placement; the LensAR system has been designed for flexibility using an ergonomic design that resulted from extensive surgeon feedback. Multiple screens allow the surgeon, scrub nurse and circulating nurse to see procedure progress as well as to have access to the system software. It enables a number of patient orientations; perpendicular to the laser head or at right angles on either side to accommodate superior or temporal approaches to the eye (Figs 11.8A to C). The laser head retracts to allow unrestricted access to the treated eye after the laser procedure, essential for use in the operating room.

- The ability to safely extract hard +3 and +4 nuclei. Several lens fragmentation patterns have been tried: the spheres pattern, developed with the intention of creating small fragments that were small enough to allow aspiration without ultrasound energy, worked well with softer nuclei but was less effective with harder nuclei where the spheres impacted the holding power of the phaco tip, lengthening the overall procedure. Other pattern algorithms such as pies, cubes and French fries were more effective in harder nuclei (Figs 11.9A to D), even cases like white, intumescent cataracts with anterior capsular fibrosis, where capsulorhexis

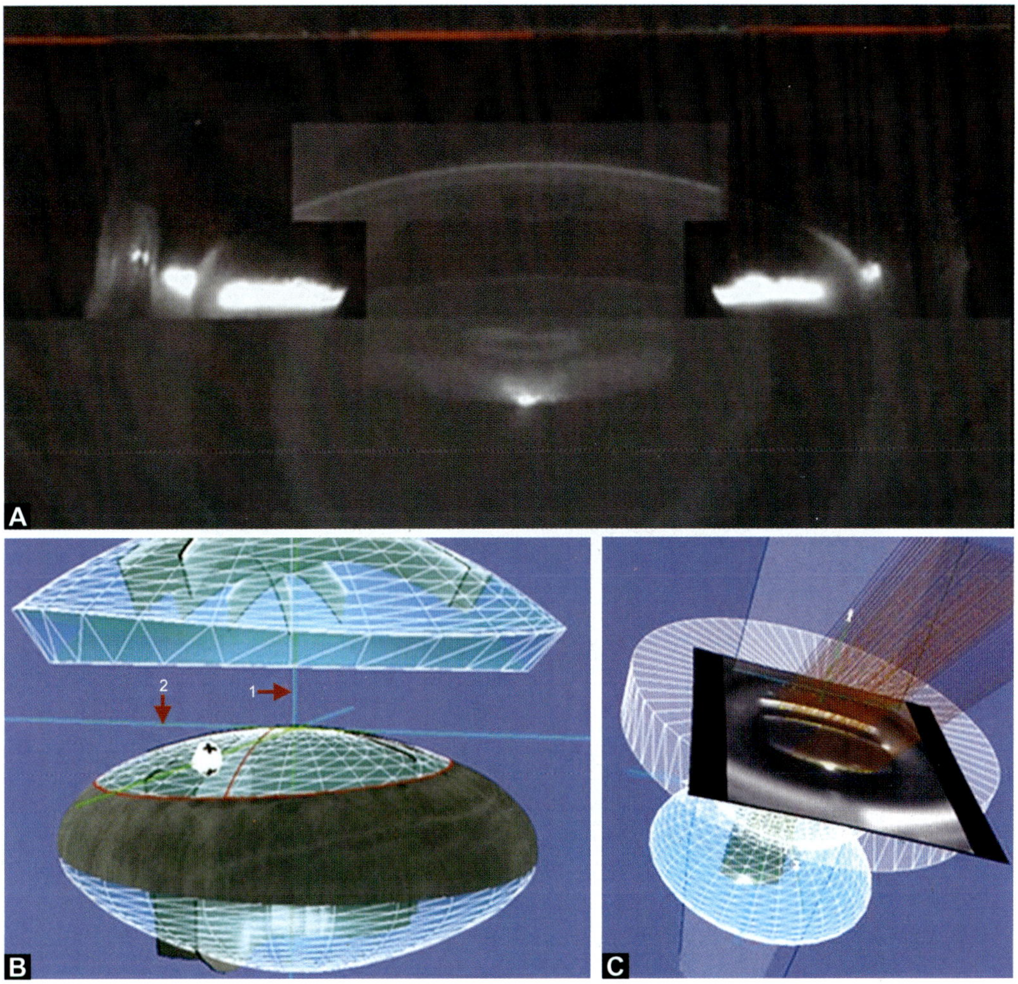

Figs 11.7A to C (A) The LensAR 3D-CSI software runs the imaging process and the auto-detection part of the program identifies the key surfaces; (B and C) The motorized Scheimpflug camera acquires different cross sections of the eye with final purpose to reconstruct a 3D model of the eye

is particularly challenging even for the experienced and gifted surgeon, can be safely faced, without using trypan blue dyeing. As the white color reflects the femtosecond light, however, the laser can only penetrate the superficial, anterior part of the lens, while it is more difficult to significantly assist in chopping up the nucleus.

In conclusion, about a decade after the introduction of the femtosecond laser into refractive surgery, we must admit that it has not yet supplanted steel blades for all patients or all surgeons, despite its stated benefits. At least in selected settings, where conventional LASIK surgery has reached an indisputable level of safety and accuracy for years, it remains controversial whether femtolaser assisting technologies are truly reducing complications and improving outcomes. It is certain, however, that they have introduced some specific drawbacks like new clinical complications, a significant burden to the procedure length and to the logistics (the extra time and personnel that it requires, the need of a room for a bulky machine) and, above all, financing issues (high cost for purchasing a very sophisticated piece of equipment and considerable per-click fees) that represent the real question mark to the future of these technologies. I feel that similar considerations can be made for the future of femtolaser-assisted cataract surgery. As the driving force of advertising-oriented patient population will probably be overwhelming, it is easy to predict that cost considerations shall be rationalized to find a viable financial

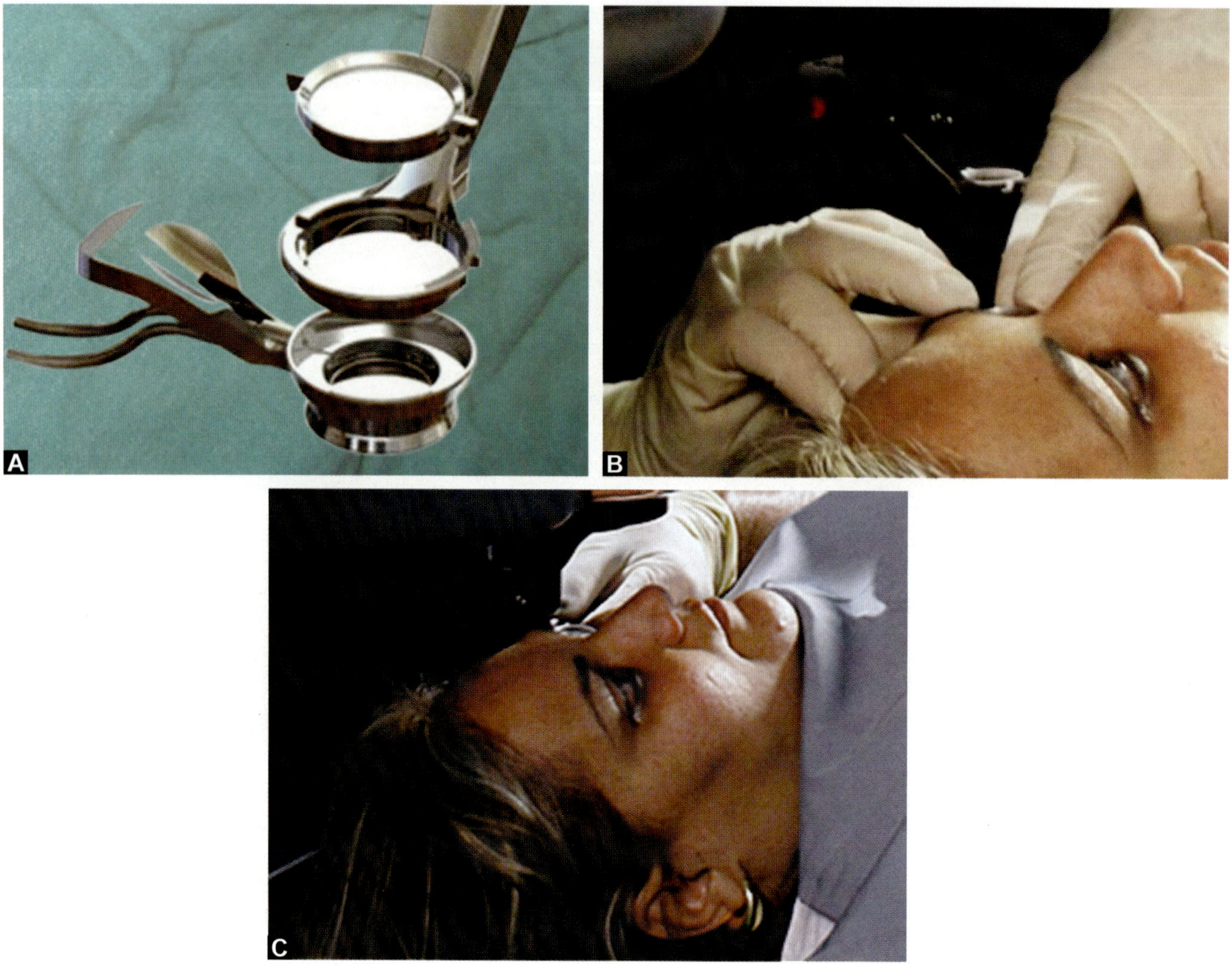

Figs 11.8A to C LensAR patient interface incorporating the suction ring, the docking arm and the glass cover plate. (A) The suction ring and glass cover plate are disposables while the docking arm is reusable after cleaning and sterilization; (B) Patient docked to the laser; (C) Laser head at final position over the patient

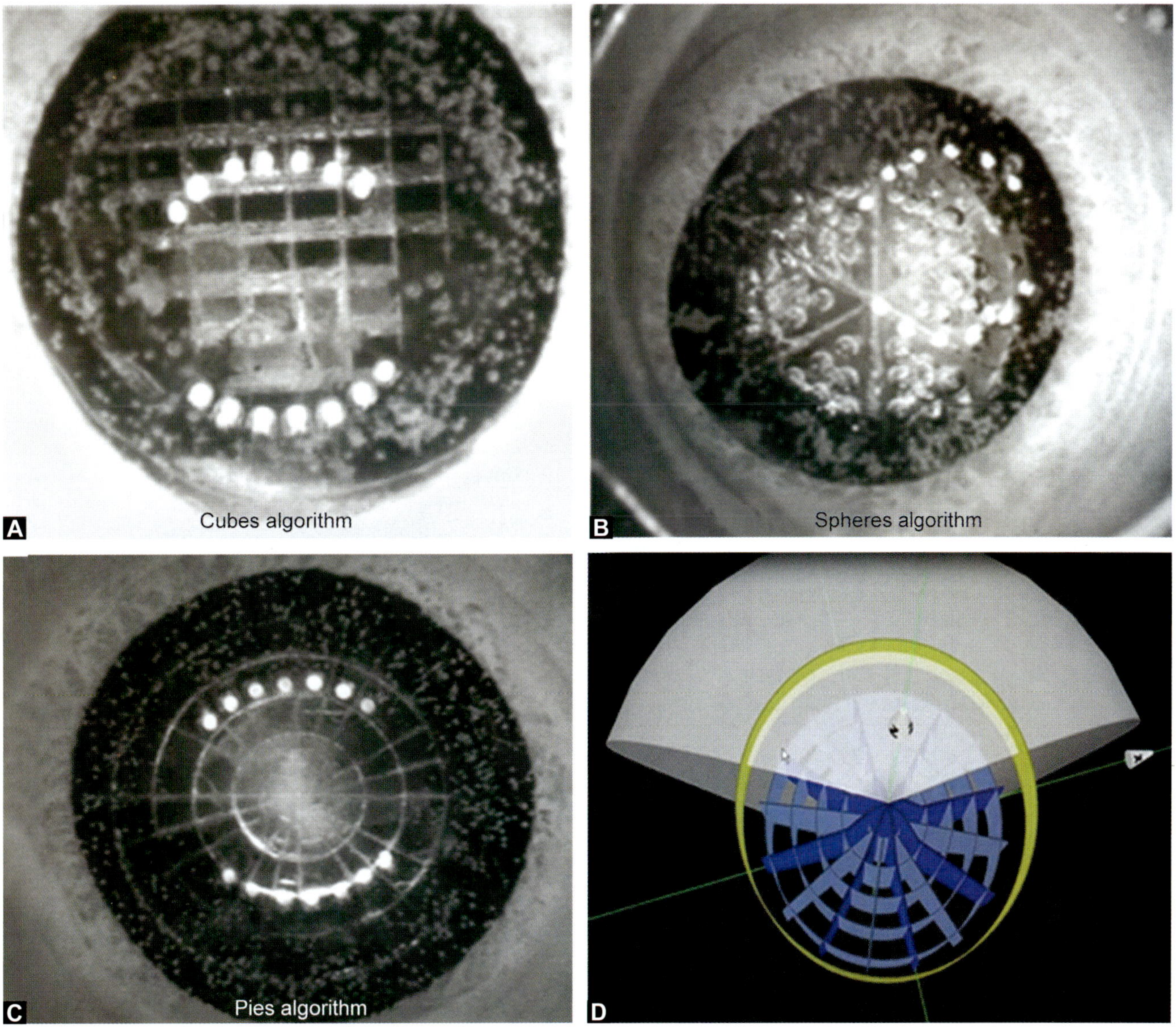

Figs 11.9A to D (A to C) Cadaver eye simulations of different algorithms applied to produce lens prechopping and softening with the LensAR; (D) Graphic of combined capsulorhexis

model proportional to the documented upstream benefit. It will take some time, probably another decade and a lot of data before anyone can say that the supposed advantages of femtosecond laser cataract surgery are real, rather than simply imagined or stated by "juke box" rented podium surgeons speaking at international meetings.

REFERENCES

1. Geusic JE, Marcos HM, Van Uitert LG. Laser oscillations in Nd-doped yttrium aluminum, yttrium gallium and gadolinium garnets. Applied Physics Letters. 1964;4(10):182-4.
2. Fankhauser F, Roussel P, Steffen J, et al. Clinical studies on the efficiency of high-power laser radiation upon some structures of the anterior segment of the eye. Int Ophthalmol. 1982;5:15-32.
3. Aron-Rosa D, Aron J, Griesemann J, et al. Use of the neodym YAG laser to open the posterior capsule after lens implant surgery: A preliminary report Am Intraocul Implant Soc 1980;6:352-4.
4. Vogel A, Capon MA, Asiyo-Vogel MN, et al. Intraocular photodisruption with picosecond and nanosecond laser pulses: Tissue effects in cornea, lens, and retina Invest Ophthalmol and Vis Sci. 1994;35(7):3032-44.
5. Chen S, Feng Y, Stojanovic A, et al. IntraLase femtosecond laser vs mechanical microkeratomes in LASIK for myopia: A systematic review and meta-analysis. J Refract Surg. 2012;28(1):15-2.
6. Prazeres TM, Souza AC, Pereira NC, et al. Intrastromal corneal ring segment implantation by femtosecond laser for the correction of residual astigmatism after penetrating keratoplasty. Cornea. 2011;30(12):1293-7.
7. Coskunseven E, Kymionis GD, Tsiklis NS, et al. Complications of intrastromal corneal ring segment implantation using a femtosecond laser for channel creation: a survey of 850 eyes with keratoconus. Acta Ophthalmol. 2011;89(1):54-7.
8. Shabayek MH, Alió JL. Intrastromal corneal ring segment implantation by femtosecond laser for keratoconus correction. Ophthalmology. 2007;114(9):1643-52.
9. Fung SS, Iovieno A, Shanmuganathan VA, et al.Femtosecond laser-assisted lock-and-key shaped penetrating keratoplasty. Br J Ophthalmol. 2012;96(1):136-7.
10. Thompson MJ. Femtosecond Laser-Assisted Half-Top-Hat Keratoplasty. Cornea. 2011;9.
11. Rousseau A, Bensalem A, Garnier, V, et al. Interface quality of endothelial keratoplasty buttons obtained with optimised femtosecond laser settings. Br J Ophthalmol. 2012;96(1):122-7.
12. Chamberlain W, Omid N, Lin A, et al. Comparison of corneal surface higher-order aberrations after endothelial keratoplasty, femtosecond laser-assisted keratoplasty, and conventional penetrating keratoplasty. Cornea. 2012;31(1):6-13.
13. Birnbaum F, Maier P, Reinhard T. Perspectives of femtosecond laser-assisted keratoplasty Ophthalmologe. 2011;108(9):807-16.
14. Cheng YY, van den Berg TJ, Schouten JS, et al. Quality of vision after femtosecond laser-assisted descemet stripping endothelial keratoplasty and penetrating keratoplasty: a randomized, multicenter clinical trial. Am J Ophthalmol. 2011;152(4):556-66.
15. Moshirfar M, Churgin DS, Hsu M. Femtosecond laser-assisted cataract surgery: a current review. Middle East Afr J Ophthalmol. 2011;18(4):285-91.
16. Uy HS, Edwards K, Curtis N. Femtosecond phacoemulsification: the business and the medicine. Curr Opin Ophthalmol. 2012;23(1):33-9.
17. Miháltz K, Knorz MC, Alió JL, et al.Internal aberrations and optical quality after femtosecond laser anterior capsulotomy in cataract surgery. J Refract Surg. 2011;27(10):711-6.
18. Masket S, Sarayba M, Ignacio T, et al. Femtosecond laser-assisted cataract incisions: architectural stability and reproducibility. J Cataract Refract Surg. 2010;36(6):1048-9.
19. Ecsedy M, Miháltz K, Kovács I, et al. Effect of femtosecond laser cataract surgery on the macula. J Refract Surg. 2011;27(10):717-22.
20. Kránitz K, Takacs A, Miháltz K, et al. Femtosecond laser capsulotomy and manual continuous curvilinear capsulorhexis parameters and their effects on intraocular lens centration. J Refract Surg. 2011;27(8):558-63.
21. Friedman NJ, Palanker DV, Schuele G, et al. Femtosecond laser capsulotomy. J Cataract Refract Surg. 2011;37(7):1189-98.
22. Mamalis N. Femtosecond laser: the future of cataract surgery? J Cataract Refract Surg. 2011;37(7):1177-8.
23. Nagy ZZ, Kránitz K, Takacs AI, et al. Comparison of intraocular lens decentration parameters after femtosecond and manual capsulotomies. J Refract Surg. 2011;27(8):564-9.
24. He L, Sheehy K, Culbertson W. Femtosecond laser-assisted cataract surgery. Curr Opin Ophthalmol. 2011;22(1):43-52.
25. Roberts TV, Sutton G, Lawless MA, et al. Capsular block syndrome associated with femtosecond laser-assisted cataract surgery. J Cataract Refract Surg. 2011;37(11):2068-70.
26. Gimbel HV, Neuhann T. Development, advantages, and methods of the continuous circular capsulorhexis technique. J Cataract Refract Surg. 1990;16(1):31-7.
27. Becker KA, Holzer MP, Reuland AJ, et al. Accuracy of lens power calculation and centration of an aspheric intraocular lens.Ophthalmologe. 2006;103(10):873-6.
28. Palanker DV, Blumenkranz MS, Andersen D, et al. Femtosecond laser-assisted cataract surgery with integrated optical coherence tomography. Sci Transl Med. 2010;2(58):58-85.
29. Lubatschowski H, Schumacher S, Fromm M, et al. Femtosecond lentotomy: generating gliding planes inside the crystalline lens to regain accommodation ability. J Biophotonics 2010;3(5-6):265-8.
30. Ackermann R, Kunert KS, Kammel R, et al. Femtosecond laser treatment of the crystalline lens: a 1 year study of possible cataractogenesis in minipigs. Graefes Arch Clin Exp Ophthalmol. 2011;249(10):1567-73.

31. Lubatschowski H, Schumacher S, Wegener A, et al. Lentotomy: presbyopia reversal by generating gliding planes inside the crystalline lens Klin Monbl Augenheilkd 2009;226(12):984-90.
32. Schumacher S, Fromm M, Oberheide U, et al. In vivo application and imaging of intralenticular femtosecond laser pulses for the restoration of accommodation J Refract Surg 2008;24(9):991-5.
33. Gerten G, Ripken T, Breitenfeld P, et al. *In vitro* and *in vivo* investigations on the treatment of presbyopia using femtosecond lasers Ophthalmologe. 2007;104(1):40-6.
34. Reggiani Mello GH, Krueger RR. Femtosecond laser photodisruption of the crystalline lens for restoring accommodation Int Ophthalmol Clin. 2011;51(2):87-95.
35. Chatoux O, Touboul D, Buestel C, et al. Crystalline lens photodisruption using femtosecond laser: experimental study J Fr Ophtalmol. 2010;33(7):472-80.
36. Tackman RN, Kuri JV, Nichamin LD, et al. Anterior capsulotomy with an ultrashort-pulse laser. J Cataract Refract Surg. 2011;37(5):819-24.
37. Straub L. Femtosecond lasers for cataract surgery. When will Laser Cataract Surgery be a routine procedure? Cataract and Refract Surg Today Europe, 2011.

12

Femtosecond–assisted Cataract Surgery

Jorge L Alio, Roberto Fernandez Buenaga (Spain)

Chapter Outline

Cataract surgery with Femtosecond lasers is approaching its practical application in modern ophthalmology. The purpose of this chapter is to update the reader in what is the state of the art of this technology in its application to the practice of cataract surgery.

Femtosecond lasers and their application to cataract surgery constitute a major innovation in modern ophthalmic surgery. These lasers, which act on the near infrared wavelength (1030 nanometers) are capable of penetrating the transparent and even opaque cornea, with the only limitation of an eventual corneal vascularization and the densly calcified plates. They are guided in their intraocular application by 3 dimensional image systems of the anatomy of the anterior segment of the eye. In this way they act in a very precise way, delivering energy to perform corneal incisions, capsulorhexis, softening or breaking the nucleus, thus enabling elimination of the cataract. The surgical procedure performed in this way, facilitates optimization of the surgical time, minimizing the surgeon's work (the least controllable variable of all those involved in cataract surgery) to the elimination of the cortical remains and the implantation of the intraocular lens.[1]

From the technological perspective, for femtosecond lasers to be applied in cataract surgery, their numeric opening should be greatly reduced with the aim of penetrating the structures of the anterior chamber. Conversely, the diameter of focus of the laser should be larger than that of corneal surgery. This combination of physical elements leads to an increase in the energy limit needed to produce the adequate tissue photodysruption. This high energetic level needed to produce the crystalline lens lysis or fragmentation, results in a limitation of the repetition index of the laser to be maintained within the safety limits, due to the heat effect that the laser causes in this tissue.

The intraocular photodysruption is guided by image systems that use optical coherence tomography systems or optical technology in real time, such as the Scheimpflug, which is able to define with precision, the structures of the anterior segment up to the posterior capsule. Given that these elements are based on optical technology, they cannot penetrate the iris or the structures located behind the iris.

What is the importance of these lasers in cataract surgery during this initial stage of their application? It should be considered that the cataract surgery that is performed today begins with a small corneal incision which enables a manual capsulorhexis of the anterior capsule to be carried out followed by the phacoemulsification and the elimination of the lens fragments. At present, these procedures, although safe and efficient, are considered to depend greatly on the skill and experience of the surgeon.

Precise performance of the incisions is one of the immediate advantages of these lasers and for reason, the concept of Microincisional cataract surgery (MICS) plays and important and relevant role. MICS, a term coined by Professor Jorge Alió and registered in 2003, embraces the principles of modern cataract surgery, which aims to perform the surgery through a minimal incision, with no refractive impact on the cornea.[2]

Given that the femtosecond lasers can today guarantee the stability, precision, length, shape and width of the corneal incisions, this step has considerably improved. The capsulorhexis, essential step for the so-called "Premium" intraocular lenses, requires an optimum size, shape and centration for its performance to have a positive effect on the lens stability and in this way contribute to the success of the accommodative, multifocal and toric lenses and also to a reduction in the risk of early development of posterior capsular opacity. The capsulorhexis is one of the steps which depends most on the ability and experience of the surgeon, and is the step in which the fewest technological advances have been incorporated into in recent years, depending totally on the skill of the surgeon.

Finally, the energy delivered by the phacoemulsification tip (cavitation, shock waves and free radicals), damages the corneal endothelium. For all these reasons, a technology that can perfectly control the performance of the incisions and the capsulorhexis and minimize the use of the phacoemulsificator inside the eye, should result in a better surgical outcome and greater control of cataract surgery.[3]

There are current studies by the different companies involved in the development of the femtosecond lasers, applied to cataract surgery, with respect to the improvement in the safety and the precision of the surgery. All these aim to avoid the collateral damage that the energy for the photodysruption may cause by controlling the length of the pulses and the levels of flow-limit of the energy delivered. In addition, the femtosecond lasers may enhance the outcome of the surgery by controlling and correcting preoperative corneal astigmatism using relaxing corneal incisions which have far greater precision and control than those that are performed with diamond knives available until now. In summary, the intergrated systems of anterior segment image diagnosis, together with femtosecond lasers, modified for use in the structures of the anterior segment of the eye can provide multiple benefits for cataract surgery.

Different femtosecond technologies are currently being developed. All are technologically similar but differ basically in the way in which they capture the image of the anatomic structures of the anterior segment, in their versitility for surgical use and speed of action.

There are currently four International companies producing femtosecond lasers that are being applied to cataract surgery: LenSx- (ALCON), LenSar, Catalys (OPTIMEDICA) and Victus (FEMTEC Technolase Perfect Vision, Bausch & Lomb).

LenSx Technology (ALCON, Aliso Viejo–California, USA)

This platform uses a femtosecond laser with a optical coherence tomography diagnostic unit with capacity for real time analysis of the intraocular structures (Fig. 12.1).

The surgical procedure with this technology begins with the fitting of a suction ring in the eye which in turn is attached to the optical system of the laser. At this moment, the OCT image captures the anterior segment in a single image, projected onto the screen with capacity for 3 dimensional manipulation of the image (Figs 12.2A and B). Using the software of the device, the incisions are located regarding their width, depth, profile and length (Fig. 12.3). In addition, the location of the capsulorhexis and the volume of the laser action inside the crystalline lens is determined by the use of the image software. With this data, determined by the software, the procedure is initiated by pressing the pedal.

The procedure is visible in real time while it is being carried out and lasts less than a minute (Figs 12.4A and B). The group that has acquired the most experience in the use of this laser is the group from the University of Budapest, directed by Dr Zoltan Nagy. The application of this laser using MICS has already been performed by this team and Professor Jorge Alió in June 2010.

In Figure 12.5 the results obtained can be observed, regarding the precision of the incisions observing the comparison between femtosecond laser surgery incisions that are sub 2 mm and 2.75 mm, in which the full impact on the cornea with respect to the induction of corneal aberrations or astigmatic changes can be seen.

This group has also studied the precision of the capsulorhexis performed with this instrument and its clear superiority with respect to the manual capsulorhexis.

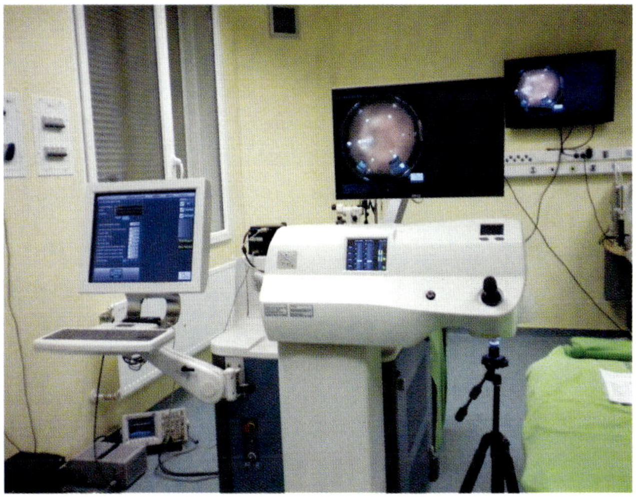

Fig. 12.1 LenSx Laser

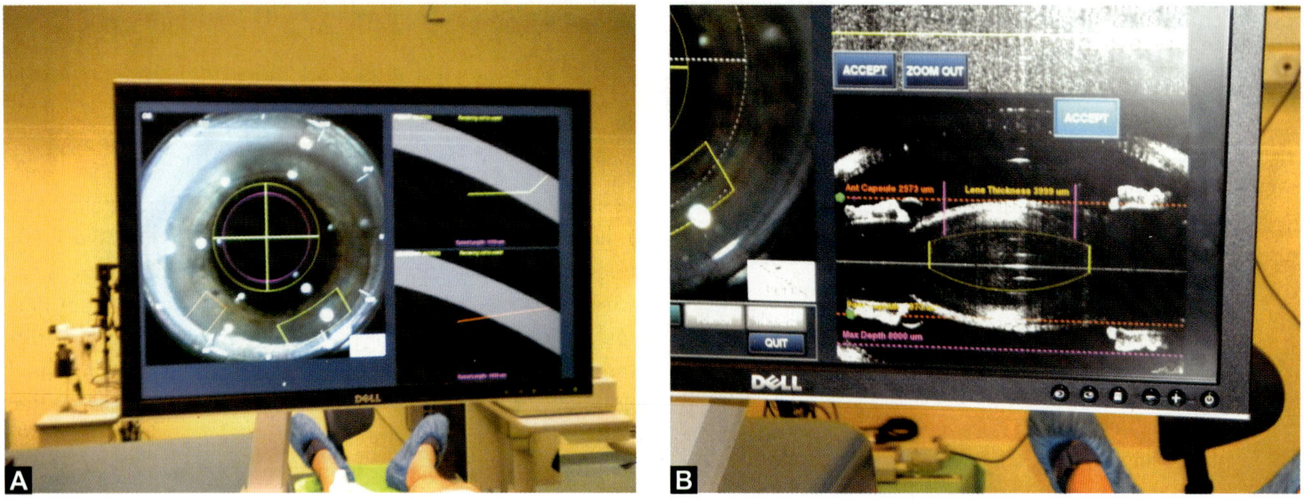

Figs 12.2A and B Images of the LenSx laser monitor

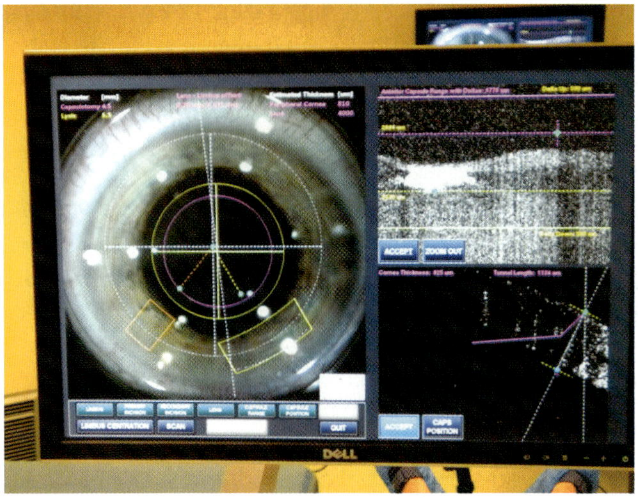

Fig. 12.3 Image of the incision drawing

The LenSx instrument can soften grade 1 and even grade 2 cataracts. Grade 2 + and 3 can be partially segmented, allowing in this way a more efficient division prior to phacoemulsification. Advanced grade 3 cataracts and over cannot be tackled at the moment by this technology.[4]

The LenSx technology will probably be available mid 2011 in Europe, once the CE mark is obtained.

LenSar Technology (Winter Park – Florida, USA)

This technology (Fig. 12.5) uses, an integrated confocal 3 dimensional platform which uses Scheimpflug technology (CSI Technology) to create the image of the anterior segment. In contrast to the optical coherence technology (OCT) the object plane, the lenticular plane and the planar image, are not parallel to each other but intersect in a joint straight line. The theoretical advantage is a greater depth of field which allows greater precision in the laser focalization. The laser is fitted to the cornea using a suction ring below. As with the previous platform, the image obtained of the anterior segment structures, which is observed in a monitor allow adjustment of the corneal incisions, the capsulorhexis and the fragmentation of the crystalline lens. This platform allows the lenticular fibers to be cut with multiple possible forms, including quadrangles, making it different from the others. This technology claims to be able to fragment even grade 4 and grade 5 cataracts, using the nuclear fragmentation in small cubes to facilitate the elimination of hard nuclei.

The CSI technology (3D Confocal Structure Illumination) using modified Scheimpflug image, according to the data provided by the company, allows much more precise nuclear fragmentation than with the other technologies. The first experiences with this instrument were carried out in May 2010, and an experimental platform is aimed to be established in Europe during 2010.

Optimedica Technology (Catalys System) (Optimedica Corporation - Santa Clara, California, USA)

This technology has developed a femtosecond laser platform for cataract surgery. It incorporates an optical coherence tomography (OCT) image to draw the image of the anterior segment anatomic structures. This is known as the Catalyst system for lens surgery. The experiences with this laser, similar to the previous ones regarding capacity to create precise corneal incisions, controlled capsulorhexis with respect to diameter, location and shape, and nucleus fragmentation, have been satisfactory in the preliminary clinical trials (Fig. 12.6).

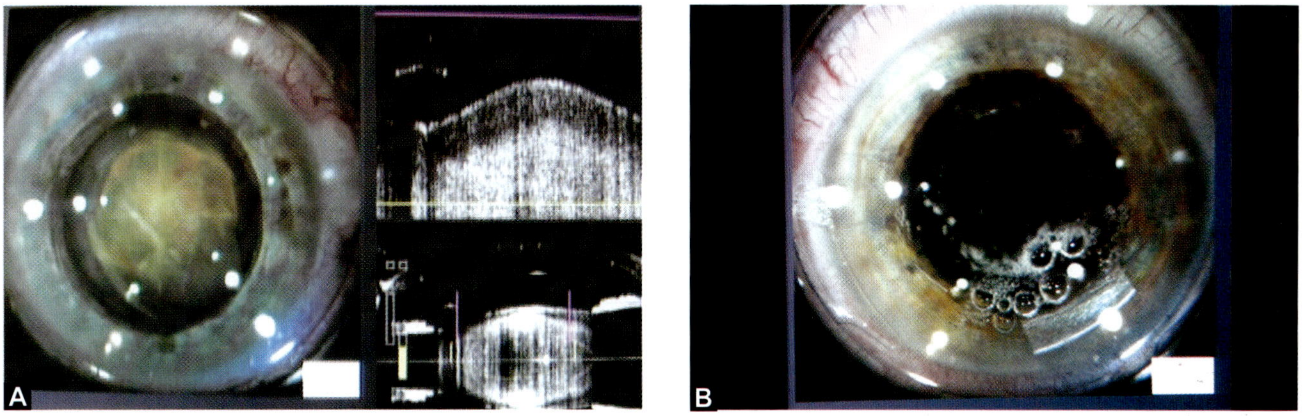

Figs 12.4A and B Final stage of the nucleus dysruption

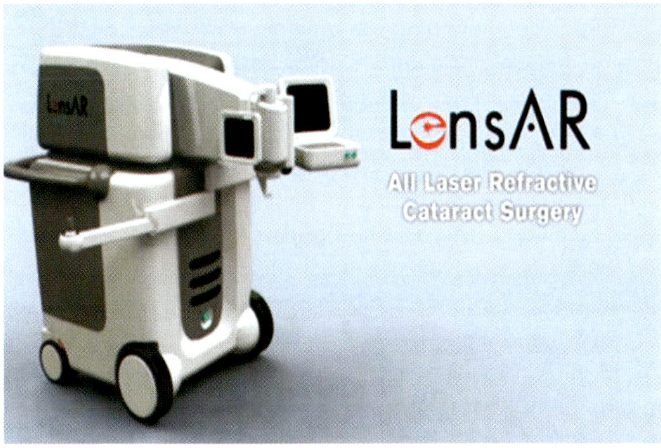

Fig. 12.5 LensAR Laser

Fig. 12.6 The Catalyst femtosecond cataract system (Optimedica)

As with the previous laser, this laser can perform multiple options for fragmentation of the nucleus, with diverse possible forms.

Victus Technolas (520F, Customlens) FEMTEC Technology (Munich – Germany) (Fig. 12.7)

This company has adapted its femtosecond corneal laser for cataract surgery (CUSTOMLENS Technology). They have initiated clinical studies in Europe, in Heidelberg University, and these are expected to be extended to other countries during this year. This is the only technology that offers a combined instrument capable to perform both corneal and intraocular surgery.

DIFFERENCES AMONG THE TECHNOLOGIES

The femtosecond lasers for use in cataract surgery, require different parameters to those used for corneal surgery. These lasers need to penetrate more deeply into the eye (approximately 7500 microns compared to 1200 microns required for cornea surgery), and the four companies LenSx, LenSar, Optimedica and FEMTEC have developed very similar with respect to the creation of an efficient photodysruption for the creation of the capsulorhexis and the softening and fragmentation of the nucleus.

The four technologies use anterior segment diagnostic technology capable of drawing the anterior segment of the eye with precision. The LenSar technology uses an optic system which requires the extraction of data for the complete reconstruction of the posterior capsule, apparantly similar to OCT technologies, which in theory can imply an limitation to its clinical application, although the initial clinical results confirm its precision.

At the moment, there are no comparative studies that can determine which of these technologies is better than the oth-

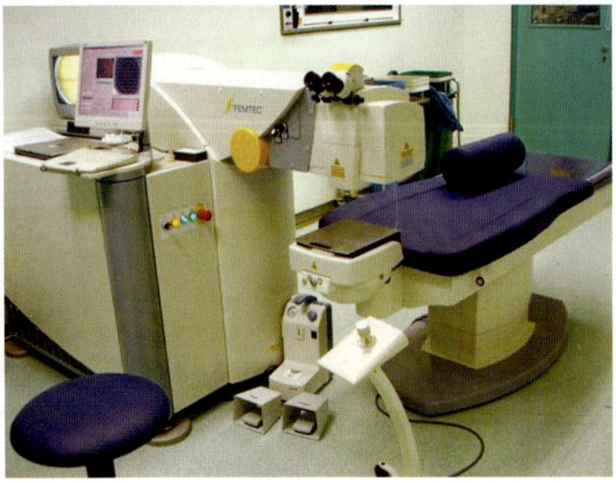

Fig. 12.7 Femtec Laser Image

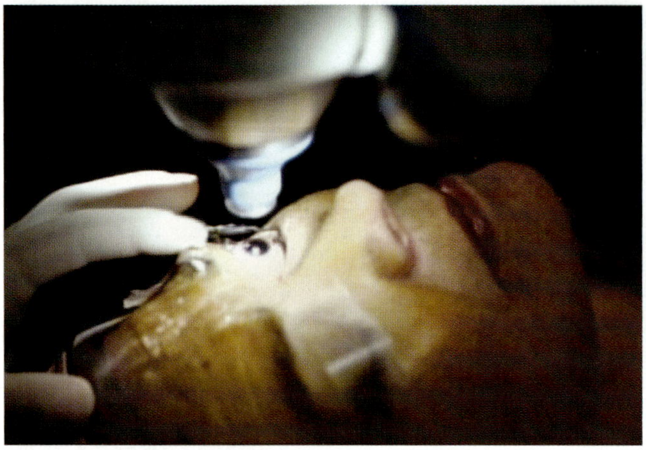

Fig. 12.8 Docking procedure

ers, although all three are different in terms of ergonomics, size and probably cost. 2011 will open a new era for the application of this technology in cataract surgery given that these three platforms will provide clinical results which will lead to their introduction, to cataract surgery, in the near or long term as a new reality. In any case, femtosecond technology will not eliminate the need for a modern phacoemulsificator, as this is necessary to emulsify hard nuclei and the cataract fragments created from grade 2 and 2+ of hardness.

An important question raised by most of the surgeons and administration interested in these technologies is cost. An increase in about 600 US$ per operated eye is expected when introducing femtosecond laser technology in cataract surgery (approx cost of the unit will be about 500,000 US$ plus 150-300 US$ for disposibles, plus taxes, loans and hidden costs). How this extra cost, very significant in today's worldwide financial environment, will be absorbed by the practice is going to be the unresolved question for the cataract surgeon.

Our First Experiences with the ALCON-LENSX Technology

For those surgeons who are used to performing corneal surgery and especially LASIK with femtosecond laser, the learning curve will be easy and short. As occurs with LASIK, the key to the success of performing cataract surgery with femtosecond laser is the docking procedure. The docking is the first step of the surgery and it determines the safety and the accuracy of the whole procedure. It is essential to achieve a perfectly centered docking in order to continue with the surgery. Once the docking is done (Fig. 12.8) and the position of the eye is checked on the screen, suction is applied by simply pressing a button. Regarding the suction, the pressure level is lower than that of the corneal surgery platforms. This is an important point because patients who undergo cataract surgery are usually elderly and a large increase in the intraocular pressure would not be appropriate.

The next step is to center the treatment plan on the screen. First of all, the incisions have to be placed. We have used both, the standard microcoaxial technique and the microincisional technique (MICS). Our preference is to perform MICS and for this purpose we use two 1 mm microincisions to complete the surgery comfortably and a third incision placed on the steep meridian of 1.8, 2.2 or 2.5 mm (depending on the IOL to implant) which is opened at the end of the surgery prior to the IOL implantation.

Once the corneal incisions are designed in location, length and width, the next step is to perform the capsulorhexis. The capsulorhexis size can be chosen according to the IOL optic diameter that we are going to implant in order to achieve an excellent anterior capsule overlapping. A limitation for the rhexis size is the iris border. Thus, those patients with a mydriasis smaller than 6 mm are not good candidates for this procedure. After determining the size, the capsulorhexis can also be perfectly centered (Fig. 12.9).

The next step is to choose the pattern of nucleus softening/photodisruption. We can choose three different patterns. The first one is the softening pattern (Fig. 12.10) for those not very hard cataracts. The laser creates several cylinders of photodisruption ("like coins") in the core of the nucleus allowing an easy removal of these soft cataracts. For the hardest cataracts we use the segmentation (Fig. 12.11) or the hybrid pattern (Fig. 12.12). The former creates 4 or more horizontal cuts in the nucleus (like a chopping) while the latter is a combination of the softening pattern creating cylinders in the core of the nucleus and the segmentation pattern producing more peripheral horizontal cuts. Using the segmentation pattern or

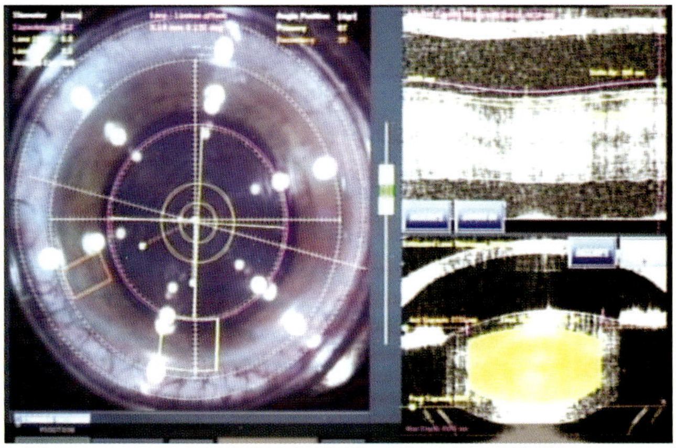

Fig. 12.9 Anterior capsulotomy plan on OCT

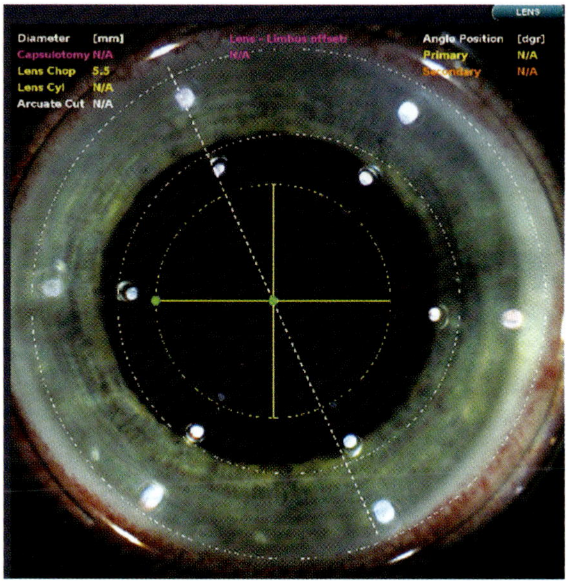

Fig. 12.11 Segmentation pattern of photodisruption

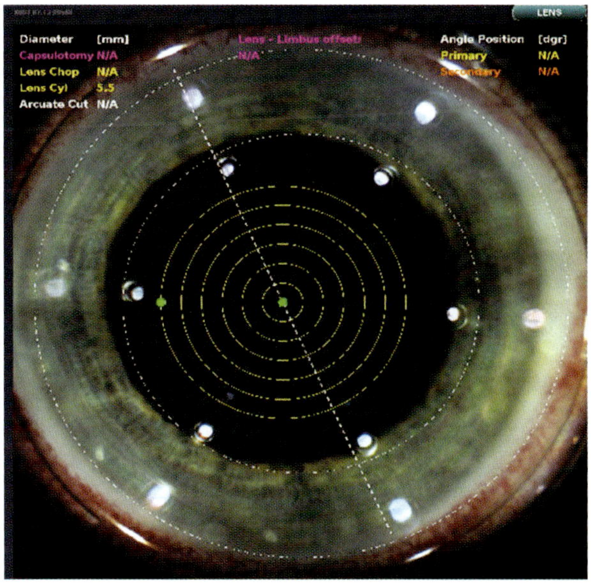

Fig. 12.10 Softening pattern of photodisruption

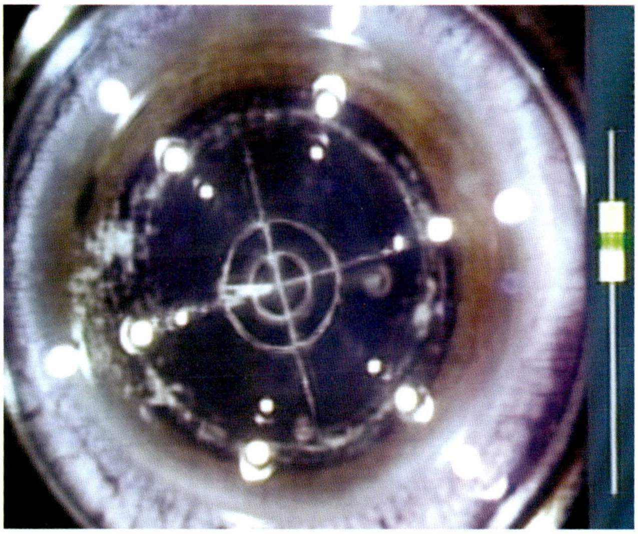

Fig. 12.12 Hybrid pattern of photodisruption.

the hybrid pattern we have been able to remove hard cataracts with a lower ultrasound energy use.

Once the pattern has been chosen, we have to set two important parameters, first the photodisruption diameter and second the laser depth of action. We determine the depth of action using the posterior capsule OCT image. We keep a safe area between the highest point of the posterior capsule and the laser action. At this moment we have not had any posterior capsule rupture due to a laser damage.

The last step prior to starting the femtosecond treatment is to determine the incision shape. The architecture of the incisions previously located has to be determined. Our preference is the "dimple up" design: there is a short first plane which is orthogonal to the epithelium followed by an inclined plane of about 1600 microns towards the endothelium and finally a third orthogonal plane that penetrates into the anterior chamber (Fig. 12.13). The angles and length of every plane can also be customized.

Finally the treatment parameters on the OCT are confirmed and the foot pedal is pressed to start the treatment. The whole procedure is completed in only 90 seconds. It starts doing the anterior capsulotomy, then continues with the photodisruption and finishes with the incision creation.

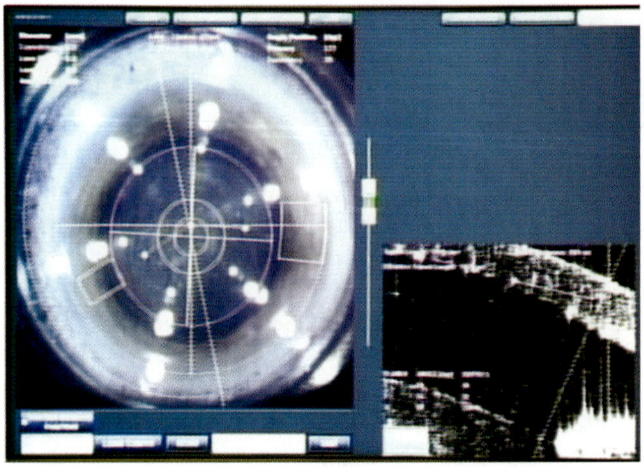

Fig. 12.13 Incisional architecture

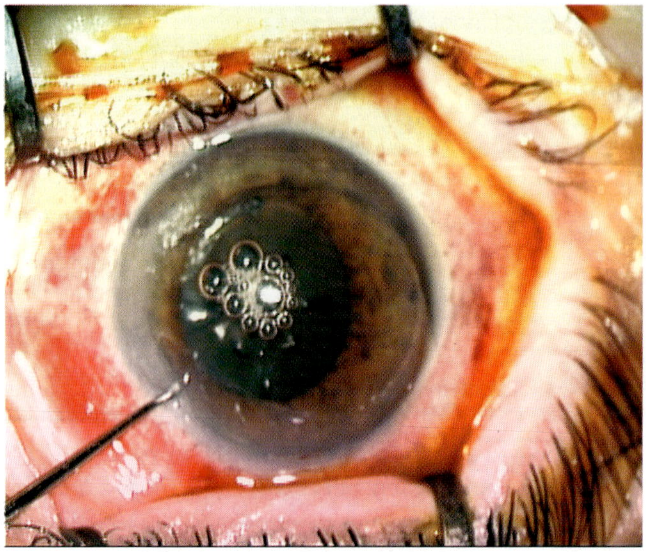

Fig. 12.14 Incision opening with a spatula

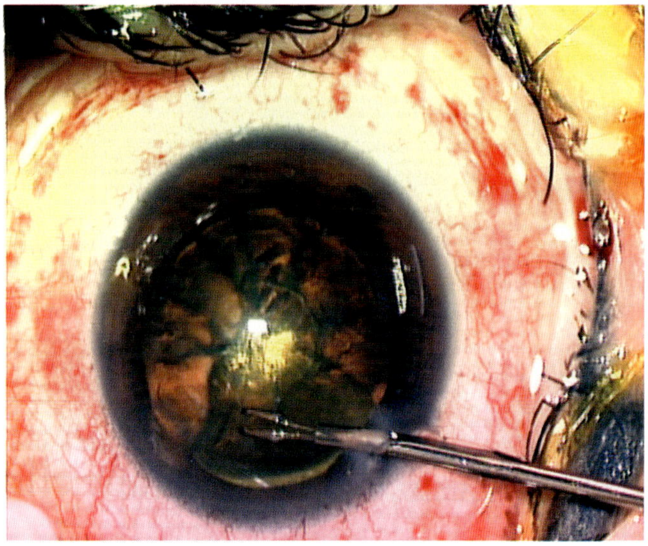

Fig. 12.15 Anterior capsulotomy flap removal with MICS forceps

We perform this procedure under ordinary corneal refractive surgery aseptic conditions. It is very important to consider that there is no risk of endophthalmitis because the eye is not opened during the femtosecond laser treatment.

Phacoemulsification after the Femtosecond Laser Treatment

Once the LENSX treatment is completed we prepare the eye under sterile conditions to perform the phacoemulsification. Alcon recommends that the time elapsed from the femtosecond treatment to the phacoemulsification should not be longer than 30 minutes. In our experience we have performed uneventful procedures with slightly longer time intervals.

The incisions may be opened using either a Sinskey hook or the Alió femtocataract spatula (Epsilon, EEUU). At this point it is very important that the incisions that we have created with the laser have a similar design to our normal manual incisions, this way the incisions will be easily opened and there will not be any problems to find the incisions planes (Fig. 12.14).

The next stage is the viscoelastic injection. We strongly recommend performing this maneuver introducing the tip of the cannula inside the eye placing it over the anterior capsulotomy flap. The viscoelastic injection placing the tip of the cannula in the corneal wound must be avoided because the viscoelastic device may go under the capsulotomy flap producing a capsulotomy tear.

Then, using a cystitome or a MICS capsulorhexis forceps (MICS Forceps Alió, Katena, USA) the anterior capsulotomy flap is carefully removed paying a lot of attention to break any potential capsular bridges (Fig. 12.15). These capsulotomy bridges are usually due to small inclinations of the eye that determine that the laser capsulotomy is done with a higher efficacy in some sectors than in others.

After the anterior capsule flap removal, the next step is the hydrodissection. There is an interesting debate about if the hydrodissection is necessary or not. In many cases, the bubbles from the photodisruption dissect the nucleus from the capsular bag making the hydrodissection maneuvers not strictly necessary. However, hydrodissection can be performed to make sure that the cataract nucleus is free. The capsular bag has an increased volume because there are usually bubbles from the photodisruption. This fact explains why it is so important to remove the viscoelastic device completely from the anterior chamber before starting the hydrodissection. If we do not take this precaution we may have a capsular block and a posterior capsule rupture.

The cataract fragments removal is easier because the nucleus already has break lines as a result of the femtosecond laser action. For this purpose we use an ultrasound energy of 2 to 4% in Burst mode and a vacuum from 300 to 350 mm Hg.

After the nucleus removal, the cortex usually looks very clear and transparent. This fact can even create confusion because the surgeon may think that there is no cortex to remove. However, if aspiration is performed using the capsulorhexis border as a guide the cortex will be removed from 360° of the capsular bag. Therefore, it is advisable that the capsulorhexis border is seen easily. We do not recommend capsulorhexis size greater than 5.7 mm. Our most frequently chosen size is about 5.4 mm.

After the regular IOL implantation, the ocular viscoeslastic is removed and the incisions are sealed. We have noticed that the femtosecond incisions are very prone to seal spontaneously even with no hydration. We think that this is explained because of the perfect incisional architecture and because with the femtosecond assistance we save intraocular maneuvers that sometimes distort the incisions.

Our first outcomes in Vissum Alicante are still preliminary but confirm that the nucleus removal is facilitated by the femtosecond laser. The incisions and anterior capsulotomy performance with the LenSx technology are excellent. The effective phaco time (EPT) and the total phacoemulsification power are diminished when using this technology.

We have not had any complications related with the femtosecond laser procedure. Our opinion is that the surgical technique will undergo important modifications in the future and only the experience will show us how to transform our current surgery into a new more efficient surgery with better results.

Early Clinical Evidence

Femtosecond laser cataract surgery may indeed provide significant advantages to operate cataracts. Part of the surgery is much more systematic as the corneal incision profiles and dimensions can be set up with much more control of the variability. Capsulorhexis should be performed on a more controlled basis also for size and dimensions. Nuclear softening should reduce the phaco power and effective phaco time used to remove the cataract, as occurs in traditional prechopping. This maneuvre has indeed been proposed by many authors but so far it is seldomly used because it is considered by many surgeons as difficult and risky. With the use of femtosecond lasers this clear advantage for cataract surgery will no longer be a real barrier for the practical surgeon.[10] There are still very few papers dealing with the real outcomes that happen when using femtosecond cataract surgery. The few reports that have been published are to be considered preliminary and the clinical evidence, although it really suggests a trend to confirm the abovementioned advantages. An adequately designed clinical comparative study, comparing the standard phacoemulsification technique and femtosecond cataract surgery is still lacking at present.

Preliminarly, there is evidence that the control of the corneal incisions in terms of size and dimensions is indeed achieved. Capsulorhexis size and shape is indeed improved[2,8,11] with a consequent improvement in the effective lens position of the implanted IOL. The consequence of this is a more precise refractive outcome that standard conventional cataract surgery.[8] The optimized control of the corneal incisions has demonstrated to lead to a better astigmatic control and less induction of higher order aberrations.[2] The total aberration pattern of the eye is indeed also improved,[2] consequently, and based on the better preservation of the corneal diopter capabilities and the better IOL position. All this preliminary evidence indicates that femtosecond cataract surgery is indeed a step forward in the quality of cataract surgery.

One more aspect to consider for femtosecond cataract surgery is that it is probably the first step towards a real robotization process in cataract procedures. In other areas of medicine, this is happening (neurosurgery, urology, tumor of the digestive tract surgery, etc.), and the trend is approaching in ophthalmology. Such process is starting and most probably with the femtosecond laser devices that we have at this moment constitute the first step of such process.

The only drawback of femtosecond cataract surgery is indeed financial, the added high cost of introducing this as a routinary procedure in cataract surgery.[1] In today's environment of cataract surgery, a reduction in the costs to increase the cost effectivity of the procedure is a global trend. How, in this environment, will a clear increase in the cost due to the benefit of an apparently small improvement, even though significant, justify such cost? This question remains unsolved. At this moment, in some environments, the cost of using femtosecond cataract surgery can duplicate the total cost of phacoemulsification and IOL implantation. In the developed countries, this is indeed also a problem even with the higher level of financial resources available. Most probably, a new business model for cataract surgery has to be developed to absorb such a cost in a comprehensive and practical way prior to the generalization of the technique.

REFERENCES

1. Uy HS, Edwards K, Curtis N. Femtosecond phacoemulsification: the business and the medicine. Curr Opin Ophthalmol 2012;23:33-9.
2. Mihāltz K, Knorz MC, Alió JL, Takács AI, Kránitz K, Kovács I, Nagy ZZ. Internal aberrations and optical quality after femtosecond laser anterior capsulotomy in cataract surgery. J Refract Surg 2011;27:711-6.
3. Roberts TV, Sutton G, Lawless MA, Jindal-Bali S, Hodge C. Capsular block syndrome associated with femtosecond laser-assisted cataract surgery. J Cataract Refract Surg. 2011;37:2068-70.
4. Ecsedy M, Mihāltz K, Kovács I, Takács A, Filkorn T, Nagy ZZ. Effect of femtosecond laser cataract surgery on the macula. J Refract Surg. 2011;27:717-22.
5. Kránitz K, Takacs A, Mihāltz K, Kovács I, Knorz MC, Nagy ZZ. Femtosecond laser capsulotomy and manual continuous curvilinear capsulorrhexis parameters and their effects on intraocular lens centration. J Refract Surg. 2011;27:558-63.
6. Friedman NJ, Palanker DV, Schuele G, Andersen D, Marcellino G, Seibel BS, Batlle J, Feliz R, Talamo JH, Blumenkranz MS, Culbertson WW. Femtosecond laser capsulotomy. J Cataract Refract Surg. 2011 Jul;37(7):1189-98. Erratum in: J Cataract Refract Surg. 2011;37:1742.
7. Mamalis N. Femtosecond laser: the future of cataract surgery? J Cataract Refract Surg. 2011;37:1177-8.
8. Nagy ZZ, Kránitz K, Takacs AI, Mihāltz K, Kovács I, Knorz MC. Comparison of intraocular lens decentration parameters after femtosecond and manual capsulotomies. J Refract Surg. 2011;27:564-9.
9. He L, Sheehy K, Culbertson W. Femtosecond laser-assisted cataract surgery. Curr Opin Ophthalmol. 2011;22:43-52.
10. Naranjo-Tackman R. How a femtosecond laser increases safety and precision in cataract surgery? Curr Opin Ophthalmol. 2011;22:53-7.
11. Kullman G, Pineda R 2nd. Alternative applications of the femtosecond laser in ophthalmology. Semin Ophthalmol. 2010;25:256-64.
12. Palanker DV, Blumenkranz MS, Andersen D, Wiltberger M, Marcellino G, Gooding P, Angeley D, Schuele G, Woodley B, Simoneau M, Friedman NJ, Seibel B, Batlle J, Feliz R, Talamo J, Culbertson W. Femtosecond laser-assisted cataract surgery with integrated optical coherence tomography. Sci Transl Med. 2010;17;2:58ra85.
13. Nagy Z, Takacs A, Filkorn T, Sarayba M. Initial clinical evaluation of an intraocular femtosecond laser in cataract surgery. J Refract Surg. 2009;25:1053-60.
14. Kullman G, Pineda R 2nd. Alternative applications of the femtosecond laser in ophthalmology. Semin Ophthalmol. 2010;25:256-64.
15. 15. Alió J, Rodríguez-Prats JL, Galal A, Ramzy M. Outcomes of microincision cataract surgery versus coaxial phacoemulsification. Ophthalmology. 2005;112:1997-2003

13 Lamellar Corneal Resection with LDV Crystal Line Femtosecond Laser after Penetrating Keratoplasty

Bojan Pajic, Slobodanka Latinovic, Farhad Hafezi, Brigitte Pajic-Eggspuehler, Michael Mrochen, Joerg Muller, Franz Fankhauser (Switzerland)

Chapter Outline

- Physics and technical approach
- Minimizing energy—maximizing cutting precision
- Technical Realization and clinical outcomes
- Patients and methods
- Results
- Discussion

PURPOSE

The purpose of this study is to describe the underlying physics of the crystal line LDV femtosecond laser (Ziemer ophthalmic systems AG, Port, Switzerland), and to demonstrate the capability of this device to perfom flap resections in postkeratoplasty corneas through scarred tissue.

INTRODUCTION

As several different models of femtosecond (fs) laser systems have become available, refractive and corneal surgeons-strive to perform all types of corneal surgery with "all-laser" procedures. Particularly, performing corneal resections through scarred tissue, as encountered in eyes that have undergone perforating keratoplasty (PKP), has remained a challenge.

As the trend in refractive and corneal surgery is moving towards all-laser-treatments, there are now different fs-laser systems for applications in corneal surgery available. All current fs-laser systems use short laser pulses to dissect corneal tissue by photodisruption. However, there are significant differences in technical detail and in clinical outcome between different devices.

PHYSICS AND TECHNICAL APPROACH

Laser—Tissue Interaction

The interaction process of all fs-laser systems for corneal surgery, e.g. preparation of corneal lenticule in LASIK procedures, is based on nonlinear absorption and consecutive disruption of the tissue, followed by formation of a cavitation bubble and a residual gas bubble. Nonlinear in contrast to linear absorption means that the tissue is transparent for the infrared laser radiation at moderate intensities, hence no absorption occurs. Only at very high intensities, which can be achieved by compressing the laser pulse in time (ultrashort) and in space (strong focusing), some photons can be absorbed co-instantaneously by the tissue. The narrowness of the region of photodisruption depends on the degree to which the beam is strongly focused and is never, in practice, a single point in z-axial space. However, nonlinear interaction gives the user the advantage of three dimensional tissue processing. The absorption process is not limited to the surface anymore.[1-4]

While photodisruption scales with pulse intensity (W/cm^2), the unwanted side-effects such as thermal heating, stress- and shockwaves, or large residual gas bubbles, scale with pulse energy (J). It is therefore desirable to minimize pulse energy while keeping the pulse intensity above the photodisruption threshold level (Fig. 13.1). Two obvious principles can be used for optimizing pulse intensity at reduced pulse energy: Compression in the time domain (pulse shortening) and in the spatial domain (strong focussing).

MINIMIZING ENERGY—MAXIMIZING CUTTING PRECISION

Shortening the pulse duration is a basic physical challenge, which is related to the spectral bandwidth of the laser medium. The pulse duration of typical fs-laser systems is around 200 to 800 fs (1 fs = 10^{-15} sec). In this range, the energy threshold for optical breakdown increases almost linearly with pulse duration (Fig. 13.2). Thus, the shorter pulse duration of Ziemer's Femto LDV (approx. 250 fs) is related to a lower pulse energy threshold than other systems, which operate at pulse widths approx. 600 to 800 fs.

The second way to decrease the energy is to decrease the focal volume of the laser spot. The focal volume of a Gaussian laser beam depends on the axial extension, the so called Rayleigh range

$$z = \frac{\pi w_0^2}{\lambda}$$

and the beam waist

$$w_0 = \frac{f\lambda}{\pi w_L}$$

where f is the focal length of the focusing lens, λ the laser wavelength and W_L the radius of the beam at the focussing

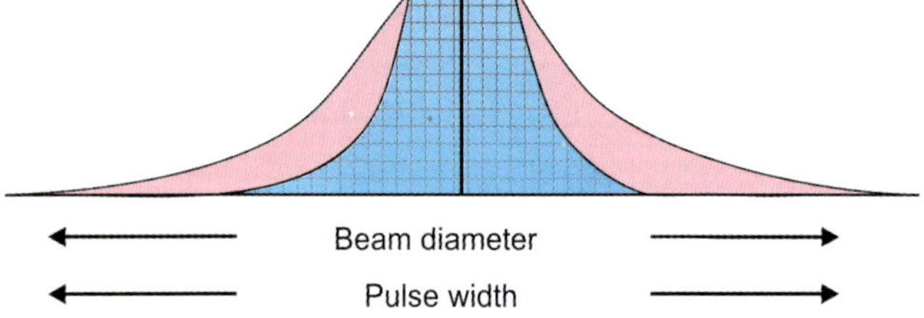

Fig. 13.1 Two laser pulses with the same pulse energy. The red labelled laser pulse is compressed in time domain (ultra short) and in space (strongly focussing with high NA). As a result its intensity increases above threshold for photodisruption

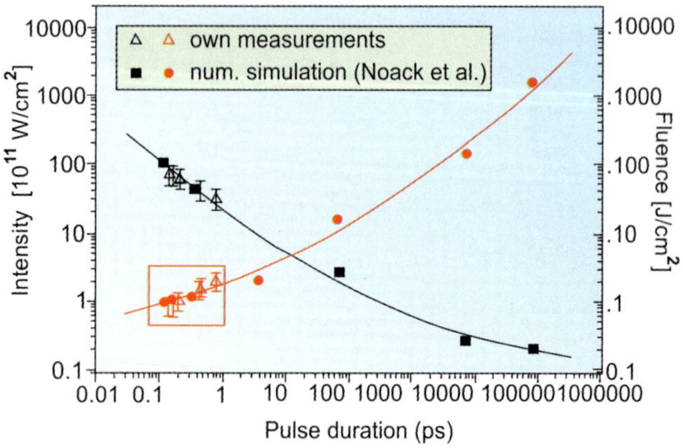

Fig. 13.2 Threshold values for photodisruption in water. The black curve shows the intensity threshold and the red curve represents the fluence of a single laser pulse. In the range of 100 fs to 1 ps (red box) the function of energy threshold is almost linear with pulse duration. [Calculated data points are from Noack[3]]

lens. In other words, the focal volume varies inversely with the fourth power of the numerical aperture

$$NA = \frac{w_L}{f}$$

of the focussing optics. The larger the NA the smaller is the focal spot and finally, the less pulse energy is needed (Fig. 13.3) to cause photodisruption. According to its definition, there are two ways to increase the NA. One possibility is to increase the beam diameter at the focusing optics, which requires large and expensive optical components. As an alternative, one can also decrease the focal length of the focusing lens, which on the other hand reduces the working distance of the laser system (Fig. 13.4). Finally, the laser's repetition rate (laser frequency) has an important influence on the pulse energy threshold. The higher the laser frequency, the less pulse energy is needed (Fig. 13.5). Thus, it is obvious that the Ziemer LDV, which operates at >5 MHz and at large NA, offers a significant advantage over other devices which operate at frequencies in the tens or hundreds of kHz MHz-Laser take advantage of this[5] and at up to 5 times larger focal widths.

TECHNICAL REALIZATION AND CLINICAL OUTCOMES

With respect to the interaction process and the beam delivery, fs surgical lasers can thus be classified into two groups: one group characterized by high pulse energy with low pulse frequency, and the other group characterized by low pulse energy with high pulse frequency. In the "high pulse energy – low pulse frequency" group, amplified laser systems are used to deliver pulse energies in the range of 1 to 5 μJ to the cornea. The typical repetition rate is about some kHz. In contrast,

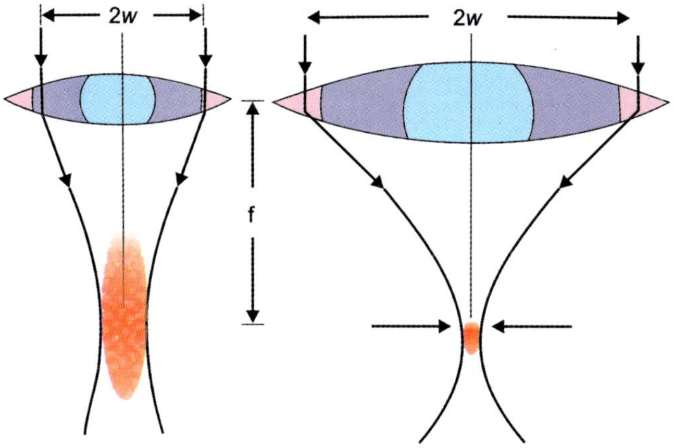

Fig. 13.3 The focal volume of a Gaussian laser beam scales with the numerical aperture NA = W_L/f of the focussing lens. The larger the NA, the smaller is the focal spot volume

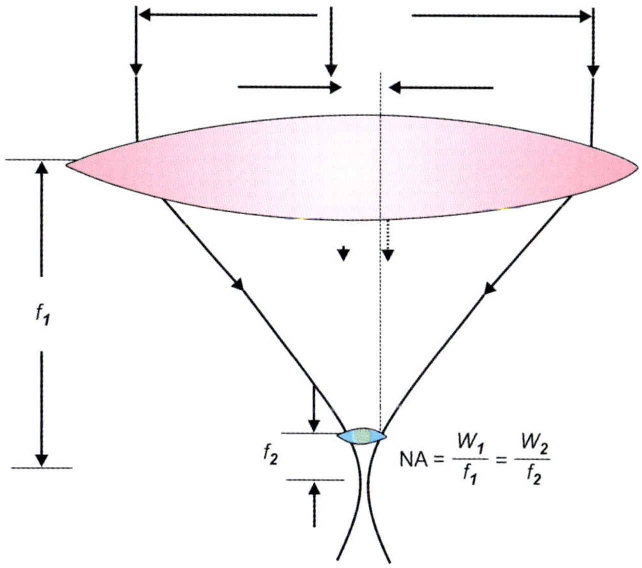

Fig. 13.4 At a constant relationship of focal length and lens diameter, a given focal diameter of the laser beam can be achieved by either large lenses and long working distances or by smaller lenses with shorter working distances

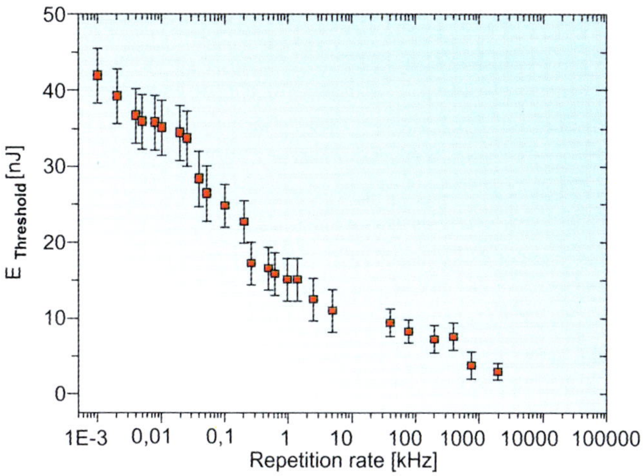

Fig. 13.5 The threshold energy in water decreases with increasing laser frequency. For MHz laser frequencies the threshold is almost half of kHz laser frequencies

the "low pulse energy-high pulse frequency" Contrary to that the Ziemer's Femto LDV, currently the only commercial device in this group, employs short 200 fs pulses, high NA optics and a very high pulse repetition rate exceeding 5 MHz. The required pulse energies in such a design are in the range of a fewless than 100 nJ (nano-Joules) which delivers only nJ pulse energies to the eye and uses MHz repetition rates clearly belongs in the "low pulse energy-high pulse frequency" group. Based on the laser parameters, the physical nature of the cutting processes of the two groups is different. In the "high pulse energy laser group", the cutting process is driven by mechanical forces which are applied by the expanding bubbles and which disrupt the tissue. This cutting process is very efficient because the radius of disrupted tissue is larger than the laser spot itself. Hence, the spot separation of the scanned laser pulses can be larger than the spot diameter (Fig. 13.6A). On the other hand, due to the lower focusing accuracy and the larger remaining gas bubbles, poorer flap qualities, OBL (opaque bubble layer), TLS (transient light sensitivity syndrome) and anterior chamber bubbles may appear. There can be an additional waiting time for the proper use of the Excimer's eye tracking system as well. Using low pulse energies, the cutting tissue disruption process is confined by to the focal spot size of the laser pulse. In fact, the actual photon interaction volume can be even smaller than the focal volume due to the nonlinear intraction requiring a certain threshold for disruption to occur. As a consequence, more pulses are needed to cut the same area. To keep the total operation procedure time at the same level, higher pulse repetition rates are required (Fig. 13.6B). Since a pure fs-laser oscillator is used, it delivers these many pulses as well as very high stability and robustness against environmental influences compared to amplifier systems. The clinical outcome shows a perfect cut even through scars.

PATIENTS AND METHODS

Nine eyes that had undergone penetrating keratoplasty 1-3 months earlier were selected to undergo lamellar keratectomy (flap creation for subsequent excimer laser ablation) with a Ziemer LDV Crystal line fs-laser (Fig. 13.7). Preoperative characteristics included an astigmatism of more than 4 D, a

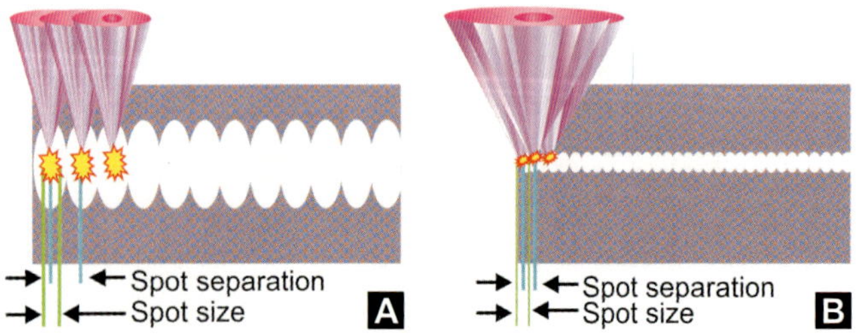

Figs 13.6A and B For smaller focusing angles with higher pulse energies (lower laser frequencies), the cutting effect is driven by mechanical forces of the increasing cavitation bubble (B). On the other hand, MHz laser frequencies can offer many more pulses that are needed for cutting with lower pulse energies and larger numerical aperture (B). In this case, the size of the cut is defined just by the focal spot size

Fig. 13.7 Crystal line LDV femtosecond laser

corneal irregularity of more than 4 D in the 5 mm zone and significant higher order aberrations (HOA). The best preoperative visual acuity range was in all cases less than 0.125. A 9.5 mm suction ring, 110 µm flap thickness and manufacturer's standard resommended laser parameters were used in all cases for performing a lamellar LASIK cut. Standard ophthalmological examination, including visual acuity examination, corneal topography and wavefront aberrometry were performed before and 6 weeks after treatment. For purposes of statistical data analysis, the SPSS program was used.

RESULTS

After performing a lamellar keratotomy and repositioning of the corneal flap we observed a significant decrease of the corneal-irregularity-index averaging -0.33±0.42 dpt in the 5 mm zone (Table 13.1). At the same time the corneal astigmatism reduction of -3.01 ± 2.14 D was slightly significant (p = 0.19), whereas the axis position changed by 14.28 ± 2.12 degrees. A highly significant decrease of HOA was detected. This resulted in an important and highly significant improvement of visual acuity by 3 up to 7 lines, (mean 5 ± 1.4). HOA analysis revealed an amplitude of 13.28 µm pre-operatively (Fig. 13.8), decreasing postoperatively to 1.82 µm (Fig. 13.9). Simultaneously, the point-spread function (PSF) decreased significantly from preoperative (Fig. 13.10) to postoperative (Fig. 13.11).

DISCUSSION

This study investigated surgical outcomes obtained by using the Ziemer Crystal line LDV Femtosecond Laser for lamellar keratectomy after PKP. The use of this fs laser technology helped to reduce or eliminate many of the minor or even severe complications associated with flap creation.[7-11] Usually, a transparent medium is necessary for cutting with a femtosecond laser. However, this study has demonstrated that the Ziemer LDV, due to its unique characteristics, is capable to perform perfect lamellar cuts and produce smooth stromal beds even in the presence of scars originating from prior PKP. In our experience, the flaps were as easily mobilized as flaps prepared with a mechanical microceratome, with no tissue bridges interfering but with the added benefit of improved flap thickness precision. The tension forces in the corneal tissue, caused by scar formation after penetrating keratoplasty, can compromise the optical characteristics of the corneal center and hence visual acuity. A lamellar keratotomy in the anterior third of the cornea leads to relevant tissue relaxation, with concomitant decrease of corneal irregularities and aberrations, leading to improved visual acuity and better visual comfort. This treatment can form the basis for subsequent refractive fine-tuning by standard refractive procedures; namely LASIK by performing excimer laser ablation in the stromal bed, by re-lifting the flap created as described in this study.

Table 13.1 Topographical corneal parameters

	MW	SD
Δ corneal-irregularity-index	-0.33	0.42
Δ astigmatism power	-3.01	2.14
Δ astigmatism axis change	14.28	2.12

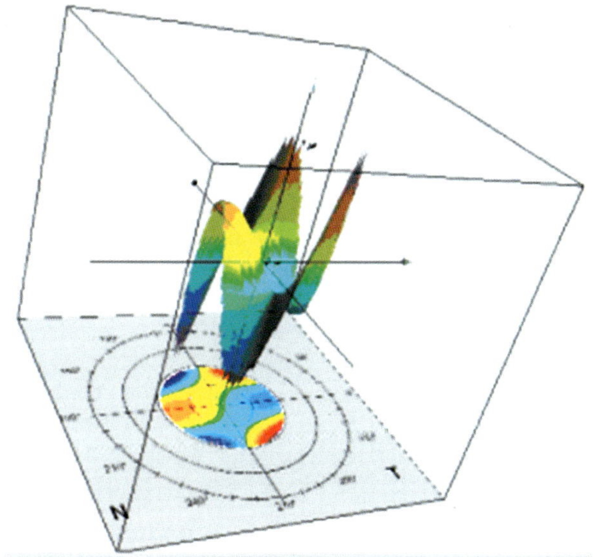

Fig. 13.8 Aberration higher order 3D before keratotomy; ampl: 13.28 µm

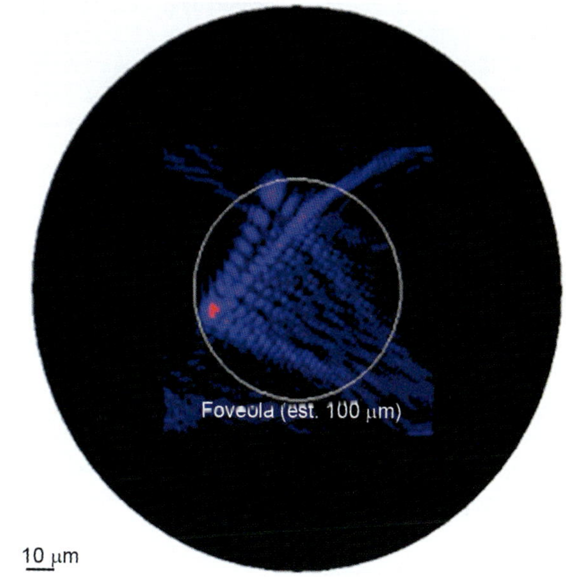

Fig. 13.10 Aberration higher order PSF before keratotomy

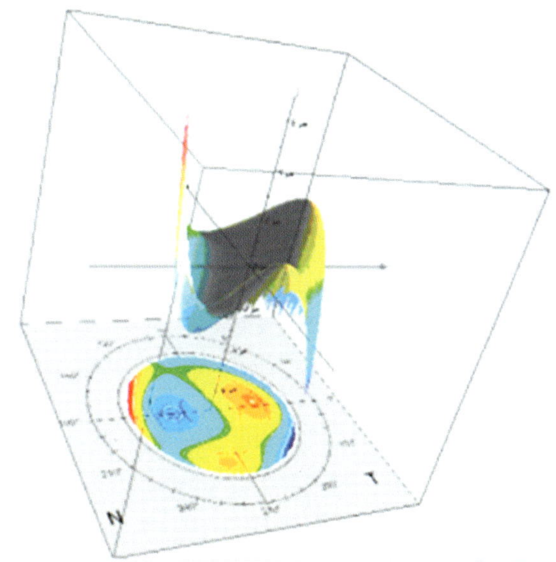

Fig. 13.9 Aberration higher order 3D after keratotomy; ampl: 1.82 µm

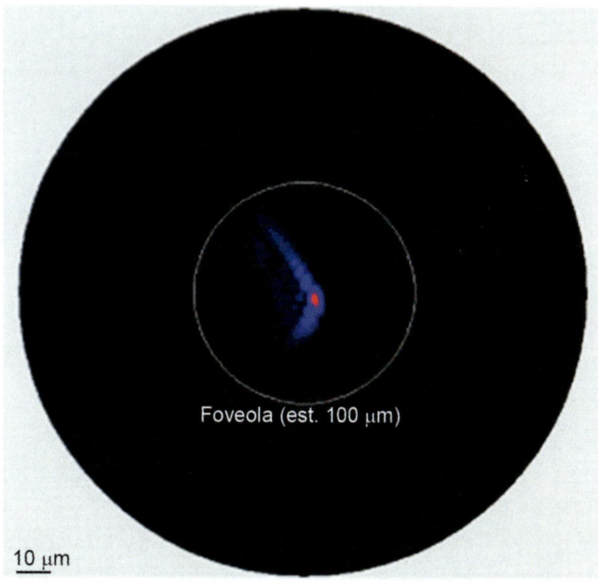

Fig. 13.11 Aberration higher order PSF after keratotomy

ACKNOWLEDGMENTS

The authors thank for dis tinct support of Drs T Ripken (Laser Zentrum Hannover, Germany), Dr and Drs C Rathjen and Dr TA Wirthlin from (Ziemer Group, Port, Switzerland) for their valuable personal support and technical advice.

REFERENCES

1. Heisterkamp A, Ripken T, Mamom T, Drommer W, Welling H, Ertmer W, Lubatschowski H. Nonlinear side effects of fs-pulses inside corneal tissue during photodisruption, Appl Phys. 2002;B,74,1-7.

2. Lubatschowski H, Maatz G, Heisterkamp, Hetzel U, Drommer W, Welling H, Ertmer W. Application of ultrashort laser pulses for intrastromal refractive surgery; Graefe's Arch Clin Exp Ophthal. 2000;238:33-9.
3. Noack J, Vogel A. Laser-Induced Plasma Formation in Water at Nanosecond to Femtosecond Time Scales: Calculation of Thresholds, Absorption Coefficients, and Energy Density; IEEE J. Quantum Electron. 1999;35(8):1156.
4. Vogel A. Noack J. Mechanism of femtosecond laser nanosurgery of cells and tissue Appl Phys. 2005;B,81(8):1015.
5. Ripken T. Anwendung von MHz-fs-lasern in der Ophthalmologie und Erarbeitung eines Therapiekonzeptes für die laserassistierte Behandlung der Alterssichtigkeit, Dissertation, Gottfried Wilhelm Leibniz Universität Hannover, 2007.
6. Nordan LT, Slade SG, Baker RN, et al. Femtosecond laser flap creation for laser *in situ* keratomileusis: six-month follow-up of initial US clinical series. J Refract Surg. 2003;19:8-14.
7. Ratkay-Traub I, Ferincz IE, Juhasz T, et al. First dinical results with the femtosecond neodynium-glass laser in refractive surgery. J Refract Surg. 2003;19:94-103.
8. Binder PS. Flap dimensions created with the IntraLase FS laser. J Cataract Refract Surg. 2004;30(1): 26-32.
9. Iran DB, Sarayba MA. Bor Z, et al. Randomized prospective clinical study comparing induced aberrations to IntraLase and Hansatome flap creation in fellow eyes: potential impact on wavefront-guided laser *in situ* keratomileusis. J Cataract Refract Surg. 2005;31:97-105.
10. Kezirian GM, Stonecipher KG. Comparison of the IntraLase femtosecond laser and mechanical keratomes for laser in situ keratomileusis. J Cataract Refract Surg. 2004;30:804-11.
11. Anderson NJ, Edelhauser HF, Sharara N. et al. Histologic and ultrastructural findings in human corneas after successful laser in situ keratomileusis. Arch Ophthalmol. 2002;120:288-93.

14

Applanation—Free Femtosecond Laser Processing of the Cornea

Bojan Pajic, Farhad Hafezi, Brigitte Pajic-Eggspuehler, Michael Mrochen, Joerg Muller, Dusan Pajic, Franz Fankhauser (Switzerland)

Chapter Outline

- The Femtosecond-laser Surgery System
- Applanation-free Surgery
- Cut Quality and the Femtosecond-parameters
- Intrastromal Lenticular Extraction

INTRODUCTION

Laser systems normally use pulse energies to "photodisrupt" the tissue with the emphasis on the word disruption. In general the longer the pulse duration is the higher is the required pulse energy. The laser pulse creates a plasma and induces a shock wave that propagates at supersonic speed. Subsequently, an expanding cavitation bubble is created that becomes much larger than the focal volume of the laser beam. Material is driven apart and high mechanical stresses are in induced at the border to the cavitation bubble. In biological tissue cells undergo large mechanical deformations. In the cellular matrix responsible for structural integrity residual deformations and damages remain which induce wound healing mechanism in living bodies. The shock wave accounts for damages on intra cellular level close to the cavitation bubble. The energy that creates the cavitation bubble finally also translates into heat which is absorbed by the surrounding tissue. Depending on the pulse duration and pulse energy thermal effects can take place as well (Besides those effect photo-chemical phenomena can be observed in particular for short wavelength).

These effects were responsible in the past that lasers with long pulse durations and therefore high pulse energies (nano and picosecond lasers) lead to unacceptable tissue reactions. This situation improved dramatically with the advent of femtosecond lasers which required much lower energy per pulse than picosecond lasers. There is gives a good introductory about laser interaction regimes form ns to fs[1]. But still side effects like TLS, DLK, OBL and rainbow glare persists. Whether OBL is only a short-term side effect or it is indirectly responsible for TLS or DLK is still an open issue. It might be possible that decomposition by-products like free radicals that are either created in the intracellular matrix or pushed into it also contribute to long lasting side effects.

How far improvements could go one day can be imagined with the following research work on cell surgery.[2] Precision cuts were performed inside cells using 80 MHz femtosecond NIR laser pulses with nJ pulse energy (known also under the terms nanosurgery or cell surgery). No thermal side effects were observed. The example below in Figure 14.1 shows a dissection of a chromosome within a living cell. The cell remained alive. This responsible process is referred to in science as photodissociation, photolysis, or photodecomposition. The process is of photochemical nature in contrast to photodisruption that always implies an optical breakdown followed by mechanical effects. Such a process in the infrared can probably best be described as multi photon dissociation. The same process was used to create flaps in rabbit eyes.[3] High repetition rate femto second tissue interaction is still a field of ongoing research. In[4] the surprising result was published that pulse trains of the same energy and number of pulses cause less tissue effects at higher repetition rates (Fig. 14.2).

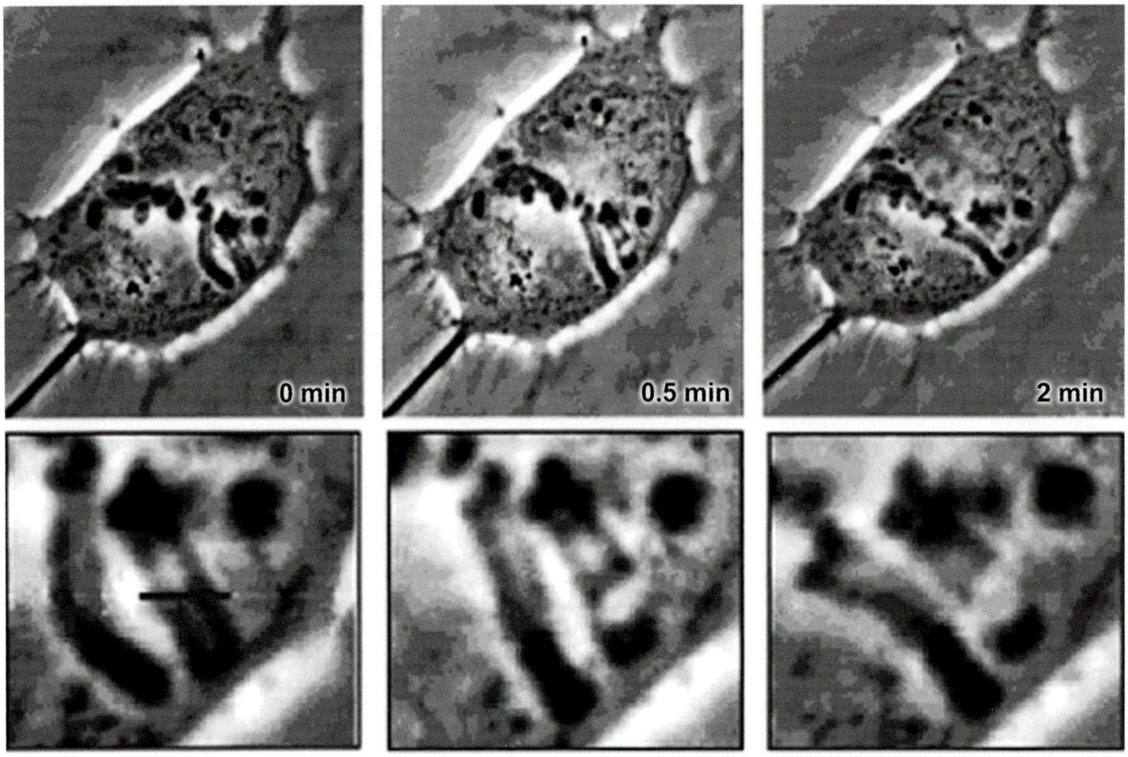

Fig. 14.1 Dissection of chromosomes in a living cell

In a very general sense, only the deposited laser energy is responsible for interactions with tissue. Using the nomenclature[5] and including the terms linear absorption and free electron generation, the following diagram depicts the energy distribution of the emitted laser energy of a surgical laser device. In a cutting process, where the laser focus is displaced only nonlinear absorbed energy interacts with tissue.

THE FEMTOSECOND-LASER SURGERY SYSTEM

Our processing setup was based on the oscillator-regenerative amplifier system Pharos (Light conversion). This laser system is operating at a wavelength of 1030 nm, providing pulses of 300 fs duration. The repetition rate can be varied from 10 to 350 kHz. For surgery in the cornea the output beam from the Pharos laser system was delivered into a xyz-scanner (Scanlab) provided with a telecentrical f-θ-lens. The optical system has a numerical aperture of 0.086 and focuses the laser to a beam waist of about 4 μm.

As described in detail in the next chapter, our approach for processing of the cornea involves no kind of applanation

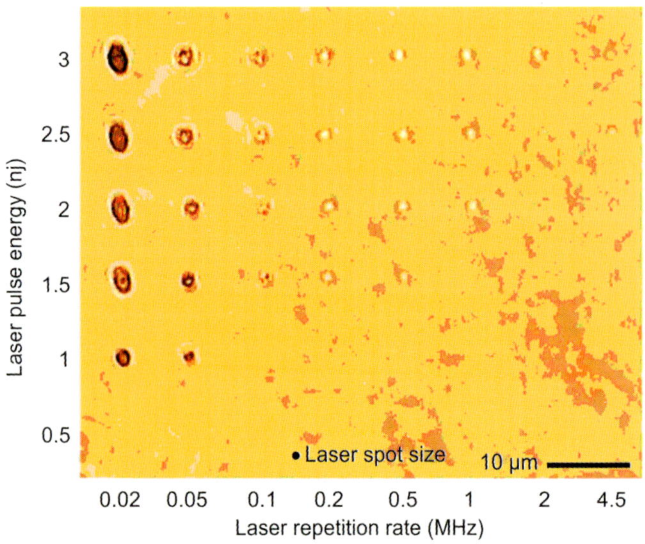

Fig. 14.2 Optical microscope images showing repetition rate dependency of intrastromal surgery in fresh pig cornea at a constant number of laser pulses of 10^6 and lambda = 800 nm. Repetition rates (20 to 4.5 MHz) and laser pulse energy (0.5 to 3 nJ) are indicated for each column and for each row respectively. Laser direction is normal to page

of the eye. The contact liquid layer used instead, however, has an individual thickness for each eye, i.e. the positioning of the laser focus can not be done by the help of a reference surface like in other fs surgery systems on the market. Therefore, we use a positioning system to find the eye vertex, which is then used as a reference point for the processing. The position of the eye vertex is measured with a confocal set-up which follows the same optical path in the system as the surgery beam.

The positioning system contains a fiber-coupled monomode laser diode at 980 nm (near the surgery laser's fundamental wavelength) with 150 mW continuous output power. It is focused on the eye and the reflection from the cornea follows the way back through the optical setup and is then detected with a photodiode connected via a monomode fiber. The laser focus is on the eye vertex when the signal of the photodiode is maximal. The resolution of this confocal system at that point merely depends on the numerical aperture of the focusing optics and on the aberrations within the system. The axial resolution of the setup is 5 μm, limited by the relatively low numerical aperture of the f-θ-lens.

The z-scanning of the laser focus is achieved by a variable telescope integrated into the xyz-scanner. Using the two mirrors of the scanner (for x- and y-coordinates respectively) and this z-scan, a 3D coordinate set which is calculated in a computer program is scanned by the laser focus. The 3D profiles in the cornea, which represent cutting masks for flap, lenticle, hinge and flap edge, are programmed based on the individual properties of the eye. These properties, in particular corneal curvature, astigmatism and the distance vertex – pupil center, were obtained from the individual eye topography measured with a Scout keratometer. The flaps were computed to follow the curvature of the cornea and the hinge could be placed wherever needed. Lenticle profiles were programmed depending on the wanted refractive correction using the Munneryll formula.[6] The cuts were created as circles starting on the eye vertex in the wanted depth increasing their diameter by changing the z-position depth in the cornea. The distance between the spots on a circle and the distance between circles could be arbitrarily chosen in the controlling software. During the whole procedure, the eye was observed with a CCD camera which gets the image of the eye through a 2 μm thin pellicle placed in the laser beam.

APPLANATION-FREE SURGERY

Most of the fs refractive surgery devices developed up to now involve either a weak or a strong applanation scheme of the eye in order to achieve a plane surface of the eye for the fs-laser entrance into the cornea.[7] The cutting directly on the curved cornea surface would induce strong coma aberrations only allowing for cuts of low quality. In the case of weak applanation on one hand, an adaption piece of almost the same curvature as the cornea is brought onto the eye. In the case of strong applanation on the other hand, a painful applanation is generated by pressing the eye down with a glass plate. The novelty of the proposed processing procedure lies in the fact that a liquid (physiological saline) with a refractive index (n=1.338) close to that of the cornea (regularly n = 1.376) is brought on top of the eye into the suction ring.[8] This liquid ensures a very good refractive index matching and thus minimizes the wave front distortion of the fs-laser pulses upon entrance into cornea. The thickness of the liquid could be varied from a few hundred μm to 1 mm. In this way, an applanation-free processing of the cornea becomes possible. A sketch of the suction ring is presented in Figure 14.3.

The required removal of the epithelial layer of the pig eyes caused a significant swelling of the cornea upon contact with the liquid. A thickness increase by up to 100 μm could be measured with the pachymeter. Therefore, processing of the eye was only started when corneas' thickness remained constant. It was verified that during the processing time which is determined by eye positioning and fs-laser processing (less than 1 min), the thickness was not changing by more than 2 to 3 μm. Another point which might influence the processing is a refractive index change of the pig cornea with thickness due to hydration, as shown in.[9] However, this problem is not relevant in case of surgery of human eyes, their epithelial layer is not removed during the surgery procedure since vital eyes would not degrade during the procedure. The pigs have cor-

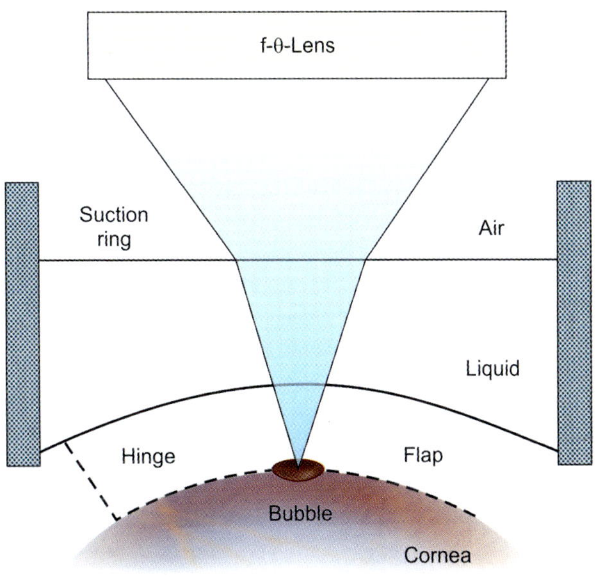

Fig. 14.3 Applanation-free geometry

neas with thickness which varies from 600 µm up to 1000 µm (depending mostly on age), accompanied by refractive index variation from 1.405 to 1.373. This change can induces errors of up to 10 µm in the calculation of the 3D profiles if the refractive index is assumed to be constant influencing the aimed cutting depth. The human cornea varies much less, from 550 µm to 600 µm inducing an error margin of only 1 to 2 µm.

Another important aspect is the nonlinear behavior of the cornea. It was recently shown that pig cornea possesses a nonlinear refractive index of

$$2.10^{-19}\,\frac{m^2}{W}$$

which is relatively high, compared to water.[10] This means that precise calculation of the focal position can only be done taking into account self-focusing effects. Of course also self-focusing in the liquid layer has to be regarded. As the nonlinear properties of the physiological saline are relatively weak (like water), it induces only a small nonlinear phase change in the laser wave fronts depending on the thickness of the liquid layer (which can be precisely measured with the positioning system). In other words, the liquid layer behaves in good approximation like a Kerr lens on top of the eye, which can therefore easily be included in calculating the 3D profiles. When all these effects are being considered, the calculation of the focal position for a certain flap thickness can be performed in a well controlled fashion.

CUT QUALITY AND THE FS-PARAMETERS

There is the misconception that the focal volume defined by Gaussian beam optics equals to the volume of processed material. In processes with nonlinear absorption like it is the case for femto second lasers this equality does not hold any longer. In particular for short pluses and a very high focussing power as, i.e. with the Ziemers LDV the photon interaction volume becomes almost spherical compared to the elliptical volume of conventional systems (Fig. 14.4).

The effect on the cavitation bubbles and mechanical forces that might be formed depending on the pulse energy is even more dramatic. The following example compares cavitation bubbles in water of a system with low focussing power on the left to a system with high focussing power on the right. The pulse energy for the high power focussing lens was about 12 times less. Although this example is only illustrative since it was made in water, it confirms the impact of the focussing power on the shape and the size of the interaction zone.

The big bubble was created with pulse energies still in use today (1350 nJ, laser light is coming from the left) (Fig. 14.5A). Whereas the small bubble was created with pulse energies (106 nJ) (Fig. 14.5B). In tissue cavitation, bubbles will

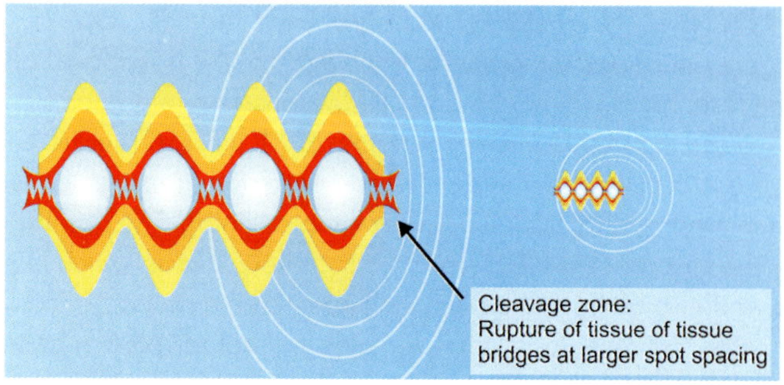

Fig. 14.4 Comparison of photon interaction zones

be smaller. Otherwise it would not possible to cut 110 micron flaps with conventional laser systems. However, the energy that inflates the bubble and pushes the water away against its inertia has to be absorbed by the surrounding tissue. This inevitably induces stresses and damage and pushes gas into intra lamellar spaces (OBL). As stated above, the smaller the bubbles are, the smaller the stresses and intrastromal gas deposit is.

Due to the horizontally oriented lamellas in the stroma, tissue separation in horizontal direction is easier if the mechanical forces of expanding large cavitation bubbles are used. Cutting can be performed partially through inter lamellar cleavage (Fig. 14.4). This mechanism only works at high pulse energies and induces nonnegligible mechanical stresses in the remaining corneal tissue. For vertical cuts cleaving through lamellas does not work. Therefore, up to 50% larger pulse energies are used with low NA femtosecond laser system than in horizontal direction.

On the contrary, due to the lower pulse energies and much lower gas pressure, i.e. Ziemers LDV does not separate tissue through mechanical forces. The LDV process is mainly based on photodissociation. Therefore, the pulse energy does not have to be increased for vertical cuts. The spacial pulse overlap in combination with the high repetition rate of the laser creates a kind of microventing channel in the opposite direction of the dissection which reduces gas pressure built up. It should be noted that the low gas bubble production can even be reduced to zero using special device tunings.

In order to have a comparable cutting speed using smaller interaction zones as conventional laser systems it is necessary to increase the repetition rate. Histological sections have been prepared in order to investigate the impact of high repetition rates. The histological section bellow shows the cutting zone (arrows) inside the cornea. Due to the very small gas production of the crystal line LDV, the dissection zone is only faintly visible. The stroma next to the cutting zone does not show any thermal side effects.

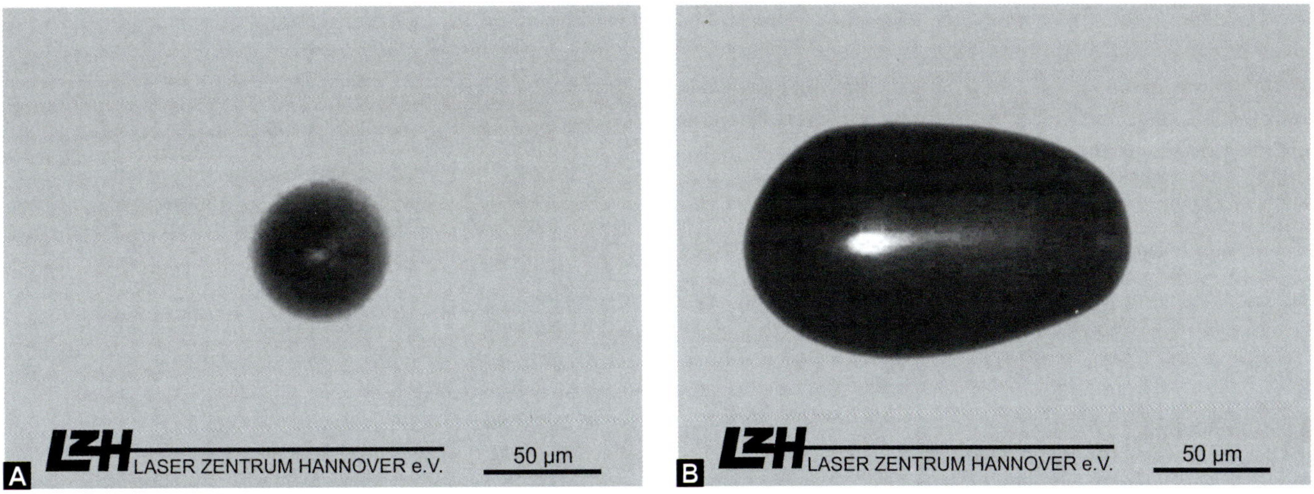

Figs 14.5A and B (A) 1350 nJ pulse at low NA compared to 106 nJ Pulse at high NA (B) 106 nJ pulse at high NA

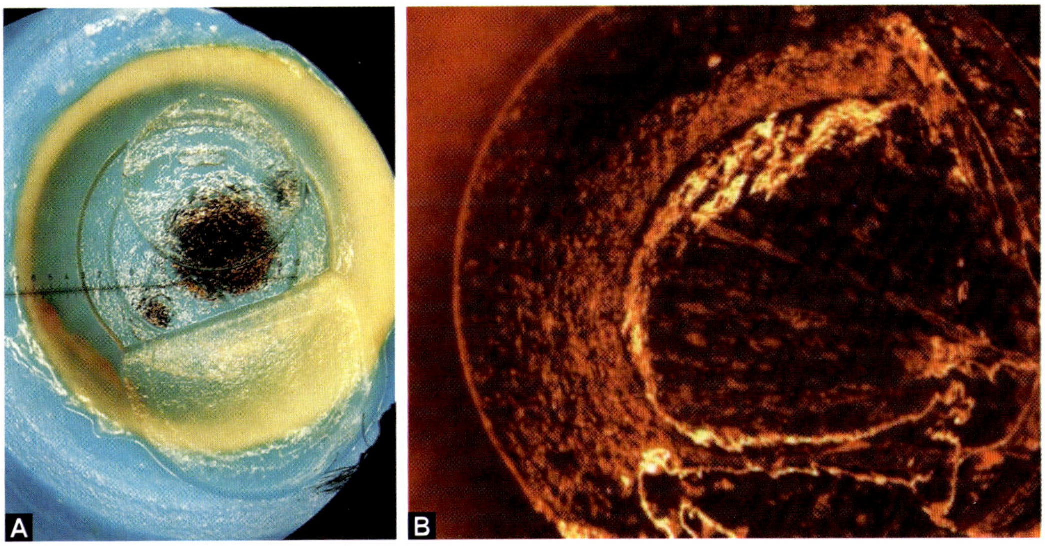

Figs 14.6A and B (A) Lenticle removal (B) Extraction of a 35 µm lenticle through a side channel

INTRASTROMAL LENTICULAR EXTRACTION

In ophthalmology, it has always been highly desired to be able to perform all steps of the refractive surgery (flap and lenticle) with one device. The procedure can create a non-refractive (flap) and a refractive (lenticle) cut by processing two surfaces with different radii of curvature. After lifting the flap, the intrastromal tissue is removed. Subsequently, the flap is repositioned. Throughout this procedure, the cornea curvature is changing and perpetually a refractive correction is performed. Moreover, the possibility to skip lifting the flap and to only extract the lenticle throughout a side channel through the cornea is a possible advantage primarily concerning the wound healing. Finally, the programmable features of this system with flexibility in the z-plane enable the creation of virtually any surface inside of the stroma, and thus to correct higher order visual defects.

Figures 14.6A and B present examples of flaps and lenticles. On the top left there is a flap in 150 µm depth with a lenticle exhibiting a central thickness of 100 µm and a diameter of 10 mm as the flap. This corresponds to a refractive correction of –5 diopters. The right case shows a very thin lenticle (a lens-like intrastromal tissue) of about 35 μm central thickness which implies a refractive correction of – 1 diopters. As can be seen from the right case, by striving for thin lenticles even weaker refractive corrections can be achieved. The bottom example shows the extraction of a 35 μm lenticle through a lateral channel.

ACKNOWLEDGMENTS

The authors thank for distinct support of Dr T Ripken, Dr C Rathjen and Dr T Wirthlin from Ziemer Group for personal and technical advice.

REFERENCES

1. Vogel A, Noack J, Nahen K, Theisen K, Birngruber R, Hammer DX, Noojin GD, Rockwell BA. Laser-induced breakdown in the eye at pulse durations from 80 ns to 100 fs. Proc. SPIE. 1998;34:3255.
2. König K. Multiphoton microscopy in life sciences. Journal of Microscopy, Vol. 200, Pt 2, November 2000, pp. 83-104/ J Microscopy. 2000;200(2):83-104.
3. Wang BG, Lohmann CP, Riemann I, Schubert H, Halbhuber KJ, König K. Multiphoton-mediated corneal flap generation using the 80 MHz nanojoule femtosecond near-infrared laser. J Refract Surg. 2008;24(8):833-9.
4. Kuetemeyer K, Baumgart J, Lubatschowski H, Heisterkamp A. Repetition rate dependency of low-density plasma effects during femtosecond-laser-based surgery of biological tissue. Applied Physics B: Lasers and Optics. 2009;97(3):695-9.
5. Vogel A, Noack J, Nahen K, Theisen D, Busch S, Parlitz U, Hammer DX, Noojin GD, Rockwell BA, Birngruber R. Energy balance of optical breakdown in water at nanosecond to femtosecond time scales. Applied Physics B: Lasers and Optics. 1999;68(2):271-80.
6. Chang AW, Tsang AC, Contreras JE, Huynh PD, Calvano CJ, Crnic-Rein TC, Thall EH. Corneal tissue ablation depth and the Munnerlyn formula, Journal of Cataract & Refractive Surgery. 2003;29(6):1204-10.
7. Faktorovich E. Femtodynamics: A Guide to Laser Settings and Procedure Techniques to Optimize Outcomes with Femtosecond Lasers, Slack Incorporated 1st edn. (2009)
8. "Auf ein Auge aufsetzbare Vorrichtung", Graener H,Lange J, Seifert G, Fankhauser F, Hollerbach T, Magnago T, Anmeldung Nr. 07115544.4-2305.
9. Kim YL, Walsh Jr JT, Goldstick TK, Glucksberg MR, "Variation of corneal refractive index with hydration", Phys Med Biol. 2004;49:859-68.
10. Miclea M, Skrzypczak U, Faust S, Fankhauser F, Graener H, Seifert G. Nonlinear refractive index of porcine cornea studied by z-scan and self-focusing during femtosecond laser processing, Optics Express. 2010;18(4):3700-7.

15

Dry Senile Macular Degeneration

Franz Fankhauser, Maria Ott, Bojan Pajic (Switzerland)

Chapter Outline

- Subjects
- Method
- Results
- Discussion

Keywords: Photon-correlation spectroscopy, quasi-elastic light scattering spectroscopy, dynamic light scattering spectroscopy, nonexudative (dry) senile macular degeneration.

ABSTRACT

Purpose

Photon-correlation spectroscopy (PCS) (quasi-elastic light scattering spectroscopy, dynamic light scattering spectroscopy) allows the determination of the macromolecular and microparticular structure of the normal retina and of pathologic alterations in the macula. It can thus be used as a diagnostic test in nonexudative senile macular degeneration and can be analyzed by the receiver operating curve (ROC).

Methods

The maculae and peripheral retina of 73 normal and of 26 eyes afflicted by an early stage of nonexudative senile macular degeneration were analyzed by autocorrelation functions that can record the sizes and diffusion coefficients of macromolecules and of mobile microparticles.

Results

The macromolecules and microparticles of eyes afflicted with nonexudative macular degeneration and of normals differ significantly.

Conclusions

Photon-correlation spectroscopy is able to assess the macromolecular and microstructural aberrations in the macular afflicted by nonexudative, senile macular degeneration. It has been demonstrated that macromolecules of this disease show abnormal kinetic behavior in the macula.

INTRODUCTION

Photon-correlation spectroscopy (PCS) has been known for more than 30 years. It is a standard technique for measuring random thermal movements, the weight and shape together with the diffusion constants of suspended particles, such as macromolecules in gels and in other complex liquid systems.[1-7] In ophthalmology, PCS has been performed in the cornea, the aqueous humor and the vitreous. Such examinations have been performed under physiologic as well as in pathologic conditions.[8-13] It has been shown that it is possible to determine the dimensions of such particles in the range of about 1 to 1000 nm. PCS seems to be very appropriate for application in ophthalmology because it is noninvasive and can make biochemical processes both on the microscopic, macromolecular and supramolecular level visible. Pathologic processes can be observed both directly, in real time and under optimal conditions in living tissues. When we assume that most pathologic processes are already present as molecular or macromolecular alterations during a certain period prior before the disease actually starts, one may be able to find such molecular preclinical pathology before the disease actually begins, perhaps at a time when therapy is perhaps more effective than afterwards.

SUBJECTS

Photon-correlation spectroscopy measurements were performed on 73 eyes (of 84 normals) and 26 eyes (of 26 patients suffering from nonexudative, senile macular degeneration [NESMD]) (Table 15.1). The average age of the normals was 60.5 years and for the abnormals 65.5 years. The experiments were performed with the written consent of the subjects and under the oversight of the local ethics committee of the University of Dresden, Germany. The declaration of Helsinki on human rights was complied with. Both the normals and the abnormals were scheduled for cataract surgery following the experiments.

The average visual acuities for the normals were 0.8 (least: 0.5) and for the abnormals 0.7 (least 0.4). The eyes were studied by a routined specialist with a Goldmann 3-mirror contact lens, the only disease accepted being a nonexudative, dry, senile, macular degeneration. The criteria were soft or hard drusen.[14,15]

METHOD

The diagnostic laser beam was emitted from a HeNe laser, working at 632.8 nm, and was operated at 1.5 to 15 μW. These values are well below the safety limits (55 μW during 20 minutes) stipulated by the American National Standards Institute (ANSI) Z-136 for the safe use of lasers, New York. The pupils were dilated with Diopine 10 and local anesthesia was induced by 0.4 percent Novesine. The focal spot was set at the macula by a hollow-beam aiming system[16] (Figs 15.1A and B) (when the focus is anterior or posterior to the macula or retina, two spots are seen). When the focus is coincident with the macula, one spot is seen, as observed with the microscope of the instrument. The ellipsoid-shaped focal volume (640 μm^3 in air) had the dimensions of 2.5 by 84 μm.

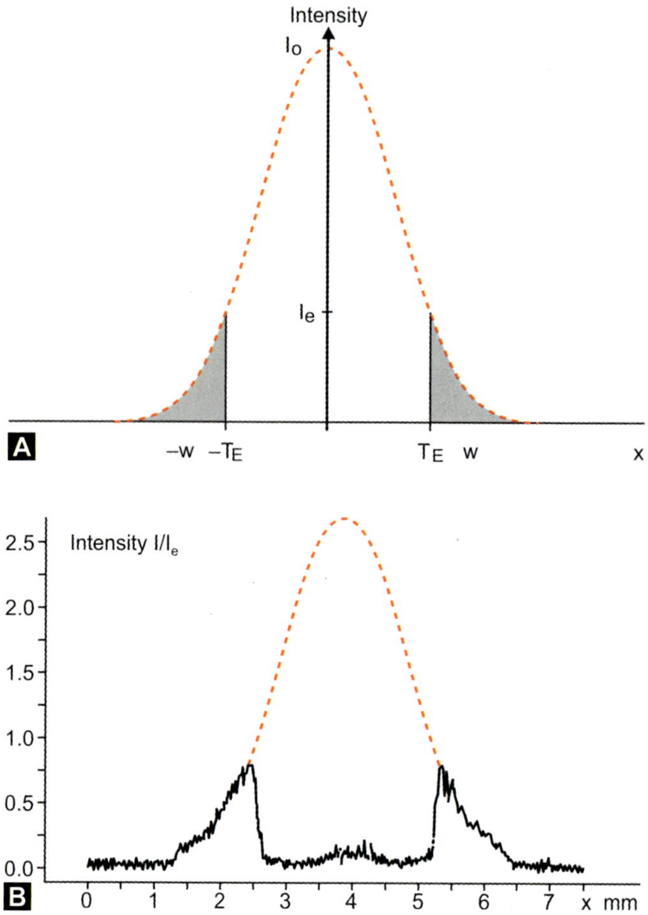

Figs 15.1A and B (A) Distribution of laser light intensity, which originally had a Gaussian shape and a maximum intensity Io. I_e is the maximum intensity of the beam that was used and Ti and w are the inside and outside radii, respectively, of the hollow beam. (B) Square shape of the hollow beam at a plane perpendicular to the beam's propagation

Table 15.1 Demography of the normal and abnormal subjects tested

	Normals	Nonexudative age-related macular degeneration
Total number of eyes	73	20
Average age (years)	60.5	65.5
Average visual acuity	0.9	0.6
Worst visual acuity	0.9	0.3
Best visual acuity	1.1	0.8
Central fixation	+	+
Other pathology	-	-
Neovascular vessels by fluorescein angiography	-	-

During operation, an auxiliary fixation spot is shown to the patient and three locations are determined: C_1 (center), C_2 (6° temporal) and C_3 (12° temporal) (Fig. 15.2). One measurement of an intensity correlation function took on the average about 10 seconds. One measurement cycle consisted of 5 measurements, which were averaged. The correlation functions were observed on-line on a monitor and, when abnormalities of the curves were seen, the curves were discarded and the measurement repeated. Such anomalies could be easily seen. In addition, the measurements were checked by a computer program which detects noise in the optical and electric system.[17]

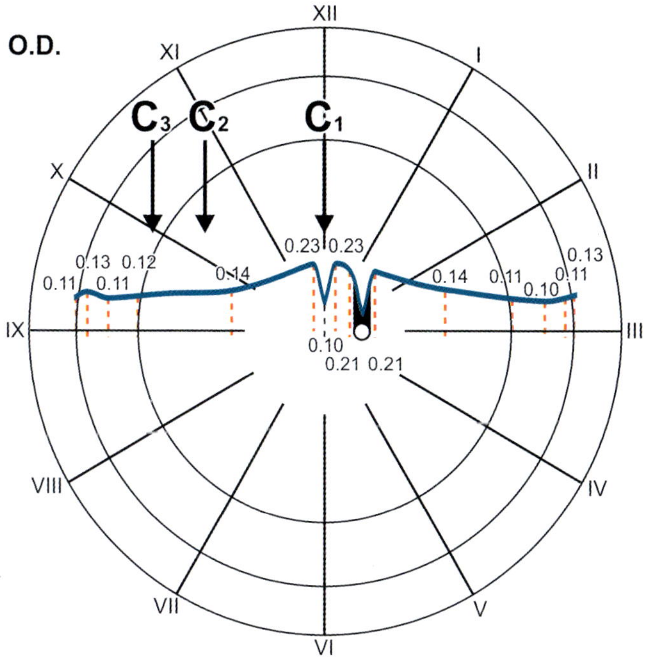

Fig. 15.2 Optical section through a human retina. The figures are thicknesses of the retina and C_1, C_2 and C_3 are the test locations

The intensity of the diagnostic stray light was measured by an avalanche diode (EG&G, SPCM-AQ-WX1-FZ,Vaudreil, Canada), which was correlated by a digital correlator (AL-5000, 5.0, ALV Germany). It contains a single photon counting diode (SPAD) fed by a stabilized high voltage supply. For the wavelength used (632.8 nm) the quantum efficiency of our avalanche diode exceeds that of a photomultiplier by a factor of about 4 and is therefore well adapted for the weak stray light signals emitted by an eye.[18] The dead time of the electronics is 30 ns and did not interfere with the measurements because the correlation time of one measurement was on the order of 1 ms. Hence, delays of $\tau > 1$ s are sufficient for the calculation of the decay constant.

The measurements were performed in the lower linear range of the photodetector, i.e. between 20 Hz and 100 Hz. Here the correction for the dead time is between 0.99 and 1.01. At 20 Hz, the background corresponds to 1.7 percent and, at 100 Hz, to 0.4 percent of the measured counting rate. The digital signal was correlated and symmetrically normalized by means of multiple schemes.[19] This kind of symmetrization optimizes the statistical accuracy for long delay times. Two monitor channels are activated, one for the direct and another for the delayed counting rates. Instead of normalizing the direct channel, the product of the mean values of the direct and the delayed were used for normalization. Because the correlation curves were displayed during the measurements they could be evaluated, and, if necessary, deleted and repeated. The criterion for a "good" correlation curve is a stable plateau not disturbed by noise at the beginning, a steep decay and a flat final plateau at $y = 1$.

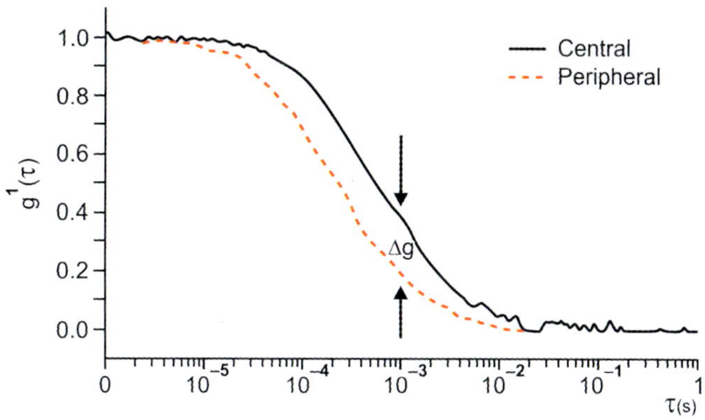

Fig. 15.3 Electric field correlation function $g^{(1)}_{(\tau)}$ at the center and the periphery of a normal subject of 50 years. Δg is the difference of the correlation curves at a delay time τ of 1 ms

RESULTS

First, the intensity autocorrelation functions $g^{(2)}_{(\tau)}$ had to be determined, which were then transformed into the electric field autocorrelation $g^{(1)}_{(\tau)}$ functions by Eq. 7. Figure 15.2 shows the test locations C_1,C_2,C_3, Figure 15.3 shows a $g^{(1)}_{(\tau)}$ autocorrelation curve at the center (macula) and in the periphery of a normal 50 years old subject. The autocorrelation function shows an exponential decay. Δg is the difference of the functions at a relaxation time τ of 1 ms. In Figure 15.4 the autocorrelation function of eyes with NESMD are shown at the center, and normal values at the macula and in the periphery.

Then, both the hydrodynamic radii of the particles and their diffusion constants were determined (Eq. 2) (Table 15.2) (Fig. 15.5) assuming for $\eta = 1.16$. Following this, a frequency distribution of the differences of the autocorrelation functions at the center and at the periphery was computed (Fig. 15.6). The determinations of Δg were performed at 1 ms and the widths of the intervals is 0.1. Because 2 measurements at each test location were available average values were used. Negative differences mean that the autocorrelation curve at the center is below the peripheral autocorrelation curve. It was found that in 46 percent of normal eyes Δg is always greater than Δg in the abnormal (NESMD) eye. In eyes with NESMD 45 percent of the Δg differences were smaller than −0.1. In order to discriminate normals from abnormals a threshold value is required, i.e. when the differences are greater than this threshold, the eye is normal and the diagnosis is negative (no disease). When this is not the case, the diagnosis is positive, i.e. there is disease. At a threshold of 0.05, the values in the

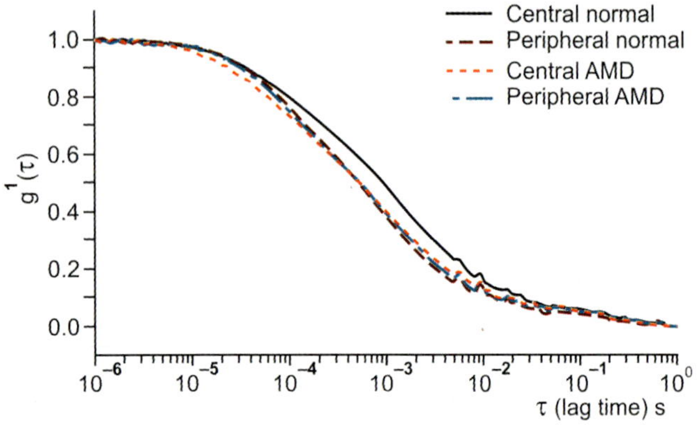

Fig. 15.4 Comparison of the electric field correlation functions at the degenerated macula ..... C_1; the normal macula ..... C_1 and the normal periphery ..... C_3

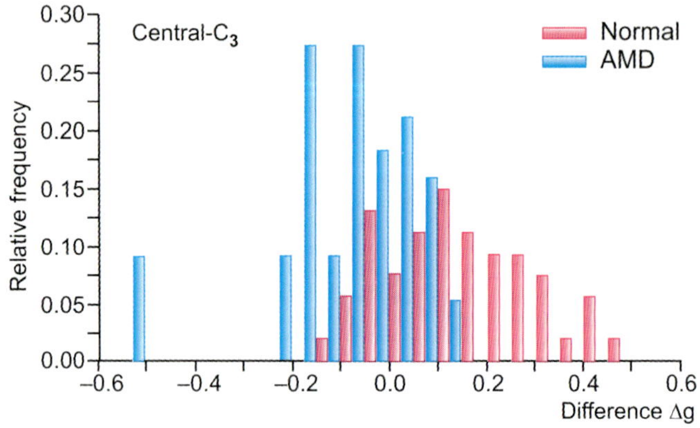

Fig. 15.6 Distribution of relative frequencies of the differences of the electric field autocorrelation functions at the center vs the periphery for the normal and an abnormal eye. The width of the intervals is 0.1

Table 15.2 Radii and diffusion constants (from Fig. 15.5)

Radius (nm)	Diffusion constant (m^2/s)
8	2.5×10^{-11}
10	2.0×10^{-11}
13	1.5×10^{-11}
83	2.4×10^{-12}
183	1.1×10^{-12}
690	2.8×10^{-12}
896	2.2×10^{-12}

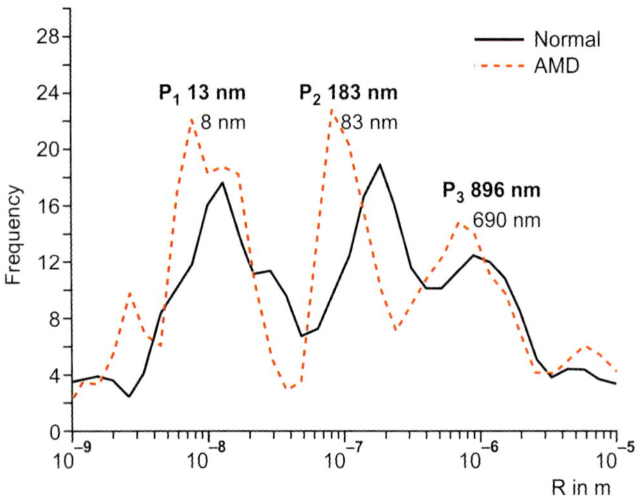

Fig. 15.5 Change of distribution of apparent sizes of particles and diffusion coefficients in the normal and diseased macula

Table 15.3 Convergence table for the measurements performed, provided their differences between central and peripheral correlation function at t = 1 ms with $\Delta g > 0.05$ are normal and those with $\Delta g < 0.05$ are abnormal

Retina condition	Diagnostic result		Total
	Positive	Negative	
AMD	0.78	0.22	1.0
no AMD	0.28	0.72	1.0

convergence table (Table 15.3) were obtained. It follows from this table that the sensitivities at a threshold of 0.05 are 78 percent, whereas the selectivity is 72 percent .

These data are displayed by a ROC curve (receiver operating curve) (Fig. 15.7).[20] It is known that a test is better the more the curve approaches 0.1. It also follows that the more false diagnoses are accepted, the more the probability of a correct diagnosis rises. It is also evident from the ROC curve that 0.00 corresponds to the value where one measurement deviates most from the correlation curve (Fig. 15.7). For one test location only 2 measurements were available, one can therefore assume that the average values of more measurements would approach the correlation curve better. Also, the area under the ROC curve is the average of the selectivities of all the sensitivity values.

DISCUSSION

Photon-correlation spectroscopy is a noninvasive optical method that is able to define molecules or small mobile parti-

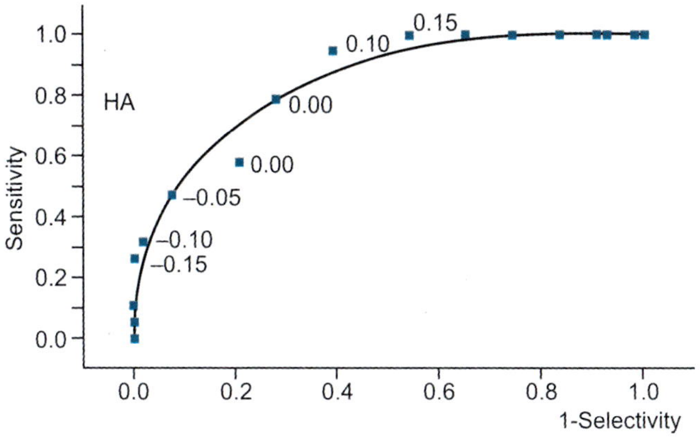

Fig. 15.7 ROC curve for the diagnosis of NESMD, realized by the differences of the autocorrelation curves $g^{(1)}_{(\tau)}$, measured at the macula and the peripheral retina at $\tau = 1$. The values, computed with the help of Table 15.2, are displayed as black squares. The adjacent figures are the threshold values. The black line is the fit-function

cles in colloidal solutions and ocular tissues. It is able to find differences among macromolecules and of mobile structures in normals as compared to those with NESMD. Photon-correlation spectroscopy computes the movements of macromolecules and of mobile microparticles and determines their sizes, their diffusion coefficients and the viscosities of the intermolecular matrices, the result being expressed as an autocorrelation function. In polydisperse, biological tissues, such as here, it measures an assembly of various molecular identities. This function eventually must be split in order to find these individual entities by means of an inversion of a Laplace function (Eq. 6) and this leads to an ill-poised mathematical problem, the result being disturbed by noise. It is for this reason that the determination of the microparticles displays some amount of noise interaction.[19] Despite such limitations the size determinations have been shown to be possible.[4-13]

This clinical study was performed on 73 normal eyes and on 20 eyes afflicted with NESMD. The measurements (i.e. the determinations of the autocorrelation functions) were performed at the macula and at two selected extramacular retinal sites. The temporal correlations of the fluctuations of the scattered light were then computed and the molecular sizes were derived from them. For this, the electrical field must be deduced from the intensity measurements of the stray light.

Despite individual differences of the correlations, the particles of the various retinal sizes could be measured and compared. In normal eyes the autocorrelation functions obtained at the center and at various peripheral retinal sites differ. When the eyes with NESMD were compared such differences vanish. This leads to the conclusion that the differences between the fovea (macula) and the periphery is caused by an alteration of the stray light characteristics of the foveal cones. Since the strongly scattering properties of melanin also add to the measured signal, degenerative alterations of the pigment epithelium and perhaps its lack may also cause this abnormal stray light behavior.

Both an abnormal stray light behavior of the receptors, of the pigment epithelium, and of other microparticles of the retina may work together. We have found macromolecular size distributions at 15, 80, and 900 nm. When possibilities oriented towards a diagnosis of a macular disease by simple means is aimed at, only one measurement at the macula and one at a peripheral site is required and an inversion of the correlation function is not necessary.

Summary

Differences of central versus peripheral stray light behavior were quantified and these were then analyzed by an ROC function that is able to correctly diagnose this disease at a given probability, here at a sensitivity of 78 percent and a selectivity of 72 percent. We have found an overlap of the difference values of the auto-correlation functions of normals as compared to abnormals. The reason for this may be the fact that the leading bioindicator for disease was an anatomical criterion, i.e., most of the time the criterion was "hard or soft drusen".[14,15] While drusen are an important indicator for the NESMD, normally functioning areas may have still existed in these maculae, while in maculae, termed "normal" pathological areas may have existed already. It is important to know that the average sensitivity of all sensitivities is 0.85. It lies in a range where other diagnostic methods in medicine are also found.[20]

For an improvement of the accuracy of the photon-correlation spectroscopy at the retina various approaches are available. These are methods related to data collection and analysis but also methods related to an improvement of the measuring system. The accuracy of the data collection procedure can be increased by increasing the number of determinations because at various retinal regions the average values differ only slightly. The determination of the autocorrelation function could be increased by a suitable filter function. Here, we have characterized the correlation functions via measurements of time. Perhaps other parameters, such as the measurement of the area under the autocorrelation function (curve) may increase accuracy. Most importantly: an optimization of the determination of the particle size distribution should be intensified. Here, also an introduction of dedicated filter functions, the development of a stray light model of the retina, the automatic detection of data which are affected by eye motions appears important. Just as important are local determinations of the refractive indices and viscosities. This would lead to an optimization of the particle measurements.

Also, when these distribution functions are better understood the sensitivity and the specificity of the inversion procedure can be improved. This could lead to a better understanding of the mechanism of the central and peripheral stray light behavior and this could increase the diagnostic power. Also the measuring system could be improved and better adapted to the specific conditions of the retina. It should be noted that with the present apparatus stray particles that are larger than the wavelength of the exiting laser beam can indeed be detected, but the detected radii are somewhat smaller than the true ones. When instead of HeNe laser light (632.8 nm) a source with a longer wavelength would be used, the measurements of the particles would be improved. According to the spectral properties (absorption) of the eye, wavelengths of 900 and 110 nm would be well suited. Here, another advantage would be to increase the maximal exposure duration/energy (BGI 2003). This would result in an augmented signal/noise ratio. This would be important when measuring cataractous eyes where a laser power of 30 μW at a wavelength of 632.8 nm is often not enough. The measuring inaccuracy and the disturbing effect of eye movements could be partly or totally inhibited by measuring central and peripheral retinal sites simultaneously.

ACKNOWLEDGMENT

This work was supported by Haag-Streit International. The author had full access to all the data in the study and takes responsibility for the integrity of the data and the accuracy of the data analysis. The authors gratefully acknowledge the editorial assistance of Arthur Funkhouser. The explanations of characteristic features of nonexudative (dry) macular degeneration by Hans E Grossniklaus have been greatly appreciated.

APPENDIX

THEORY OF PHOTON CORRELATION SPECTROSCOPY

When light interacts with material, scattering particles alters the direction of its propagation.[21] The angle distribution of the scattered light—with respect to the incoming light beam—depends thereby on the sizes and positions of the scatterers. Hence, the scattered light carries detailed information on the molecules or microstructures from which it is scattered. The information of the scatterers is encoded in temporal intensity fluctuations I(t) of the scattered laser light and can be analyzed by calculating the autocorrelation function of the detected light intensity $g^2(\tau)$:

$$g^2(\tau) = 1 + \frac{\langle I(t)I(t+\tau)\rangle}{(\langle I\rangle\tau)^2} \quad \text{Eq. 1}$$

Intensity fluctuations originate, thereby, mainly from the motions of the particles with respect to each other. Furthermore, rotational movements of nonspherical particles or changes in the observed volume will also lead to fluctuations.

The random movements of unbound particles can be described by the mathematical model of Brownian motion which characterizes the motions by a translational diffusion coefficient D.

This diffusion coefficient gives access to the sizes of the scatterers, since it depends inversely on the hydrodynamic radius (R_h) of the particle as given by the Stokes-Einstein equation:

$$D = \frac{k_B T}{6\pi\eta R_h} \quad \text{Eq. 2}$$

where η corresponds to the viscosity of the surrounding solvent, T is the temperature, and k_B is the Bolzmann constant.

Relatively small particles with large diffusion coefficients show fast molecular movements which lead to only briefly correlated fluctuations of the scattered light.

To characterize the decay of the correlation function it is helpful to define a relaxation rate Γ (or decay constant τ_c) by:

$$\Gamma = \frac{1}{\tau_c} = q^2 D \quad \text{Eq. 3}$$

which also accounts for the spatial distribution of the scattered intensity described by the scattering vector q:

$$q = \frac{4\pi n}{\lambda}\sin(\theta/2) \quad \text{Eq. 4}$$

for a scattering angle of θ.

For monodisperse solutions the relaxation rate is linked to the autocorrelation function $g^2(\tau)$ by:

$$g^{(2)}(\tau) = I_0(1 + \exp\{-2\Gamma\tau\}) \quad \text{Eq. 5}$$

In case of a polydisperse solution $g^{(2)}(\tau)$, the determination of the distribution of $G(\Gamma)$ becomes an ill-posed problem and has to be derived from the correlation function of the electric field, $g^{(1)}(\tau)$.

Provided there is no interaction by parasitic, static stray light and there is no particle-particle interactions, the electric field correlation function $g^{(1)}(\tau)$ and the measured intensity correlation function $g^{(2)}(\tau)$ are related by Siegert's relation:

$$g^{(2)}(\tau) = 1 + |g^{(1)}(\tau)|^2 \quad \text{Eq. 6}$$

The size distribution of the particles yields the intensity autocorrelation function $g^{(2)}(\tau)$ which consists of superimposed exponential functions with decay times corresponding to the distribution's r distribution. Polynomial fit procedures have been described before.[22]

$$g^{(1)}(\tau) = \int_0^\infty G(\Gamma)\exp(-\Gamma\tau)d\Gamma \quad \text{Eq. 7}$$

$G(\Gamma)$ must be analyzed by the regularization program CONTIN.[23]

Heterodyning

Two correlation functions should be discriminated, i.e. the correlation function of the light intensity $g^{(2)}(\tau)$ and the correlation function which measures the electric field $g^{(1)}(\tau)$. One can measure light intensity but not, however, the electric field. Hence the two quantities must be correlated by means of Eq. 9. Principally speaking, two possibilities for such a correlation exist. The first is the homodyne method. Here the intensity of the scattered light is measured and its correlation function is determined. In the other, the heterodyne version, the diagnostic HeNe laser light is split into two. One of the beams is directed to the sample. The resulting backscattered light is then superimposed with the nonscattered light and directed to the detector.

This same effect is seen when under nonideal conditions the heterodyne portion is superimposed upon the diagnostic (homodyne) beam. This is called partial heterodyning,[24] falsifying the information in the autocorrelation function. This effect is observed in nonideal conditions such as *in situ* experiments in the eye. Here, nasty side effects may occur and one has to expect interferences of unknown amounts of parasitic stray light. Such stray light may be generated, for instance, at the surface of a contact lens but also, e.g. by collagenous fiber strands or membranes in proliferative, diabetic vitreoretinopathy. Especially problematic is the case of weak heterodyning in which merely a small decrease of the amplitude of the correlation function (= the coherence factor) is observed while the more important distortion of the correlation function may escape attention. For example, with a 45 percent local oscillator, the amplitude decreases by only 15 percent but the error in the estimation of the hydrodynamic radius of a particle is 40 percent.

For purely homodyne experiments the Siegert relation holds:

$$g^{(2)}(\tau) = 1 + \beta\,|\,g^{(1)}(\tau)\,|^2 \qquad \text{Eq. 9}$$

β is the coherence factor $0 < \beta < 1$. In monomode fibers, such as are used here, β is 1. It decreases when parasitic stray light (e.g. an incoherent background light) intrudes into the measuring system, but also when, e.g. dead-light or saturation effects of the detector are observed. This is what could happen (to a larger or smaller extent) in the eye. Such global noise effects can be detected, e.g. by the program of Flammer and Ricka.[22] Due to heterodyning, Eq. 9 is no longer valid and the heterodyne parameter h must be introduced:

$$h = I_l/I_s \qquad \text{Eq. 10}$$

where I_l is the homodyne (diagnostic) signal and I_s the heterodyne (parasitic) light. Now Eq. 9 must be corrected and replaced by:

$$g^2(\tau) = 1 + \frac{h}{(1+h)^2}\,\mathrm{Re}\{g^{(1)}(\tau)\} + \frac{h}{(1+h)^2}\,\mathrm{Re}|\,g^{(1)}(\tau)\,|^2 \qquad \text{Eq. 11}$$

where the parameter h can be obtained from Eq. 12:

$$h = \frac{1}{g^{(2)}(0) - 1} - 1 \qquad \text{Eq. 12}$$

We are now in the position to compute $g^{(1)}(\tau)$ from $g^{(2)}(\tau)$ and can perform an analysis of our data.

APPARATUS

As a consequence of space limitations due to the pupil size, the optical system used has the excitation beam (into the eye) and the beam carrying the scattered light (out of the eye) in common. This system has been used before[18] and is described elsewhere.[18] Hence, we shall restrict ourselves to some key elements.

Radiation Source

The radiation source used for these experiments is a HeNe laser (Uniphase 1103P, California, USA) working at a power of 20 mW. The emitted wavelength is 638.2 nm, 20 percent of which is absorbed on its way to the retina. The wavelength used is also from a range that is least toxic for the receptors because, according to Planck's law, photon energy decreases with increasing wavelength. As a consequence, the retinal metabolism is least disturbed. For light conduction, a monomode optical fiber (TEM-mode > 95%) is used, being coupled into the system by a coupler. Monomode fibers, in contrast with multimode fibers, have the advantage that higher modes are suppressed and that the exciting beam has a Gaussian profile.

Optical Diagram

The Gaussian intensity distribution of the exciting beam is:

$$I(x) = I_0 e^{-2(x-x_0)/w^2} \qquad \text{Eq. 13}$$

where w is the distance at which the intensity maximum at the x_0 has fallen to e^{-2} (= 13.5%) of the maximum. A Haag-Streit slit-lamp (900BQ) is part of the apparatus. In addition various optical supplementary parts were shown to be necessary for these experiments (Figs 15.1A, 15.8 and 15.9) had to be supplemented. The laser beam was given a hollow intensity distribution (Fig. 15.8). In this, the excitation and diagnostic beams pass through the same optical elements (Fig. 15.8). In this optical system, the focal planes of the observation, diagnostic and excitation beams are the same. Hence, an exact localization of these beams during the experiment is possible. In figure 15.8, the principle ray paths during one measurement at the retina are shown. The red beam is the excitation beam and the blue one is the measurement, i.e. the scattered, diagnostic beam. For computations, the scattering angle, instead of using all possible scattering angles, an average scattering angle = 6° corresponds to an angle of 4.5° in water and to 4.1° in the retina (n = 1.35). The scattering angle is $\theta = 180° - \alpha_s$. For this situation a correlation time τ_c = 0.31 ± 0.15 ms can be computed for latex spheres (100 nm diameter) for 100 nm = 0.3 ms when a constant scattering vector is used. For all possible scattering vectors, a correction of only 0.1 percent would be required. From this it follows that for high precision, a constant angle of θ_s suffices for a precise description of the light scattering.

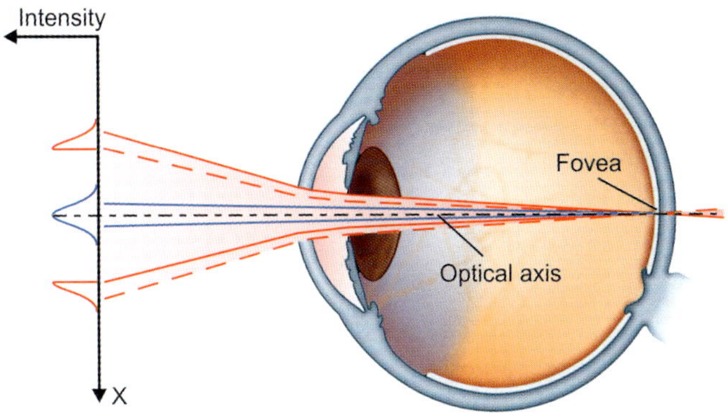

Fig. 15.8 Schematic optical scheme. The hollow beam (red) is used for focusing. Blue part of the stray light beam reaching the detector

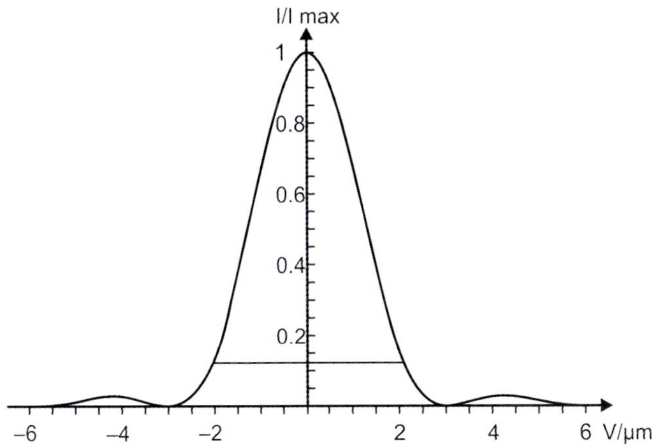

Fig. 15.9 Intensity distribution perpendicular to the focal plane

Intensity Distribution in the Focus and in the Scattering Volume

We are interested in the intensity distribution of the hollow beam and its focus (Fig. 15.9), the focal depth and the scattering volume. Taking advantage of Huygen-Fresnel's principle, the intensity distribution near the focus, as shown in Figure 15.8, is obtained. The integrals to be computed were solved numerically by program MATHEMATICA 4.0. When the focus of the excitation beam is at the retina, this corresponds to a numerical aperture of 15 mm for a pupil diameter of 6 mm. Figure 15.9 displays the intensity distribution along the optical axis and in the radial direction. For both the excitation and the scattered volumes, the isoline being used is where the intensity has decreased to 50 percent of the maximal value.

Scattering Beam

The focal volume V of the exciting beam is stimulated by its electric field. This is given by an overlapping integral of the stimulating and the scattering beam.

$$V = \int dr\, I_e(r)\, I_s(r) \qquad \text{Eq. 14}$$

In air this volume amounts to 640 m^3. The isoline where the intensity has decreased to 50 percent of the maximum is an ellipsoid having a width of 2.5 µm and a length of 84 µm in air. At the retina this amounts (n = 1.35) to a width of 3.4 µm and to a length of 154 µm. The resulting resolution suffices to have at least 2 photoreceptors in the focal volume. The focal length of 154 µm also suffices to have both the receptor layer and the pigment epithelial in focus simultaneously.

Limitations of the Laser Power

In order to avoid thermal and photochemical damage to the retina by high laser power, the excitation power of the laser beam is decreased by a thermal filter. Damage to the retina depends both on the duration and the power of the laser beam. According to the specifications of ANSI Z 1361 (1993) a maximal laser power of 55 µW (measured at the cornea) and an irradiation duration of 20 minutes avoids damage. The laser power was measured before each examination with a Silica detector. In addition, maximum laser power was used in order to adjust the counting rate of the detector. In patients with reasonably clear lenses, an examination could usually be performed with 20 µW. When the opacity of the lens was considerable, not even 30 µW were able to generate a signal.

Detector and Correlator

The intensity of the stray light was measured with a single-photon counting module (SPAD), having a high voltage, voltage-stabilized supply. The quantum efficiency for the wavelength is better than 4 times that of a photomultipler and is therefore ideally suited for registering the weak stray light signals of the retina. Because the autocorrelation time of the retina has an order of magnitude of 1 ms, the dead time of 1 µs therefore suffices to measure the decay constant. The measurements were performed in the lower range of the photon detector, i.e. between 20 kHz and 100 kHz. The correction for the dead time amounts here to 0.99 and 1.01. The amount of the background radiation corresponds, at a frequency of 20 kHz, to 1.7 percent and, at 100 kHz, to 0.4 percent of the counting rate. The digital measuring signals were correlated by means of a Tau scheme[19] and were symmetrically normalized. This kind of symmetrization optimizes the statistical accuracy, particularly for measurements with large delay times. Two monitor channels, one for the direct and one for the delayed counting rates, were used. The product of the average values of the direct and the retarded channels[19] is used for normalization. By a display of the autocorrelation curves during the measurements on the screens, these could be evaluated already during the measurements and hence could be repeated, if necessary. The criterion of a "good" autocorrelation curve is when a stable initial plateau, undisturbed by noise, contains a steep slope and a final plateau at y = 1.

EYE MOVEMENTS

Three types of involuntary eye movements exist: (i) drifts, (ii) microsaccades, (iii) microtremor. Drifts are the slowest movements, having amplitudes of 2.5 angular minutes and a speed of 2 to 8 angular minutes per second. Microsaccades have amplitudes of 3 to 50 angular minutes and a linear maximal velocity which depends on the amplitude of the motion. Microtremor, which is a kind of correcting movement, has an amplitude of less than 12 angular minutes, a velocity of 10 angular minutes per second, and a frequency of about 30 Hz.

The diffusion constant of spherical particles is indirectly proportional to its hydrodynamic radius. Hence, in order to assess the influence of involuntary eye movements, the moved retina can be treated as static particle for which diffusion constants and the hydrodynamic radii can be computed. Consequently, the above mentioned drifts can be compared with the apparent movements of static particles that lie between 1×10^{-11} m/second and 5×10^{-11} m/second. The apparent movements of static particles are therefore 1 to 2 orders of magnitude higher than the diffusion constant of a 100 nm diameter particle. According to the Stokes-Einstein relationship, the apparent diffusion constants of the eye drifts correspond to hydrodynamic radii between 5 and 20 nm. As a result, static particles that are apparently moved by eye drifts would be assigned hydrodynamic radii of 5 to 20 nm, microsaccades and microtremors would yield values of 5 nm and 20 nm, respectively. Hence, disturbances due to involuntary eye movements are negligible. However, slow movements due to voluntary movements can be highly disturbing and should be suppressed as far as possible. The Goldmann contact lens is helpful in this respect, because it inhibits eye movements. Negative suction contact lenses would inhibit eye motions altogether.

REFERENCES

1. Fankhauser F II. Dynamic light scattering in ophthalmology. Results of in vitro and in vivo experiments. Technol Health Care. 2006;14(6):521-35.
2. Berne B, Pecora R. Dynamic Light Scattering. New York: Plenum Press; 1985.
3. Brown. Dynamic Dynamic Light Scattering. Oxford: Clarendon Press; 1993.
4. Chu B. Laser Light Scattering. New York, Academic Press; 1974.
5. Chu B. Laser Light Scattering. Basic Principles and Practice, Boston: Academic Press; 1991.
6. Chu B. Selected papers on quasi-elastic elastic light scattering by macromolecular, supramolecular and fluid systems. Bellingham: The International Society for Optical Engineering; 1990.
7. Schmitz KS. Dynamic Light Scattering by Macromolecules. San Diego: Academic Press; 1990.
8. Bursell SE, Serur JF, Haughton H, Nipper JN, Sinclair JR, Spears JR, Paulin S. Cholesterol levels assessed with photon correlation spectroscopy. Proc SPIE. 1986;712:175-82.
9. Cummins HZ, Pike ER. Photon Correlation and Light Beating Spectroscopy. New York: Plenum Press; 1974.
10. Farrell RA, Bargeron CB, Green WR, McCally. Collaborative biomedical research on corneal structure. Johns Hopkins APL Techn Digest 4. 1983;65-79.
11. Hart RW, Farrell RA. Light Scattering in the cornea. J Opt Soc Am. 1969;59:766-74.
12. Jedzhiniak JA, Nicoli DF, Baram H, Benedek GB. Quantitative verification of the existence of high molecular weight protein aggregates in the intact normal human lens by light-scattering spectroscopy. Invest Ophthalmol Vis Sci. 1978;17:51-7.
13. Latina M, Chylack P, Fragerhom P, Nishio I, Tanaka T, Palmquist BM. Dynamic light scattering in the intact rabbit lens. Invest Ophthalmol Vis Sci. 1987;28:175-83.
14. Green WR, McDonnel PH, Yeo JH. Pathologic features of senile macular degeneration. Ophthalmology. 1985;92:615-27.
15. Green WR, Enger C. Age-related macular degeneration. Histological studies. The 1992 Lorenz Zimmermann Lecture. Ophthalmology. 1993;100:15-39.
16. Rovati L, Fankhauser II, Ricka J. Design and performance of a new ophthalmic instrument for dynamic light-scattering measurements in the human eye. Rev Sci Instr. 1996;67:2615-20.
17. Flammer I, Ricka J. Dynamic light scattering with single mode receivers: partial heterodyning regime. Appl Opt. 1997;36: 7508-17.
18. Daulet H, Deschamp P, Dians B, McGregor A et al. Photon counting techniques with silicon avalanche photodiodes. Appl Opt. 1983;32:3894-900.
19. Schätzel K. Noise in photon correlation data. I. autocorrelation functions. Quantum Opt. 1990;2:287-305.
20. Zou XH, Obuchowski NA, McClish DK. Statistical Methods in Diagnostic Medicine. New York: Wiley. 2002.pp.15-56.
21. Cummins HZ, Knable, Yeh Y. Observation of diffusion broadening of Rayleigh scattered light. Phys Rev Lett. 1964;12:150-3.
22. Koppel DE. Analysis of macromolecular polydispersity in intensity correlation spectroscopy: The method of cumulants. J Chem Phys. 1972;57(11):4814-21.
23. Provencher SW. CONTIN: a general purpose constrained regularization program for inverting noisy linear algebraic and integral equations. Comp Phys Comm. 1982;27:229-42.
24. Siegert AJF. Radiation Laboratory Report 465,1943.

16 Analysis of the Vitreous of Diabetic Patients with Dynamic Light Scattering Spectroscopy: Solutions and Problems Related to Photon Correlation Spectroscopy

Franz Fankhauser, Bojan Pajic (Switzerland)

Chapter Outline

- Macromolecule Size and Shape
- Limits of the Model for Size of Particles
- Heterodyning
- Subjects
- Method
- Disturbances Due to Eye Movements
- Results

Keywords: Dynamic light scattering spectroscopy, quasi-elastic light scattering spectroscopy, photon-correlation spectroscopy, diabetes, diabetic vitreo-retinopathy.

ABSTRACT

Objective: To explore the molecular alterations of the vitreous by dynamic light scattering spectroscopy (quasi-elastic light scattering spectroscopy, photon-correlation spectroscopy) in normals and in patients afflicted with various degrees of non-proliferative and with proliferative diabetic retinopathy.

Research Design and Methods: Dynamic light scattering spectroscopy was employed to analyze the vitreous of normal and of diabetic patients non-invasively in order to estimate both the sizes and diffusion coefficients of mobile macromolecules and/or microparticles.

Results: Abnormal molecular behavior of vitreous molecules was observed in diabetic patients afflicted with various degrees of diabetic vitreo-retinopathy. In the non-proliferative (background) retinopathy, both the diameters of the microparticles increase and the diffusion constants decrease significantly and progressively as the diabetic disease progresses. In the proliferative phase a significant trend in the direction of smaller particles and greater diffusion constants is obvious. These behaviors could also be interpreted as an increase of the viscosity of the intermolecular substance in the first case and as a decrease in the second.

Conclusion: The vitreous in normals and even more so in diabetics with diabetic vitreo-retinopathy is optically a highly non-isotropic, multi-dispersive structure, making an optical analysis difficult. Advanced, but available models and technology, however, permits a major step forward in the optical analysis of the normal and the diseased vitreous.

INTRODUCTION

Attempts to analyze the structure of the vitreous with optical methods date back to Bettelheim and Balasz.[1] These experiments were conducted with methods of static light scattering. Dynamic light scattering (DLS) in in-vitro experiments was first performed both by Sebag et al.[2-4] and Ansari et al.[5] In-vitro experiments were conducted by Rovati et al.[6] and Fankhauser II et al.[7] Ansari et al.[5] Fankhauser II et al.[7] and Sebag et al.[8] found evidence of heterogeneity in particle sizes with larger particles sizes within the vitreous from diabetic subjects as compared to non-diabetic controls.

Dynamic light scattering (DLS) is a strong function of the structure and composition of the vitreous which varies as a function of age,[9,10] pathology[8,11] and of numerous chemical, respectively metabolic processes.[12] Particle sizing in colloidal systems by means of DLS is a commonly known technique over nearly 30 years.[13-22] The technique, however, demands a fast signal acquisition capability. The Brownian motions of suspended particles causes intensity fluctuations of scattered light when a stray light volume is illuminated by an incident coherent light beam. The signal fluctuations are acquired by counting single photons and are analyzed with an autocorrelator that calculates the self-similarity of recorded intensity signals. The achieved intensity fluctuations autocorrelation function $g^{(2)}(\tau)$ can be described by an exponential decay. The time constant of the exponential function corresponds to the relaxation properties of the investigated system.

Theory of Dynamic Light Scattering and the Autocorrelation Function

The Stokes-Einstein equation (Eq. 1) links the hydrodynamic radius R^H of a spherical suspended particle, and D, the diffusion constant together with the viscosity η, and the temperature T of a suspension. K is the Bolzmann constant:

$$D = \frac{K_B T}{6\pi\eta R_H} \qquad \text{Eq. 1}$$

When the number of the illuminated centers of scattering particles is high enough, the light intensity fluctuations are Gaussian distributed and the autocorrelation function, $g^{(2)}(\tau)$, describes a single exponential decay added to a constant baseline term B which (without heterodyning) is 1.

$$g^{(2)}(\tau) = B + \beta \exp \Gamma(\tau) \qquad \text{Eq. 2}$$

where

$$\Gamma = 20\, q^2 \qquad \text{Eq. 3}$$

and t represents the sampling time structure of the correlator, so that β equals the dynamic part of the amplitude. The baseline term and the amplitude are often referred to as intercept and the lag time of the investigated suspensions so that τ_c is derived. The intercept represents a quality criterion for the autocorrelation acquisition: $B + \beta = 2$ in Eq. 2 (without heterodyning). The scattering vector is defined as:

$$q = \frac{4n}{\lambda} \sin\left(\frac{\theta}{2}\right) \qquad \text{Eq. 4}$$

where λ is the wavelength of incident light in vacuum, n is the refractive index of the particle suspension and θ is the scattering angle. The relaxation time of the suspension can be expressed as:

$$\tau_c = \frac{1}{\Gamma} = \frac{1}{2Dq^2} \qquad \text{Eq. 5}$$

where D is the translational diffusion coefficient of the particle. The hydrodynamic radius of the particle R_H is easily calculated through the Stokes-Einstein equation (Eq. 1). For polydisperse particles, or particles with sizes approaching the wavelength of the incident beam, the autocorrelation function is a sum of contributions from the various particle sizes and diffusional modes, and the interpretation of D is less straightforward.

$$g^{(1)}_{(\tau)} = \int_0^\infty G(\Gamma) \exp(-\Gamma_t) d\Gamma \qquad \text{Eq. 6}$$

There exist many programs for fitting autocorrelation functions to particle size distributions; probably the best known is the cumulant method[23] and the program CONTIN.[24] Stock and Ray present a clear explanation and comparison of several of these methods.[25]

For dilute solutions, the Rayleigh ratio $R_s(q)$ is related to molecular properties as

$$\frac{K_c}{R_s(q)} = \frac{1}{P(q) \langle M \rangle_w} + 2B_C \qquad \text{Eq. 7}$$

where $K_c = 4\pi^2 n^2 (dn/dc)^2 / N_A \lambda_o$, dn/dc is the refractive index increment, c is the solute mass concentration, NA is Avogadro's number, $\langle M \rangle_w$ is the average molecular weight of the particles in solution, *P(q)* is a particle shape factor - formulas for which are available for a number of particle shapes,[26] and B_c is the second virial coefficient. There are various ways of plotting scattered intensity data as a function of scattering angle and particle concentration, of which the most commonly used is the Zim plot.[26] Thus, these data yield information both on particle molecular weight, $\langle M \rangle_w$, particle size and shape, *P(q)*, and particle-particle interactions, B_c.

MACROMOLECULE SIZE AND SHAPE

Small particles act as point scatterers so there is no angular dependence of the scattered light intensity (and hence no particle shape information, since $P(q) = 1$ at all q) if $qR_g \ll 1$, where R_g is the radius of gyration. For biological macromolecules of this size (which would correspond to "typical" globular proteins), the key measurable parameter by static light scattering is therefore molecular weight. An alternative approach is to combine *D* from dynamic light scattering with sedimentation coefficients from ultracentrifugation to yield information on molecular weight and shape.[27] As the particle size increases, the particles are no longer point scatterers: scattered light from two different parts of the same molecule will therefore interfere constructively or destructively. For particles with length scales of the order of $qR_g \sim 1$, the approximation $R(q)^{-1} = 1 + q2Rg^{2/3}$ is quite good and a radius of gyration can be reasonably accurately determined from the angular dependence of the scattered light. If R_h (R_h being the hydrodynamic radius) is measured by dynamic light scattering, the ratio *Rh/Rg* (*Rg* being the radius of gyration) can be used to ascertain whether the particle in solution is globular, rod-like, or intermediate in shape.

As the characteristic particle dimensions get even larger, the angular dependence of the scattered light becomes a more significant feature. This phenomenon brings with it more information on particle shape, but also increases the difficulty in interpreting dynamic light scattering data. Similarly, as particle size increases, diffusional modes other than translational (e.g. rotational and flexing) may contribute to the autocorrelation function. Again, this carries with it the potential for more information, but also carries an increased complexity of data fitting. Two classes of biological macromolecules that fall into this category are DNA and polysaccharides. For DNA, two studies are noted.[28,29]

LIMITS OF THE MODEL FOR SIZE OF PARTICLES

The model of non-interacting globular particles (as is used here) has been adopted for the calculation of the hydrodynamic radius, Rh and is the prerequisite for equations 1-7. In order to be valid, Rayleigh-Gans-Debye`s approximation must be satisfied:

$$\frac{4\pi}{\lambda} Rh \mid m - 1 \mid << 1 \qquad \text{Eq. 8}$$

where m is the molecular mass. Eq. 8 is only satisfied when the particles are small as compared to the wavelength λ of the stimulating light (632.8 nm for HeNe laser light) and when their refractive indices deviate only little from that of the solvent. When Eq. 8 is violated, solutions derived from the Mie theory for globular particles are available. Also, when particles approach or exceed the wavelength of the stimulating laser light, they scatter light predominantly in the forward direction at the cost of the backward scattered light. For particles > 632.8 nm the sensitivity decreases. At the lower end of the measurement range, the sensitivity decreases because the scatter is a power function of the radius of the particles (particle radius^6). This drop could be compensated by increasing the power of the laser light but this is problematic for DLS applications in the eye.[30]

HETERODYNING

Two correlation functions should be discriminated, i.e. the correlation function of the intensity $g^{(2)}_{(\tau)}$ and the autocorrelation function which measures the electric field $g^{(1)}_{(\tau)}$. One can only measure intensities but not, though, the electric field. Hence the two quantities must be correlated. Two possibilities exist for this correlation. The first is the homodyne method. Here, the intensity of the scattered light is measured and the correlation of the two quantities is determined. In the other, the heterodyne version, the diagnostic HeNe laser light is split into two beams. One is directed to the sample and the back-scattered light is then superimposed on the non-scattered light – the heterodyne portion – and directed to the detector. The same effect is seen when under non-ideal conditions, such as in measurements of the eye, the heterodyne light is superimposed upon the diagnostic homodyne beam. This is partial heterodyning, falsifying the information in the autocorrelation function. This effect is seen in non-ideal conditions such as *in situ* experiments in the eye. Here, nasty side effects may occur and one has to expect the interference of an unknown amount of parasitic stray light.[31,32]

Such stray light may be generated, for instance, at the surface of a contact lens but also, e.g. by collagenous fiber bundles or strands or membranes in the proliferative, diabetic vitreo-retinopathy. Especially problematic is the case of weak heterodyning in which merely a small decrease of the amplitude of the correlation function (the coherence factor) is observed while the more important distortion of the correlation function may escape attention. For example, with a 40 percent local oscillator, the amplitude decreases by only 15 percent, but the error in the estimation of the hydrodynamic radius of a particle is 40 percent.[33]

For purely homodyne experiments, the Siegert relation holds:

$$g^{(2)}(\tau) = 1 + \beta \mid g^{(1)}_{(\tau)} \mid^2 \qquad \text{Eq. 9}$$

where β is the coherence factor $\beta > 0$. In monomode fibers, such as are used here, β is 1. It decreases when parasitic stray light (e.g. an incoherent background signal) intrudes into the measuring system (partial heterodyning), but also when dead-light, or saturation effects of the detector are observed. This is what could happen to a larger or smaller extent in the eye. Such global noise effects can be detected, e.g., by the program of Flammer and Ricka.[33] Because of heterodyning, Eq. 9 is no longer valid and the heterodyne parameter h must be introduced:

$$h = I_1 / I_s \qquad \text{Eq. 10}$$

I_1 is the homodyne (diagnostic) signal and I_s is the heterodyne (parasitic) stray light. Now Eq. 9 must be corrected and replaced by:

$$g^{(2)}_{(\tau)} = 1 + \frac{h}{(1+h)^2} R_e\{g^{(1)}_{(\tau)}\} + \frac{1}{(1+h)^2} \mid g^{(1)}_{(\tau)} \mid^2 \qquad \text{Eq. 11}$$

while the parameter h can be obtained from Eq.12

$$h = \frac{1}{g^{(2)}_{(\tau)}(0) - 1} - 1 \qquad \text{Eq. 12}$$

We can now determine $g^{(1)}_{(\tau)}$ from $g^{(2)}_{(\tau)}$ and can analyze our data.

SUBJECTS

Fifty normal eyes and 49 eyes of diabetics were tested (Table 16.1). These two populations were matched for age by a Mann-Whitney test and the average age was found to be the same in both categories at p: 0.001. The diabetics were subdivided into categories of non-proliferative diabetic retinopathy I, II, IIIa and b and the proliferative retinopathy (Table 16.2). The normals had a routine ophthalmic status. Their visual acuity was not less than 0.8 and no refractive errors > 3 diopters were accepted. The diabetics also had a general ophthalmic check-up and no eye disease except diabetes was accepted. Their visual acuities ranged from 1.0 to 0.2. Both the vitreous body and the retina were inspected with the Goldmann 3-mirror contact lens and a fluorecein angiography was performed in each case (Table 16.2). Forty one percent of the diabetics had a posterior vitreous detachment while this was only observed in two percent of the normals. A variable degree of liquefaction of the vitreal structure was seen in 85 percent of the diabetics as compared to only 10 percent in the normals. The classification of the diabetics (Table 16.2) was according to the Diabetic Retinopathy Study Group (1991): Classification of diabetic retinopathy from fluorescein angiograms (Ophthalmology 1991;98:807-833).

Table 16.1 Demographic and clinical characteristics of the diabetic patients included in study (n=49). DM: Diabetes mellitus, DR: diabetic retinopathy, NPDR: Non-proliferative (background) diabetic retinopathy, PDR: Proliferative diabetic retinopathy

	DM without DR	NPDR I	NPDR II–III	PDR
Total no. of patients	21	8	12	8
Male	9	7	7	4
Female	12	1	5	2
Duration of DM (years)	1-25	3-18	4-23	4-27
(mean duration)	(8.5)	(9.1)	(12.3)	(14.2)

Table 16.2 The various states of diabetic retinopathy tested. From the Early Treatment Diabetic Retinopathy Study Group (1991). Classification of diabetic retinopathy from fluorescein angiograms. ETD Report. Ophthalmology 1991;98: 807-33

Level	Severity	Definition
I	Very mild NPDR	Microaneurysms only
II	Mild NPDR	Microaneurysms plus hard exudites, cotton wool spots, and/or mild retinal hemorrhages
IIIa	Moderate NPDR	More extensive intraretinal microvascular abnormalities, severe retinal hemorrhages, or venous bleeding in one quadrant
IIIb	Severe NPDR	Severe retinal hemorrhages in four quadrants, or venous bleeding in at least two quadrants, or moderately severe intraretinal microvascular abnormalities in at least one quadrant
IV	PDR	Fibrous and fibro-vascular membranes, retina as in IIIb

METHOD

The pupils of the patients were dilated with Diopine 10 percent and local anesthesia was induced by 0.4 percent Novesine. The intercept of the excitation laser beam with the vitreous can be observed both with the microscope of the DLS instrument and with an auxiliary microscope.[34] The focal spot is elliptic and measures 2.5/84 μm in air. The diagnostic laser beam has a hollow intensity-configuration (Fig. 16.1) and is split in two. Therefore, except when it is in focus, two spots are seen. The one spot/two spot criterion results in high precision. Other details concerning the focusing system are given elsewhere.[34] The position of the focal spot was set at the approximate center of rotation of the eye (taking into account that here rotational and translational motions are least).

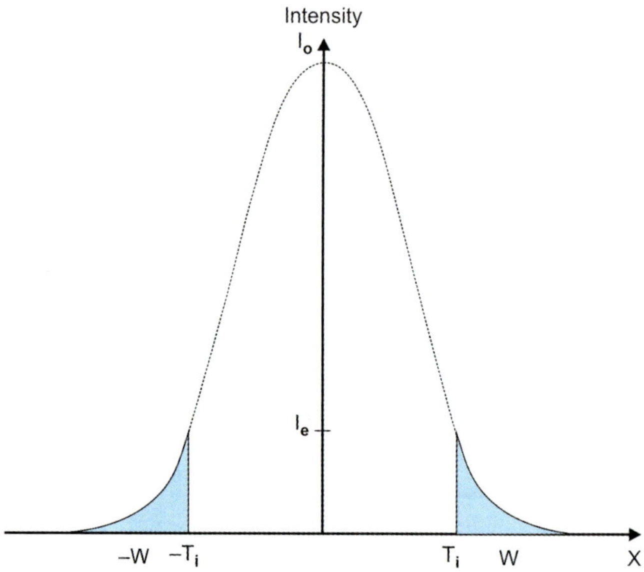

Fig. 16.1 Intensity distribution of 632.8 nm HeNe laser excitation beam. I_o: maximum of the original laser beam and I_e: the intensity of the hollow laser beam that is actually used. T_i and w are the inner and outer radii, respectively, of the Gaussian portion of the beam

The diagnostic laser beam was operated at powers from 1.5 to 15 μW and at exposure durations of 10 seconds. These values are below the safety limit (55 μW) during 20 seconds, stipulated by the National Standards Science (ANSI) Institute for safe use of lasers, Z-136, New York, 1976 and the procedure was in agreement with the Declaration of Human Rights, Helsinki (1975). Written consent was obtained in each case and also an agreement of the local ethics committee was obtained. As a rule, one measurement of $g^{(2)}_{(\tau)}$ took 10 seconds.

One measurement cycle consisted of five measurements, two seconds each, which were averaged. When the shape of a single curve deviated from the shapes of the standard, determined curves, the measurement was interrupted and discarded and this measurement was repeated. Gross deviations of the shapes of the determined curves (functions) could be easily seen because this was an online experiment. Such deviant (abnormal) correlation curves, as compared to normals, had a flat gradient or the amplitude (coherence factor) was less than 1. Eye movements were considerably inhibited by the contact lens and also because the measurements were performed at the approximate center of rotation of the eye.

The intensity of the scattered light was measured by an EG and G avalanche diode, SPCM-AQ-MX1-FZ, Vaudreuil, Canada, and was correlated by a digital correlator AL-5000, 5.0, Germany. This element contains a single photon counting diode (SPAD) that is fed by a stabilized high voltage supply. For the wavelength used (632.8 nm), the quantum efficiency of the diode exceeds that of a photomultiplier by many times and is therefore well adapted for the weak stray light signals re-emitted by the eye. The dead-time of the electronics was 30 ns. This value does not interfere with the measurement because the correlation time of a measurement is of the order of 1 ms. Hence, delays of $\tau > 1$ μs are sufficient for calculating the decay constant. Measurements were performed in the lower linear range of the photodetector, i.e. between 20 and 100 Hz. Here, the correction of the dead-time is between 0.99 and 1.01. At 20 Hz the amount of the background radiation amounted to 1.7 percent and at 100 Hz it was 0.4 percent of the measured counting rate.

The digital signal was correlated with a digital correlator. (ALV-5000, Langen, Germany). Following correlation the signal was symmetrically normalized by means of a multiple scheme. This kind of symmetrization improves the statistical accuracy for long delay times. Two monitor channels were activated, one for the direct, and the other for the delayed counting rates. Instead of normalizing, the product of mean values of the direct and the delayed correlation curves (functions), the product of the means values of the direct and the delayed value of the second was used for normalization. The criterion of a "good" correlation curve (function) is a stable plateau of the correlation function (Fig. 16.4) at the beginning, not disturbed by noise, a steep decay and a flat, final plateau at y = 1. All functions, together with the whole electronic system, were also controlled by a computer program.[33]

DISTURBANCES DUE TO EYE MOVEMENTS

The effects of involuntary eye movements, when measurements are performed at the retina, were analyzed by Wunderlich.[30] She found that involuntary eye movements would be of no relevance for measurements at the retina. This would then be even less disturbing for measurements at the center of the eye, such as is the case here. Voluntary eye movements have a detrimental effect.

The Parameters of the Correlation Function

An analytical expression must be found for the autocorrelation function for computation and in particular for statistics. Eq. 13 is an analytical description of the auto-correlation function:

$$y = \frac{A_1 + A_2}{1+(X/X_0)^E} + A_2 \qquad \text{Eq. 13}$$

This equation has been shown to well represent our autocorrelation functions (Fig. 16.4). The variable X_a is a value (on the abscissa) corresponding to the initial value of the correlation function, measured at the ordinate corresponding to 3 dB, and X_0 is the value where this value has decreased to 50 percent. The ordinate value A_1 corresponds to the initial value of the autocorrelation function and A_2 to the end of the correlation function. Both average values and their confidence limits (based on 50 normals) are given in Table 16.3.

RESULTS

On Figure 16.2, the logarithms of the decay times (τ) are plotted for both 50 normals and 49 diabetic eyes as a function of age. The values for X_a are plotted for the same populations on Figure 16.3. On Figure 16.5 the electrical fields of the autocorrelation functions $g^{(1)}_{(\tau)}$ are plotted as a function of τ and, the difference values of the autocorrelation functions, setting the function of the normal eye to zero.

It follows from the scatterplot of Figure 16.2 that there is a statistically non-significant overlap of the logarithms of the

Table 16.3 Average values + confidence limits (1 standard deviation) for normals (n=50) of parameters E and x_o

	Normal
E	0.866 ± 0.1
x_o	0.47 ± 0.3

Table 16.4 The values of E for normals and of various states of diabetic retinopathy were compared by an analysis of variance (ANOVA)

	E	
Population 1 vs Population 2	p-value (ANOVA)	
Normal	DM without DR	0.31
Normal	NPDR I	0.05
Normal	NPDR II – III	0.005
Normal	PDR	0.01
DM without DR	PDR	0.01
DM without DR	NPDR II – III	0.01
NPDR II - III	PDR	0.001

Table 16.5 The values of x_o for normals and for various states of retinopathy were compared by an analysis of variance (ANOVA)

	x_o	
Population 1 vs Population 2		p-value (ANOVA)
Normal	DM without DR	0.75
Normal	NPDR I	0.05
Normal	NPDR II – III	0.01
Normal	PDR	0.01
DM without DR	PDR	0.001
DM without DR	NPDR II – III	0.05
NPDR II - III	PDR	0.005

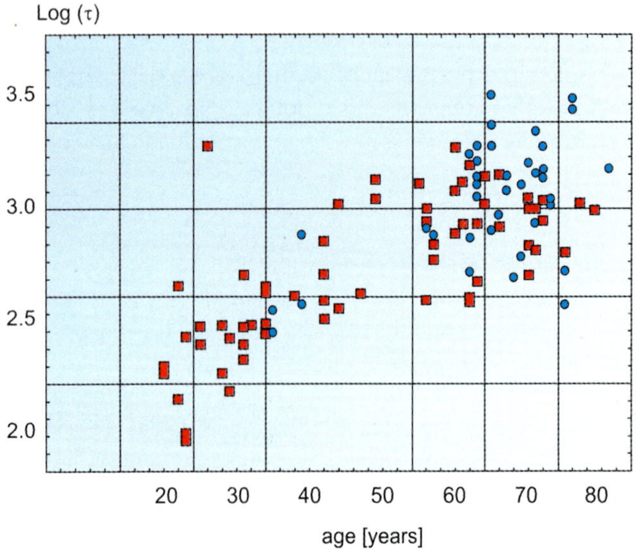

Fig. 16.2 Scatterplot of normals vs diabetics. Log τ (decay time) as a function of age

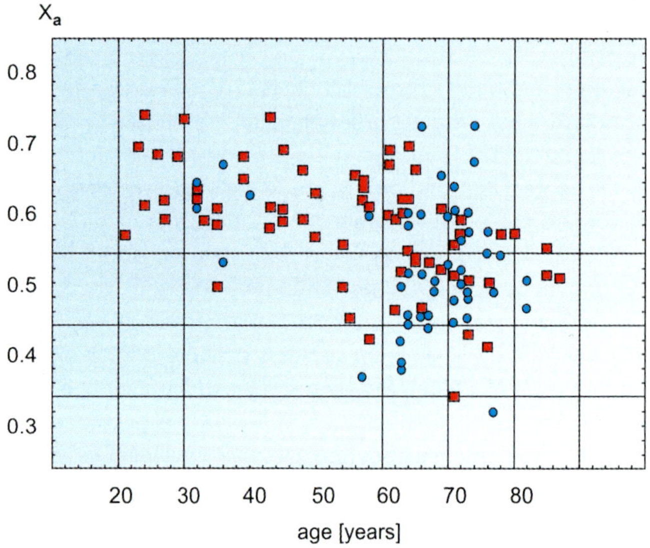

Fig. 16.3 Scatterplot of normals vs diabetics. x_o as a function of age

relaxation times τ of the normals vs the diabetics, in particular in the age-groups of more than 60 years, while in the younger age group, probably due to the paucity of cases, such a trend is less obvious. From the scatterplot of the quantity X_o of Figure. 16.3, a similar, statistically non-significant trend (overlap) of normals vs diabetics is obvious for age groups more than 60 years, while, for younger age groups, this trend is not obvious, again, probably, due to the paucity of the statistical material.

On the autocorrelation functions (Fig. 16.5) for normals and for various stages of the non-proliferative diabetic vitreoretinopathy, a shift towards "the right", i.e. towards larger particles and smaller diffusion constants is obvious (see Eqs. 1 and 2). This could also be interpreted as an increase of viscosity of the intermolecular (or inter-particular) substance without or with an enlargement of the microparticles. When the diabetic vitreo-retinopathy changes to the proliferative phase, a reverse trend is obvious: the particles decrease in size and the diffusion constants increase, implying that the viscosity of the intermolecular substance decreases. These comparisons were subjected to an analysis of variance (ANOVA) (Mann-Whitney test) and the significances of the pairs were compared (Tables 16.4 and 16.5).

The increase of the relaxation times (log τ) on Figure 16.2, both as a function of age and of diabetic disease, suggests an increase of the particle sizes and/or a decrease of the diffusion constants. On Fig. 16.3, the values X_a (i.e. the amplitude of the function or the coherence factor) decrease both with age and diabetes. The decrease of X_o on Figure 16.3 is compatible with an increase of heterodyning, i.e. prob-

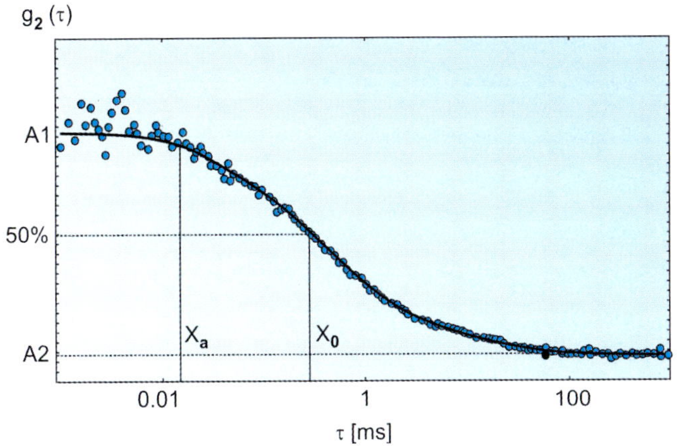

Fig. 16.4 Scatterplot of intensity correlation curve $g^{(2)}_{(\tau)}$. Best fit. A1: Starting value, A2: End value (ordinate value). x_a: Starting value, x_o (abscissa value) where intensity has fallen to 50%

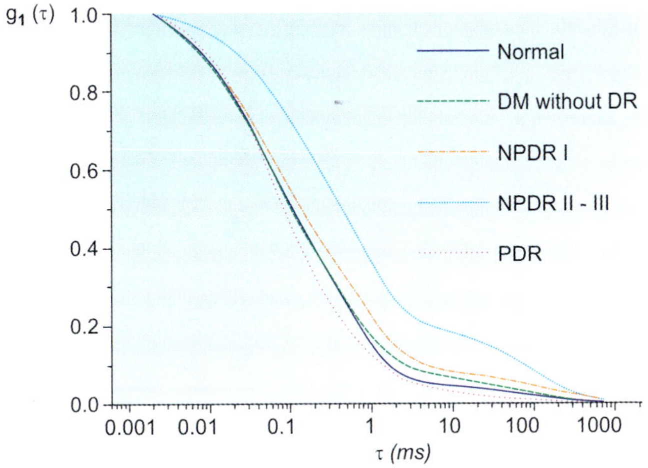

Fig. 16.5 $g^{(1)}_{(\tau)}$ autocorrelation curves for normals, diabetic phases NPDR I to NPDR II and III and PDR

ably by the appearance of enlarged, static microstructures. The overlap in Figures 16.2 and 16.3 suggests that age and diabetic mechanisms may have, at least in part, a metabolic mechanism in common. However, such a trend, within the limitations of this experiment, is not significant, in particular because this is a global comparison, integrating over various diabetic stages and we as yet do not know the behaviors of the various stages of diabetic retinopathy.

In the comparisons of the autocorrelation functions in Figure 16.5 and their differences (setting the autocorrelation function of normals to zero), the parameters E and X_o were deduced and displayed (Tables 16.4 and 16.5) for normals and for various diabetic stages (normals: N; diabetics: D; without diabetic retinopathy: DM; diabetics with non-proliferative retinopathy: NPDR I, II, III; and proliferative diabetic retinopathy: PDR) and were compared by an analysis of variance (Mann-Whitney test). In this statistical analysis we have combined phases NPDR IIIa and IIIb (Table 16.2) and have called this combination NPDR III (Table 16.2). When the normals are compared with NPDR I-II-III or, even worse, when they compared with the proliferative phase PDR, differences in E of high significance are found (p: .005 and p: .001), illustrating the intensity of the progression of the metabolic aberration. Principally analogous results are seen, when considering parameter X_o.

DISCUSSION

In *in-vitro* and *in-vivo* experiments on human eyes afflicted with non-proliferative diabetic retinopathy, Ansari et al.,[5] Fankhauser II[7] and Sebag et al.[8,11] have found an increase of microparticles in the vitreous. Sebag[8] has shown a difference between the particle sizes of age-matched diabetics and normals. Our results point in the same direction (Figs 16.2 and 16.3). In addition, the autocorrelation functions (Fig. 16.5) in our experiment also show a statistically significant, progressive increase in particle sizes during the non-proliferative phases (I, II, III). However, we have also shown a significant reverse trend when the proliferative diabetic stage starts. During this stage the autocorrelation functions could be falsified by static stray light effects due to fibrous structures. This heterodyne stray light therefore ought to be eliminated[33] but ought to be measured because its increase could signal an increase of the fibrotic process.

We have not found a statistical difference between normals and diabetics without retinopathy (Tables 16.4 and 16.5). This is important because it may mean that the metabolic aberration has not yet started or it may also mean that there may be indeed a difference, but we did not find it. This could have various reasons; in particular this could indicate insufficient statistical material. Hence these comparisons should be repeated on more extended material.

Sebag's[8,11] and Ansari's[5] determinations also differ from ours because these authors have solved equations (Eq. 6) and we have not. This Laplace equation represents a collective (polydisperse) answer which must be inverted to find single, molecular components. The solution, however, is an ill-posed mathematical problem and the inversion of equation 6 could result in an infinite number of solutions. Solutions for this problem exist.[25] Remarkably, both approaches (ours and [8–11]) have resulted in the same solution. However, they may be in part speculative and could be impaired by noise effects. Our results significantly display particles that are neither defined by shape nor by weight. The problem of polydispersity, a common problem in all tissue determinations, dead or living, remains and, for a structural or molecular diagnosis, it ought to be attacked by supplementary optical or other methods. Importantly, the effects of heterodyning should be strictly avoided in experiments on the eye since otherwise errors due to heterodyning are preprogrammed.

At this time good solutions can be found by comparing the results of dynamic light scattering with other informative sources, such as fluorography[35,36] and Raman spectroscopy.[8] Other analytical optical approaches should also be considered. Such methods are low-coherence reflectometry,[37] heterodyning[38] or ultrafast gating and detection techniques.[38-41] Also, the technique of diffuse-wave spectroscopy (DWS) appears as a promising solution.[42-44] Such apparative technology, however, is not available in ophthalmology at this time.

ACKNOWLEDGMENT

This work was supported by Haag-Streit International. The author would also like to thank M ott-Wunderlich for scientific help and A Funkhouser for editorial assistance.

REFERENCES

1. Bettelheim FA, Balasz EA. Light scattering patterns of the vitreous humor. Biochem Biophys Acta. 1968;158:309-12.
2. Sebag J, Dunker S, Suh K, Ansari AA. Dynamic light scattering measurements in vitreous. Invest Ophthalmol Vis Sci. 38 (ARVO). 1997;p.662.
3. Sebag J, Ansari RR. Dynamic Light Scattering of Vitreous in Aging and Diabetes. The Club Jules Gonin, Edinburgh, Scotland, 1998.
4. Sebag J, Kenneth M, Yee P. Vitreous: From biochemistry to clinical relevance. In: Tasman W and Jaeger EA (Eds.) Duane's Foundations of Clinical Ophthalmology. Lippincott, Williams and Wilkins, Philadelphia, 2006; Vol. 1, Chapter 16.
5. Ansari RR, Suh K, Dunker S, et al. Quantitative molecular characterization of bovine vitreous and lens with non-invasive dynamic light scattering. Exp Exe Res. 2001;73:859-66.
6. Rovati L, Fankhauser F II, Docchio F, Van Best J. Diabetic retinopathy assessed by dynamic light scattering and corneal autofluorescence. J Biom Optics 1998;3:357-63.
7. Fankhauser F II. Dynamic light scattering in ophthalmology. Results of in vitro and in vivo experiments. Technology and Health Care 2006;14:521-35.
8. Sebag J. Diabetic vitreopathy. Ophthalmology 1996;103: 205-06.
9. Sebag J. Ageing of the vitreous. Eye 1987;1:254-62.
10. Sebag J. Age-related changes of the human vitreous structure. Graefe's Arch Clin Exp Ophthalmol; 1987;225(2):89-93.
11. Sebag J, Ansari RR, Dunker S, Suhe K. Dynamic light scattering in diabetic vitreopathy. Diabetes Technology & Therapeutics 1999;1:169-76.
12. Sebag J. Biochemistry of the vitreous. In: Sebag J. The Vitreous: Structure, Function, and Pathology. Springer Verlag, New York. 1989;pp 17-26.
13. Cummins HZ, Pike ER.Photon Correlation and Light Beating Spectroscopy. Plenum Press. New York, 1974.
14. Ford NC Light scattering apparatus. In: R Pecora (Ed.) Dynamic Light Scattering: Applications of Photo Correlation Spectroscopy. Springer, New York, 1985;pp.7-57
15. Joosten JGH, Gelade TF, Pusey PN. Dynamic light scattering by nonergodic media: Brownian particles trapped in polyacrylamide gels. Phys Rev A, 1990;42(4):2161-75.
16. Schaefer DW, Berne BJ Light scattering from non-Gaussian concentration fluctuations. Phys Rev Lett 1972;28(8):475-78.
17. Ansari RR, Dhadwal HS, et al. (1992) In: Proceedings of OE and LASE, 19-25, Biomedical Optics. Fiber optic probe for ophthalmic refractive diagnosis
18. Benedek G. The molecular basis of cataract formation. CIBA Found Symp. 1984;pp.237-47.
19. Bursell SD, Baker RS, Weiss JN, Haughton, Rand LI Clinical photon correlation spectroscopy: Evaluation of human diabetic lenses. Exp Eye Res. 1989;49:241-58.
20. Benedek G. Quantitative detection of molecular changes associated with early cataractogenesis. 1987;6:1421-32.
21. Bursell S. The effect of acute changes in blood glucose on lenses in diabetic subjects using quasi-elastic light scattering. Curr Eye Res. 1975;8:821-34.
22. Tanaka T, Benedek G. Observation of protein diffusivity in intact human and bovine eye lenses with application to cataract. Invest Ophthal Vis Sci. 1975;6:449-56.
23. Koppel DE. Analysis of macromolecular polydispersity in intensity correlation spectrosopy: The method of cumulants. J Chem Phys. 1972;57:4814-20.
24. Provencher SW. A constrained regularization method for inverting data represented by linear algebraic or integral equations. Comp Phys Comm. 1982;27:213-27.
25. Stock RS, Ray WH. Interpretation of photon correlation spectroscopy data: a comparison of analysis methods. J Polym Sci B-Polym Phys. 1985;23:1393-447.
26. Burchard W. Static and dynamic light scattering from branched polymers and biopolymers. Adv Polym Sci. 1983; 48:4-124.
27. Rodgers KK, Bu Z, Fleming KG, Schatz DG, Engelman DM, Coleman JE. A zinc-binding domain involved in the dimerization of RAG1. J Med Biol 1996;260:70-84.
28. Ma C, Sun L, Bloomfield VA. Condensation of plasmids enhanced b-Z-DNA conformation of d(CG)a inserts. Biochemistry. 1995;34:3521-28.
29. Fishman DM, Patterson GD. Light scattering studies of supercoiled and nicked DNA. Biopolymers 1996;38:535-52.
30. Wunderlich M. Dynamische Lichtstreuung in der Netzhaut des Auges. Diplomarbeit. Mathematisch naturwissenschaftlich-technische Fakultät. Fachbereich Physik. Martin Luther Universität, Halle, Deutschland, 2005.

31. Gisler T, Rüger H, Egelhaaf SU, Tschumi J, Schurtenberger P, Ricka J. Mode-selective dynamic light scattering: Theory vs Experimental realization. Appl Opt. 1995;34:3546-53.
32. Schätzel K. Noise in correlation data: I. autocorrelation functions. Quantum Opt. 1990;2:287-05.
33. Flammer J, Ricka J. Dynamic light scattering with single-mode receiver: partial heterodyning regime. Appl Opt. 1997;36:7508-17.
34. Rovati L, Fankhauser F II, Ricka J. Design and performance of a new ophthalmic instrument for dynamic light-scattering measurements in the human eye. Rev Sci Instrum 1996;67:2615-20.
35. Docchio F. Ocular fluorometry: principles, fluorophores, instrumentation. Lasers Surg Med. 1989;9:515-32.
36. Rovati L, Docchio F. Autofluorescence methods in ophthalmology. J Biom Optics. 2004;9:9-21.
37. Schmidt JM, Knüttel A, Bonner RF. Measurement of optical properties of biological tissues by low-coherence reflectometry. Appl Opt. 1993;32:6032-42.
38. Yoo KM, Xing QR, Alfano RR. Imaging objects hidden in highly scattering media using femtosecond second-harmonic generation cross-correlation time gating. Opt Lett. 1991;16:1019-21.
39. Yoo KM, Alfano RR. Determination of the scattering and absorption lengths from the temporal profile of a backscattered pulse. Opt Lett. 1990;15:276-8.
40. Yoo KM, Alfano RR. Biological materials probed by the temporal and angular profiles of the backscattered ultrafast laser pulses. J Opt So Am B. 1990;7:1685-93.
41. Toida M,Kondo M, Ishimura T, Inaba H. Two-dimensional coherent detection imaging in multiple scattering media based on the directional resolution capability of the optical heterodyne method. Appl Phys B. 1991;52:391–4.
42. Rovati L, Cattini N, Zambelli N, Viola F, Staurenghi. In vivo diffusing-wave-spectroscopy of ocular fundus. Optics Express 2007;15:4030.
43. Maret G, Wolf PE. Multiple light scattering from disordered media. The effect of Brownian motion scatterers. Z. Phys B -Condensed Matter 1987;65:409-13.
44. Pine DJ, Weitz DA, Zhu JX, Herbholzheimer E. Diffusing-wave spectroscopy: dynamic light scattering in the multiple scattering limit. J Phys France 1990;51:2101-27.

17

Femtosecond Laser Flap Complications

Aylin Kilic (Turkey)

Chapter Outline

- Flap Thickness
- Corneal Biomechanics and Ectasia
- Stromal Wound Healing and Inflammation
- Opaque Bubble Layer
- Gas Bubbles Anterior Chamber
- Vertical Gas Breakthrough
- Transient Light-Sensitivity Syndrome
- Rainbow Glare
- Dry Eye
- Corneal Abberation Changes
- Flap Centralization

A high-quality flap is critical to laser *in situ* keratomileusis (LASIK) and to prevent flap-related complications. Femtosecond laser-assisted flap creation has rapidly gained popularity. This technology offers many advantages over the microkeratome, better predictability and increased safety of the procedure.[1,2] However, potential problems with the femtosecond laser are possible.

With the femtosecond laser, free flaps, irregular flaps, microperforations, decentered flaps, epithelial defects, and abrasions are significantly reduced or eliminated.

FLAP THICKNESS

Reports[3,4] show a 20 to 40 microns deviation from the intended corneal flap thickness when a microkeratome is used. Exposure of the cornea, dehydration, the intraocular pressure when cutting, the accessibility of the cornea to the blade, and other variables concerning the microkeratome.[5]

Kim et al[6] compared ultrasound pachymetry, Orbscan, and optical coherence tomography (OCT) measurements of the central corneal thickness before LASIK and evaluate the reproducibility of flaps created with the femtosecond laser using OCT. According to this report, the femtosecond laser created highly reproducible flaps that corresponded with the preoperative intended thickness. Recent studies[7] show that the femtosecond laser produced flaps with less variability in flap thickness. Most of these studies reported that most eyes had a flap thickness within ±20 microns of the intended results. Unpredictable flap thickness can lead to vision-threatening complications such as iatrogenic keratectasia.[8] Unexpectedly thinner or thicker flaps sometimes occur with the femtosecond laser and can lead to complications.[9] A thicker than expected flap results in insufficient residual stromal thickness, which can lead to iatrogenic keratectasia.[8] An extremely thin flap can lead to flap tears and irregular astigmatism.[10] Several studies show a narrower range of flap thicknesses and a smaller flap thickness standard deviation with the FS laser than with current microkeratomes[1,11] Choi et al[9] used of an femtolaser resulted in much thinner than expected flap. It is difficult to discriminate the flap from the true epithelial flap during flap lifting. The very thin nature of the flap could have resulted in vision threatening complications such as flap wrinkling and irregular astigmatism during epithelial removal.[12]

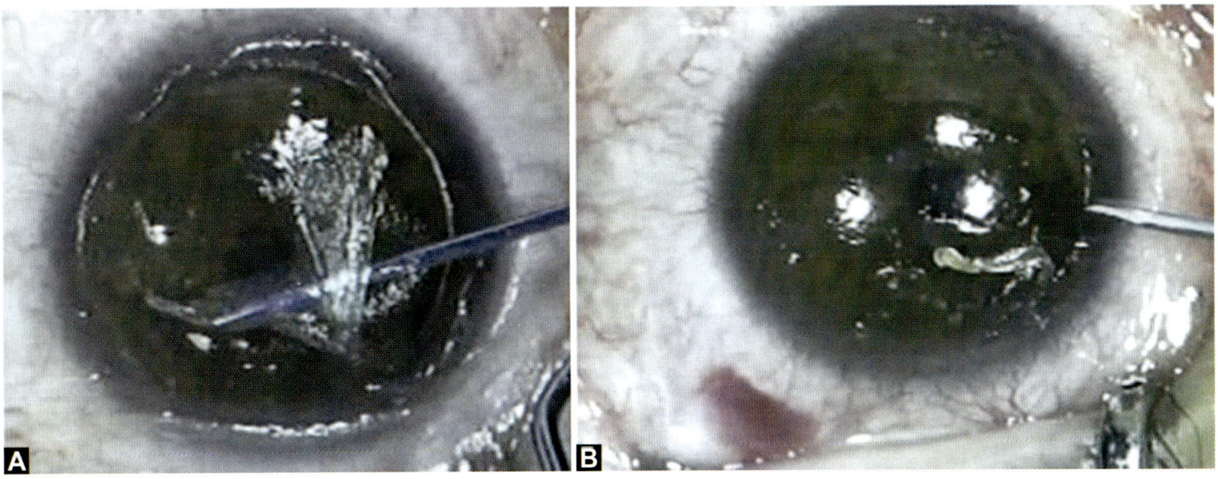

Figs 17.1A and B (A) The flap of the right eye created using intraLase was so thin it was difficult to differentiate it from an epithelial flap; (B) The flap in the left eye was even thinner than that in the right eye. During flap lifting, a tear occurred in the supranasal area and the procedure was abandoned[9]

CORNEAL BIOMECHANICS AND ECTASIA

According to a proposed model that characterizes the biomechanical response to corneal lamellar procedures, the corneal structure is a series of layered lamellae, at various orientations, that stretch limbus to limbus and are loaded by the intraocular pressure.[13-17] The severing of these lamellae results in reduction of tension in the peripheral lamellar segments, which leads to peripheral expansion and increasing radially outward force that is transmitted to the posterior stroma via crosslinking between lamellar layers. This results in central flattening of the anterior cornea with midperipheral steepening.[17] In cases of lamellar corneal refractive laser surgery, the response can be ascribed to the creation of the lamellar flap, photoablation, as well as an interaction between the two.[18]

The creation of a lamellar flap with an IntraLase laser reduces corneal stiffness, as measured by the amplitude reduction of the first infrared signal peak, Peak 1. However it does not affect the magnitude of the viscoelastic parameters, CH or CRF. Subsequently, laser ablation of the stromal bed reduces CH and CRF, along with a much greater reduction of Peak 1 than occurred with flap creation alone. This is consistent with a reduction in corneal stiffness overall. Therefore, the major biomechanical consequence of intraLASIK is in the ablation, rather than the flap creation. However, the component added by the flap creation is still important in the overall biomechanical response.[19]

Compared with published results of mechanical microkeratomes, femtosecond laser reduced the standard deviation of flap thickness as well as the achieved range. It eliminated physical complications associated with mechanical flap creation, and the impact of preoperative pachymetry and corneal power, thereby permitting more myopia to be corrected without risking deep ablations.[20]

Mechanical microkeratomes tend to create meniscus-shaped corneal flaps that are thinner in the center and thicker in the periphery, whereas the femtosecond laser tends to create more uniform planar flap. A more uniform flap has the theoretical advantage of being more mechanically stable.[20]

Knorz and Vossmerbaeumer[21] evaluated and compared the adhesion strength of flap created with a mechanical microkeratome and a femtosecond laser. They found that flap adhesion was stroger with a femtosecond laser flap. This likely reduces the risk for flap dislocation associated with trauma and, possibly, the risk for corneal ectasia.

STROMAL WOUND HEALING AND INFLAMMATION

Netto et al[22] evaluated early postoperative wound healing in rabbit corneas that had LASIK flaps formed with three different models (15 kHz, 30 kHz and 60 kHz) of a femtosecond laser compared with flaps formed with a microkeratome. They found that flap cuts with the 15 kHz femtosecond laser triggered more cell death, stromal cell proliferation, and inflammatory cell infiltration than flaps cut with the 30 kHz or 60 kHz laser or microkeratome. All corneas that had flap formation with the femtosecond laser or microkeratome had stromal cell death in the anterior central stroma and the peripheral stroma at 24 hours after the procedure. Cellular necrosis promotes inflammation. Therefore, procedures that trigger more keratocyte necrosis are likely to cause greater inflammation.

Diffuse lamellar keratitis (DLK), also known as sands of Sahara syndrome or interface keratitis, is a noninfectious condition in which white blood cells infiltrate the space between the flap and the stromal bed shortly after LASIK.[23] Gil-Cazorla et al[24] found a higher incidence of DLK in the femtosecond laser group than in the mechanical microkeratome group despite a more frequent postoperative topical corticoid regimen in the femtosecond laser group. In the 2 studies reporting the incidence of DLK after femtosecond laser surgery, the authors suggest that a decrease in the energy parameters for cutting the flap seemed to reduce the incidence of postoperative DLK.30 kHz and 60 kHz of the femtosecond lasers not only allows for much faster treatment, it also permits decreasing the energy use for cutting the flap.[24] More recent models, such as the 60 kHz IntraLase, induce a small inflammatory response that is similar to that with microkeratomes.[22]

According to animal studies, haze generation in these eyes likely results from a combination of basement membrane damage and proximity of the myofibroblast progenitor cells to the epithelial cell source of transforming growth factor B and other cytokines that drive modulate myofibroblast development.[25] In a recent study, Rocha et al[26] found that the generation of ultrathin (90 microns) femtosecond flap was associated with greater risk for haze development after LASIK. The authors propose that stromal haze after thin-flap LASIK is attributable to epithelium and basement membrane damage caused by the proximity of the laser-induced photodisruption of tissue.

Hainline et al[27] reported central flap necrosis after microkeratome and femtosecond laser created flaps. Central lamellar flap necrosis appears to differ from diffuse lamellar keratitis because the location of stromal inflammation is not in the flap interface but rather in the flap anterior stroma (Fig. 17.2).

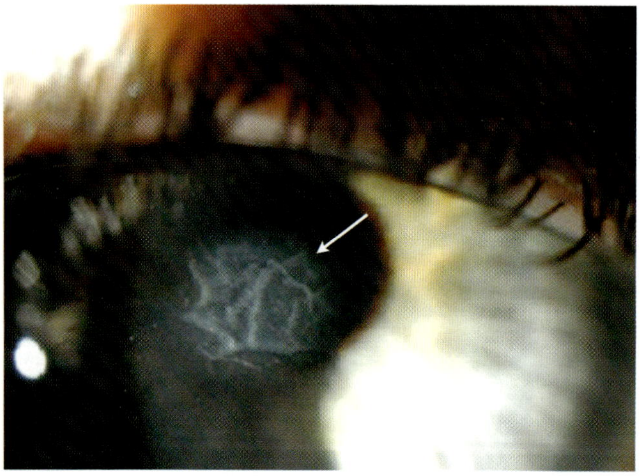

Fig. 17.2 Central flap necrosis[27]

OPAQUE BUBBLE LAYER

The femtosecond laser creates a corneal resection by delivering laser pulses of 1 micron diameter at a predetermined depth in the cornea, which expands to 2 to 3 microns. These pulses create microphotodisruption or an expanding bubble of gas (carbon dioxide) and water, which in turn cleaves the tissue and creates a plane of seperation.[28] Kaiserman et al[29] described 2 types of opaque bubble layer. An early (or hard) opaque bubble layer appears when gas bubbles spread into the tissue anterior or posterior to the plane at which the laser places additional pulses. A late (diffuse) opaque bubble layer occurs when gas travels into the intralamellar spaces after the raster pattern has passed through a particular portion of the resection. This type of opaque bubble layer generally looks more transparent, patchy and builds as more gas is introduced into the corneal tissue. They found that hard opaque bubble layers cover a significantly larger area of the flap than diffuse layers. An excessive opaque bubble layer might also interfere with or fool some excimer laser trackers, especially when the layer is located at or near the pupil. The opaque bubble layer might also interfere with intraoperative pachymetry measurements because the gas bubbles are very reflective to acoustic signals.

An opaque bubble layer occurred relatively frequently with the 15 kHz IntraLase laser. Thicker corneas and smaller flap diameters were associated with an increased risk for opaque bubble layer[29] (Fig. 17.3).

GAS BUBBLES ANTERIOR CHAMBER

Gas bubbles infrequently appear in the anterior chamber after application of the femtosecond laser. Gas bubbles are probably associated with formation of the pocket and expansion through posterior corneal or stromal layers into the anterior chamber.

Lifshitz et al[30] speculated that many small bubbles coalesced and created larger bubbles that migrated through the posterior stroma and endothelium without being absorbed by the endothelial pump, and these bubbles appeared in the anterior chamber. Theoretically, the lasershockwave might push the air bubbles posteriorly, and if stromal lamellae are weak or endothelial junctions are not too tight, the bubbles can appear in the anterior chamber.

Large bubbles could also be created near the pocket and then would migrate to the posterior stroma and anterior chamber. Another mechanism might be a retrograde passage of the bubbles from the perilimbal area to the trabecular meshwork into the anterior chamber. They may interfere with the eye tracker of the excimer laser if they are multiple or large.

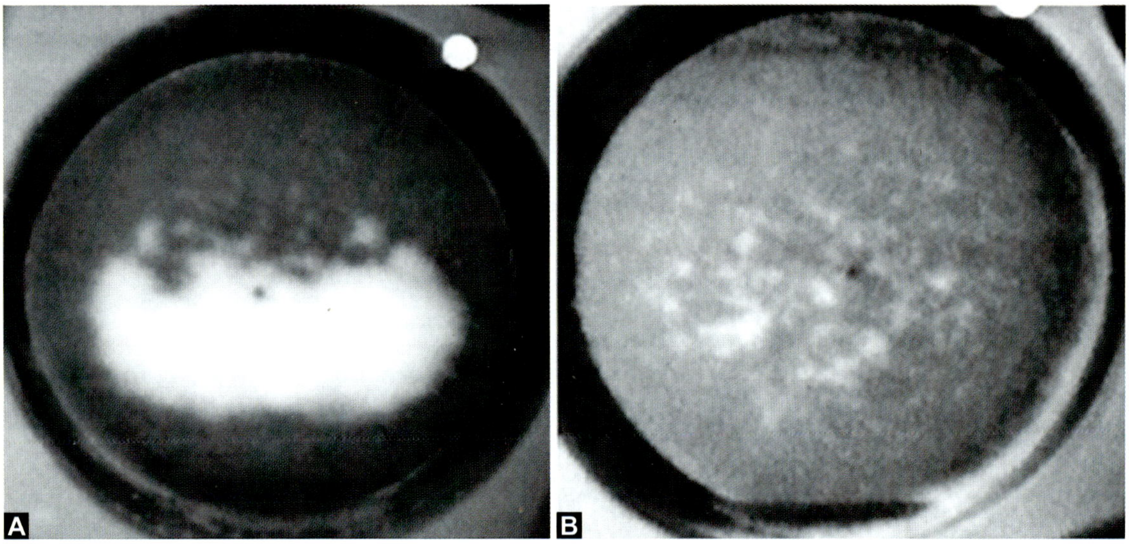

Figs 17.3A and B Opaque bubble layer after flap creation with femtolaser[29]

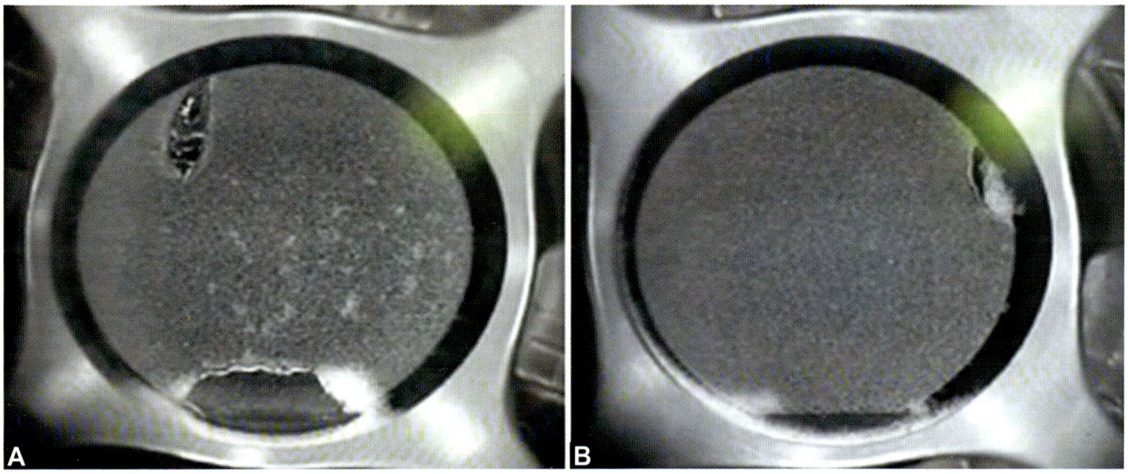

Figs 17.4A and B Vertical gas breakthrough

VERTICAL GAS BREAKTHROUGH

Each laser pulse generates a small amount of microplasma without affecting anterior tissues, which results in an interface of microscopic gas bubbles.[31] This effectively creates a dissection plane for flap lifting. These microbubbles are presumed to be placed to be placed under significant pressure due to the suction ring and glass applanating the corneal surface, although no formal analysis has been done. The microbubbles may break vertically through the corneal epithelium during flap creation, especially if the cornea weakened. Femtosecond laser cannot photo disrupt corneal stroma in a small portion of the interface or if there is resistance within the interface from scar tissue, incomplete flaps may result[32] (Fig. 17.4)

The real cause is uncertain, although some authors propose that local defects in the Bowman membrane or altered epithelium[30] contribute to it because act as a low-resistance pathway fort he bubbles to expand. Accordingly, fibrotic corneal stroma may increase the resistance to an effective plane of seperation and may be associated with this complication.[32]

TRANSIENT LIGHT-SENSITIVITY SYNDROME

The term transient light-sensitivity syndrome (TLSS) has been used to describe a clinical condition characterized by unusual photosensitivity with normal visual acuity several weeks after otherwise uneventful LASIK with the femtosecond laser that typically responds to topical treatment with steroids or cyclosporine.

Although the etiology of TLSS is unknown, there are clinical clues that point toward an inflammatory origin. Proposed mechanisms include inflammation caused by necrotic cell debris or subproducts of the gas bubbles, cytokines migrating from the flap interface to the perilimbal sclera and iris base, or activated keratocytes in the interface.[33] The combined use of the lowest possible energy and an intensive postoperative regimen of topical steroids fort he first few days after femtosecond laser flap creation helped decrease the incidence of TLSS. The differential diagnosis of TLSS should include dry-eye syndrome and DLK.[34]

TLSS is a unique side effect of the femtosecond laser that is primarily associated with earlier 6 kHz and 15 kHz models of the IntraLase laser but has been noted rarely with the 30 kHz and 60 kHz models.[35]

RAINBOW GLARE

Microscopic irregularities anywhere within the visual axis can produce forward and/or backward light scattering, which is usually insignificant. However, in rainbow glare, the raster (grating) pattern of femtosecond laser pulses, which remains on the back surface of the flap, summates the scattered rays into a peripheral light spectrum.[36] The cause of the rainbow glare was reported as the diffraction of light from the grating pattern created on the back surface of the LASIK flap after femtosecond laser exposure.[37]

Bamba et al[38] reported the incidence of and factors associated with rainbow glare after LASIK flap creation with 60 kHz femtosecond laser. In this study, they found that rather than faster pulse frequencies or lower raster energies, it is the quality of the beam and the dimension of the numerical aperture of the focusing optics that seem to be the most important factors in reducing this symptom. Thus, it is extremely important to have regular laser maintenance to ensure beam quality and alignment and avoid this side effect.

DRY EYE

A condition that is thought to play an important role in dry eye after LASIK is LASIK-induced neurotrophic epitheliopathy (LINE), a term suggested by Wilson[39] and Ambrossio[40] et al to describe the neurotrophic component of LASIK dry eye and that results from damage to the nerves during flap formation and stromal ablation.[39,40]

Corneal innervation is an integral component of the lacrimal gland-ocular surface functional unit, which coordinates basal and stimulated tear production, lid blinking, tear spreading, and tear clearence.[39-41]

Salomao et al[42] compared LASIK-associated dry eye and the need for postoperative cyclosporine A treatment after flap creation with a femtosecond laser and a mechanical microkeratome. In this study eyes with femtosecond flaps had a lower incidence of LASIK-associated dry eye and required less treatment for the disorder. In addition to neurotrophic effects from corneal nerve cutting, other factors may be important because no correlation was found between flap thickness (or ablation depth) and the incidence of LASIK-induced dry eye.

The introduction of femtosecond laser technology into the LASIK procedure has the potential to improve the long-term postoperative incidence of dry-eye symptoms. Golas et al[43] examined the effects on postoperative dryness of wavefront-guided LASIK performed in fellow eyes, using a mechanical keratome in 1 eye and a femtosecond laser in the other eye. They found no statistically significant differences in dry-eye symptoms between the mechanical and femtosecond laser keratotomy techniques.

The effects of LASIK on the reported postoperative dry-eye symptoms showed an initial increase at 1 month with a return to baseline by 6 months.[43]

CORNEAL ABBERATION CHANGES

Standard (non-wavefront guided) LASIK is routinely performed, and it would be beneficial to evaluate the effect of femtosecond laser flap creation in decreasing the optical side-effects and high order abberations (HOA) reported after LASIK with mechanical microkeratome flap creation. Calvo et al[44] showed that corneal optical HOAs after femtosecond LASIK and microkeratome LASIK stable over time. Calvo et al[44] concluded that there were no differences in HOAs between the mechanical and femtolaser methods and that the HOAs remained stable through 36 months postoperatively. Munoz et al[45] results confirm their conclusion and show that the stability continues for 12 more months. Tran et al[46] found that lamellar keratotomy with a mechanical microkeratome without excimer ablation produced significant increases in trefoil, quadrafoil, and total HOAs, whereas lamellar keratotomy with a femtosecond laser did not increase or decrease HOAs. The authors found a statistically significant increase in coma after excimer laser ablation using the Hansatome microkeratome and suggest that it might be related to the difference in hinge angle between the microkeratome flap and the femtosecond flap.

It is important that the ablation surface be smooth to limit the generation of HOAs.[47] Several studies have correlated better visual and refractive outcomes with more regular bed surfaces.[48] Recent studies report less induction of HOAs with the femtosecond laser than microkeratomes during LASIK. Mederios et al[49] compared the Moria-M2 and Hansatome microkeratomes with the 15 kHz and 30 kHz Intralase femtosecond lasers and found that, on average, the femtosecond laser group had significantly less induced aberrations than the microkeratome group.

FLAP CENTRALIZATION

It is now generally accepted that the line from the fovea through the center of the pupil should be used in calculating the aberrations of the eye.[50,51] However, if there are large changes in the location of the center of the pupil with changes in pupil size, this axis becomes less valuable because its location depends on the measurement conditions. Although there are many advantages of channel creation using af emtosecond laser, we observed decentration after Intacs placement. Before the centration process with the IntraLase device, the suction ring is inserted, after which the asymmetric and relatively thin keratoconic cornea is flattened with a glass lens. When the applanation is completed, the pupil dilates slightly. Pupil dilation may be the result of the compression applied by the suction ring or flattening of the cornea with the glass applanation lens. It has been observed that during the centration process, the pupil becomes dilated and the asymmetric cornea flattens. The cornea and pupil return to their natural positions after the applanation, and the geometric center of cornea shifts. There are several studies of the change in pupil centration with changing pupil size.

The latest tendency has been to try to create flaps with the largest possible diameter, attempting to compensate for any decentration.[52]

In our study[53] we suggest that flap centration is also influenced by patient and surgeon factors. Improper patient head position may lead to inadequate ring placement and applanation. Applanation point should be very well centered and that the glass lens should not be moved once the cornea is touched because the friction between the surface may cause decentration.

In summary, less than 1 percent of eyes had direct or indirect complications due to femtosecond laser flap creation. Haft et al[54] evaluated complications associated with femtosecond laser-assisted flap creation in LASIK surgery. Included in the study were 4772 eyes, of which 44 (0,92%) eyes had direct or indirect complications due to flap creation. Thirty-two eyes had indirect complications - 20 (0,42%) eyes developed DLK and 122 (0,25%) eyes had transient light sensitivity syndrome. Twelwe (0,25%) eyes had direct femtosecond laser-related complications - 8 (0,17%) eyes had prematüre breakthrough of gas breakthrough the epithelium within flap margins, 3 (0,06%) eyes had incomplete flaps due to suction loss, and 1 (0,02%) eye had irregular flap due to prvious corneal scar.

Table 17.1 Complications in 4772 eyes that underwent LASIK with femtosecond laser flap creation

Complications	No. Eyes (%)
Total	44 (0.92)
Direct (Flap)	12 (0.25)
Epithelial gas breakthrough	8 (0.17)
Incomplete flap/suction loss	3 (0.06)
Irregular flap due to corneal scar	1 (0.02)
Indirect	
Diffuse lamellar keratitis	20 (0.42)
Transient light sensitivity	12 (0.25)

REFERENCES

1. Nordan LT, Slade SG, Baker RN, Suarez C, Juhasz T, Kurtz R. Femtosecond laser flap creation for laser in situ keratomileusis: six months follow up initial US clinical series. J Refract Surg 2003;19: 8-14.
2. Stonecipher K, Ignacio TS, Stonecipher M. Advances in refractive surgery: microkeratome and femtolaser flap creation in relation to safety, efficiacy, predictability, and biomechanical stability. Curr Opin Ophthalmol. 2006;17:368-72.
3. Shemesh G, Dotan G, Lipshitz I. Predictability of corneal flap thickness in laser in situ keratomileusis using three different microkeratomes. J Refract Surg. 2002;18:347-51.
4. Seiler T, Koufala K, Richter G. Iatrogenic keratectasia after laser in situ keratomileusis. J Refract Surg. 1998;14:312-17.
5. Buratto L, Brint S, Ferrari M. Surgical instrument. In:Buratto L, Brint SF, (Eds.) LASIK: Principles and Techniques. Thorofare, NJ, Slack. 1998;pp35-68.
6. Kim JH, Lee D, Rhee K-II. Flap thickness reproducibility in laser in situ keratomileusis with a femtosecond laser: Optical coherence tomography measurement. J Cataract Refract Surg. 2008;34:132-6.
7. Kim JH J Cataract Refract Surg 2008;34:132-136) (Sutton G Hodge C. Accuracy and precision of LASIK flap thickness using the IntraLase femtosecond laser in 1000 consecutive cases. J Refract surg. 2008;24:802-6.)
8. Binder PS. Ectasia after laser in situ keratomileusis. J Cataract refract Surg. 2003;29:2419-29.
9. 9 Choi SK, Kim JH, Lee D, Oh SH, Lee JH, Ahn MS. Creation of an extremely thin flap using IntraLase femtosecond laser. J Cataract Refract Surg. 2008;34:864-7.
10. Maldonado MJ, Juberias JR, Pinero DP, et al. Flap tearing during lift-flap laser in situ keratomileusis retreatment. J Cataract Refract Surg. 2005;31:2016-8.

11. Kezirian GM, Stonecipher KG. Comparison of the IntraLase femtosecond laser and mechanical keratomes for laser in situ keratomileusis. J Cataract Refract Surg 2004;30:804-11.
12. Sharma N, Ghate D, Agarwal T, Vajpayee RB. Refractive outcomes of laser in situ keratomileusis after flap complications. J Cataract Refract Surg. 2005;31:1334-7.
13. Smolek MK, McCarey BE. Interlamellar adhesive strength in human eye bank corneas. Invest Ophthalmol Vis Sci. 1990; 31:1087-95.
14. Smolek MK. Interlamellar cohesive strength in the vertical meridian of human eye bank corneas. Invest Ophthalmol Vis Sci. 1993;34:2962-9.
15. Roberts C. "The Cornea is Not a Piece of Plastic." J Refract Surg. 2000;16:407-13.
16. Roberts C. "Future Challenges to Aberration-Free Ablative Procedures" J Refract Surg. 2000;16:S623-S629.
17. Dupps WJ, Roberts C: "Effect of Acute Biomechanical Changes on Corneal Curvature After Photokeratectomy" J Refract Surg 2001,17:658-69.
18. Roberts C, Dupps WJ Jr. Corneal biomechanics and their role in corneal ablative procedures. In: MacRae SM, Krueger RR, Applegate RA, (Eds.) Customized Corneal Ablation; the Quest for Supervision. Thorofare, Nj, Slack. 2001;pp109-31.
19. Uzbek (Kiliç) AK, Kamburoglu G, Mahmoud AM, Roberts CJ. Change in biomechanical parameters after flap creation using the IntraLase femtosecond laser and subsequent excimer laser ablation.Curr Eye Res. 2011;36(7):614-9.
20. Binder PS. One thousands consecutive IntraLase laser in situ keratomileusis flaps. J Cataract Refract Surg. 2006;32:962-9.
21. Knorz MC, Vossmerbaeumer U. Comparison of flap adhesion strength using the Amadeus microkeratome and the IntraLase iFS femtosecond laser in rabbits. J Refract Surg. 2008;24:875-8.
22. Netto MV, Mohan RR, Medeiros FW, Dupps WJ, Sinha S, et al. Femtosecond Laser and Microkeratome Corneal Flaps: Comparison of Stromal Wound Healing and Inflammation. J Refract Surg. 2007;23:667-76.
23. Smith RJ, Maloney RK. Diffuse lamellar keratitis; a new syndrome in lamellar refractive surgery. Ophtalmology. 1998;105:1721-26.
24. Gil-Cazorla R, Teus MA, Benito-Llopis L, Fuentes I. Incidence of diffuse lamellar keratitis after laser in situ keratomileusis associated with the IntraLase 15 kHz femtosecond laser and Moria M2 microkeratome. J Cataract Refract Surg. 2008;34:28-31.
25. Netto MV, Mohan RR, Sinha S, Sharma A, Dupps W, Wilson SE. Stromal haze, myofibroblast and surface irregularity after PRK. Exp Eye Res. 2006; 82:788-97.
26. Rocha KM, Kagan R, Smith SD, Krueger RR. Thresholds for interface haze formation after thin-flap femtosecond laser in situ keratomileusis for myopia. Am J Ophthalmol. 2009;147:966-72.
27. Hainline BC, Price MO, Choi DM, Price FW. Central flap necrosis after LASIK with microkeratome and femtosecond laser created flaps. J Refract Surg, 2007;23:233-42.
28. Kurtz RM, Liu X, Elner VM, et al. Photodisruption in the human cornea as a function of laser pulse width. J Refract Surg. 1997;13:653-8.
29. Kaiserman I, Maresky HS, Bahar I, Rootman DS. Incidence, possible risk factors, and potential effects of an opaque bubble layer created by femtosecond laser. J Cataract Refract Surg, 2008;34:417-23.
30. Lifshitz T, Levy J, Klemperer I, Levinger S. Anterior chamber gas bubbles after corneal flap creation with a femtosecond laser. J Cataract Refract Surg, 2005;31:2227-9.
31. Maatz G, Heisterkamp A, Lubatschowski H, et al. Chemical and physical side effects at application of ultrashort laser pulses for intrastromal refractive surgery. J Opt A: Pure Appl Opt, 2000;2:59–64.
32. Seider MI, Ide T, Kymionis GD, Culbertson WW, O'Brien TP, Yoo SH. Epithelial breaktrough during IntraLase flap creation for laser in situ keratomileusis. J cataract Refract Surg, 2008;34:859-63.
33. Stonecipher KG, Dishler JG, Ignacio TS, Binder PS. Transient light sensitivity after femtosecond laser flap creation: clinical findings and management. J Cataract Refract Surg. 2006;32:91-4.
34. Munoz G, Albarran-Diego C, Sakla HF, Javaloy J, Alio JL. Transient light-sensitivity syndrome after laser in situ keratomileusis with the femtosecond laser: Incidence and prevention. J Cataract Refract Surg. 2006;32:2075-9.
35. Stonecipher K, Ignacio TS, Stonecipher M. Advances in refractive surgery: microkeratome and femtosecond laser flap creation in relation to safety, efficacy, predictability, and biomechanical stability. Curr Opin Ophthalmol. 2006;17:368-72.
36. Krueger RR, Thornton IL, Xu M, Bor Z, van den Berg TJTP. Rainbow glare as an optical side effect of Intralasik. Ophthalmology. 2008;115:1187-95.
37. Sonigo B, Iordanidou V, Chong-Sit D, Auclin F, Ancel JM, Labbe A, Baudouin C. In vivo corneal confocal microscopy comparison of IntraLase femtosecond laser and mechanical microkeratome for laser in situ keratomileusis. Invest Ophthalmol Vis Sci. 2006;47:2803-11.
38. Bamba S, Rocha KM, Ramos-Esteban JC, Krueger RR. Incidence of rainbow glare after laser in situ keratomileusis flap creation with 60 kHz femtosecond laser. J Cataract Refract Surg. 2009;35:1082-6.
39. Wilson SE. Laser in situ keratomileusis-induced (presumed) neurotrophic epitheliopathy. Ophthalmology 2001;108:1082-7.
40. Ambrosio R, Tervo T, Wilson SE. LASIK-associated dry eye and neurotrophic epitheliopathy. Am J Ophthalmol. 2001;132:405-6.
41. Stern ME, Gao J, Siemasko KF, Beuerman RW, Pflugfelder SC. The role of the lacrimal functional unit in the pathophysiology of dry eye. Exp eyer. 2004;78:409-16.
42. Salomao MO, Ambrosio R, Wilson SE. Dry eye associated with laser in situ keratomileusis: mechanical microkeratome versus femtosecond laser. J Cataract Refract Surg. 2009;35:1756-60.

43. Golas L, Manche EE. Dry eye after laser in situ keratomileusis with femtosecond laser and mechanical keratome. J Cataract Refract Surg. 2011;37:1476-80.
44. Calvo R, McLaren JW, Hodge DO, Bourne WM, Patel SV. Corneal aberrations and visual acuity after laser in situ keratomileusis: femtosecond laser versus mechanical microkeratome. Am j Ophthalmol. 2010;149:785-93.
45. Munoz G, Albarran-Diego C, Ferrer-Blasco T, Lazaro SG, Exposito AC. Long-term comparison of corneal aberration changes after laser in situ keratomileusis: mechanical microkeratome versus femtosecond laser flap creation. J Cataract Refract Surg. 2010;36:1934-44.
46. Tran DB, Sarayba MA, Bor Z, Garufis C, Duh Y-J, Soltes CR,Juhasz T, Kurtz RM. Randomized prospective clinical study comparing induced aberrations with IntraLase and Hansatome flap creation in fellow eyes; potential impact on wavefront guided laser in situ keratomileusis. J Cataract Refract Surg. 2005; 31:97-105.
47. Porter J, MacRae S, Yoon G, Roberts C, Cox IG, Williams DR. Separate effects of the microkeratome incision and laser ablation on the eye's wave abberation. Am J Ophthalmol. 2003;136:327-37.
48. Vinciguerra P, Azzolini M, Airaghi P, Radice P, De Molfetta V. Effect of decreasing surface and interface irregularities after photorefractive keratectomy and laser in situ keratomileusis on optical and functional outcomes. J Refract Surg. 1998; 14:S199-S203.
49. Medeiros FW, Stapleton WM, Hammel J, Krueger RR,Netto MV, Wilson SE. Wavefront analysis comparison of LASIK outcomes with the femtosecond laser and mechanical microkeratomes. J Refract Surg. 2007;23:880-7.
50. Thibos LN, Applegate RA, Schwiegerling JT, Webb R. Report from the VSIA taskforce on standards for reporting optical aberrations of the eye; VSIA Standards Taskforce Members.J Refract Surg. 2000;16:S654-S655.
51. Mrochen M, Eldine MS, Kaemmerer M, et al. Improvement in photorefractive corneal laser surgery results using an active eye-tracking system. J Cataract Refract Surg. 2001; 27:1000-6.
52. Slade SG. The use of the femtosecond laser in the customization of corneal flaps in laser in situ keratomileusis. Curr Opin Ophthalmol. 2007;18:314-7.
53. Ertan A, Karacal H. Factors influencing flap and INTACS decentration after femtosecond laser application in normal and keratoconic eyes. J Refract Surg. 2008;24:797-801.
54. Haft P, Yoo S, Kymionis GD, Ide T, O'Brien TP, Culbertson WW. Complications of LASIK Flaps made by the IntraLase 15-and 30 kHz Femtosecond Lasers. J Refract Surg. 2009;25:979-84.

18

Femtosecond Laser Astigmatic Incisions

Eric D Donnenfeld, Allon Barsan (USA)

Chapter Outline

- Limbal Relaxing Incisions (LRI) Nomograms

LIMBAL RELAXING INCISIONS (LRI) NOMOGRAMS

Cataract surgery is one of the most commonly performed procedures worldwide and has become a refractive procedure in which patients have increasingly high expectations for both functional visual improvement as well as spectacle independence.[1,2]

Management of astigmatism is the single most important aspect of the ability to provide good surgical outcomes, quality of vision, and satisfaction for patients requesting refractive cataract surgery and optimal postoperative uncorrected visual acuity. Astigmatism decreases visual acuity through meridional blur, in which one axis of the cornea is steeper than the other, causing distortion of the visual image. Residual astigmatism after cataract surgery of 0.50 diopters (D) or even less may result in glare, symptomatic blur, ghosting, and halos.[3] As a result, greater emphasis has been placed on treating corneal astigmatism at the time of cataract surgery with limbal relaxing incisions (LRIs). A recent study of 4,540 eyes of 2,415 patients showed that corneal astigmatism is present in the majority of patients undergoing cataract surgery, with at least 1.50 D measured in 22.2 percent of study eyes.[4] Approximately 38 percent of eyes undergoing cataract surgery have at least 1.00 D of pre-existing corneal astigmatism, and 72 percent of patients have 0.5 D or more.[5]

Astigmatism management is important regardless of the IOL that is used. Patients implanted with presbyopia-correcting IOLs, however, are significantly more sensitive to even minor refractive errors than those implanted with conventional monofocal IOLs. Refractive IOL surgeons must be willing and able to treat small amounts of postoperative astigmatism in order to achieve happy postoperative patients.

Astigmatic incisions are corneal incisions that are used to relax and flatten the steep axis of regular corneal astigmatism while steepening the flat axis. The procedure reduces corneal tension and allows the cornea to heal into a more spherical shape (Figs 18.1A and B). Incisions placed adjacent to the limbus are LRIs, and incisions placed further from the limbus and closer to the visual axis are astigmatic keratotomies. Because there are several advantages of LRIs over astigmatic keratotomies, LRIs are performed more frequently. These advantages include a reduced tendency to cause axial shift, less irregular astigmatism, a 1:1 coupling ratio, a decreased

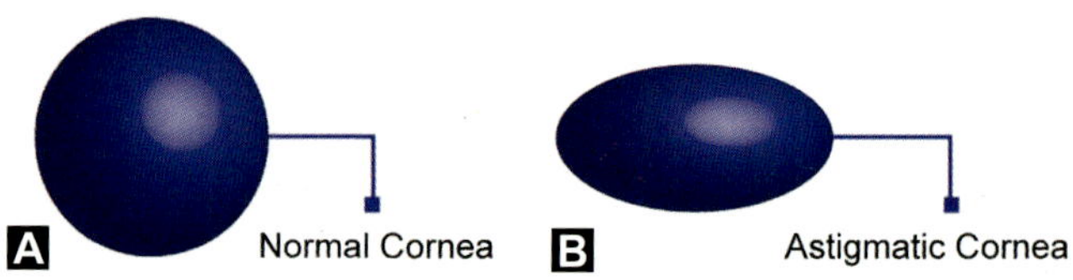

Figs 18.1A and B In a normal cornea (A), shaped like a basketball, both axes are equal; in an astigmatic cornea (B), shaped like a football, one axis is steeper than the other

chance of visually significant scarring, and a reduced likelihood of perforation.

Unfortunately, the majority of cataract surgeons are uncomfortable performing manual incisional astigmatic incisions. In a survey of 233 surgeons, only 73 routinely performed LRIs.[6] In general, irregular astigmatism (Fig. 18.2) should not be treated with LRIs. There are several different options for treating astigmatism, including excimer laser photoablation, toric IOL implantation, and conductive keratoplasty. These options, however, may not be available to all cataract surgeons and patients. Correction of pre-existing corneal astigmatism is not a covered benefit of Medicare and most other insurance plans. Therefore, the patient who is the beneficiary of a noncovered service has the option to pay for the procedure and the additional physician time and testing that accompanies it.

Once the required location of the incision has been ascertained, the incision length and depth are determined by the surgeon. Sometimes, a nomogram may be consulted. There are a number of LRI nomograms for correcting small amounts of cylinder, and many studies evaluating LRIs have been performed.[7-21] In one study on the effectiveness of LRIs, there was a 60 percent average reduction of cylinder,[22] with 79 percent of patients corrected to less than 1.00 D of cylinder and 59 percent corrected to less than 0.5 D of cylinder. The 60 percent reduction in cylinder compares favorably with the results achieved using toric IOLs, which result in a mean 58.4 percent reduction in cylinder.[23]

Many LRI nomograms are adjusted for age and cylinder axis, making them detailed and complex and giving the impression that the procedure is extremely precise and unforgiving. In my opinion, however, this simply is not the case. The LRI procedure has numerous chances for error, and the many variables that help determine how the incision is performed can each be measured incorrectly. The incision itself, if performed manually, is only as precise as the surgeon making the incision and the accuracy of the blade. Any of these errors can compound and result in suboptimal visual acuity. Manual LRIs are therefore as much an art as a science.

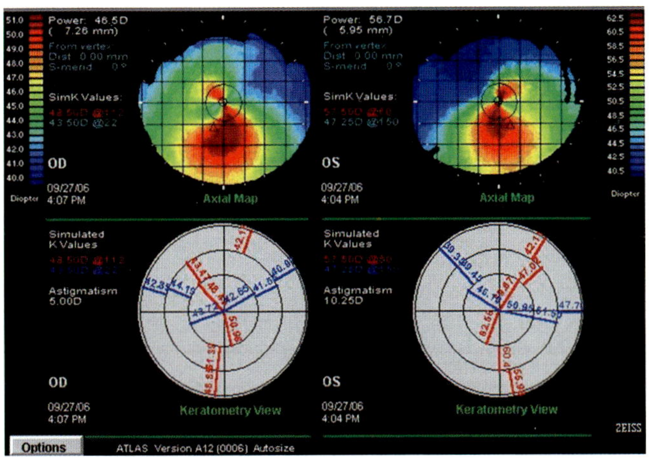

Fig. 18.2 Irregular astigmatism should generally not be treated with incisional surgery

In addition, the phacoemulsification incision must be factored into the preoperative corneal astigmatism to determine the final postoperative residual cylinder. Using vector analysis, the surgeon should determine the proper axis to perform an astigmatic incision. All of these factors can be calculated online at www.LRIcalculator.com (Fig. 18.3). The online LRI calculator uses vector analysis to calculate where to make LRI incisions based on preoperative patient keratometry and the surgeon's induced astigmatism; it also uses the Donnenfeld and Nichamin nomogram and provides a visual map of the axis and length of incisions that should be performed. A printout of the LRI calculator can be brought to the operating room and used as a guide when marking the cornea and performing LRIs.

An LRI is performed on all patients who, based on topography readings and surgical incision, are likely to end up with visually significant cylinder. Surgeons who make their incisions superiorly need to be aware that additional against-the-rule cylinder will be induced. For a patient who has 0.5 D of against-the-rule cylinder, it would be appropriate to perform an LRI at 180° preoperatively. Conversely, for a patient who has pre-existing 0.5 D of with-the-rule cylinder, this astigmatism will be corrected by the surgical technique of a superior incision. For oblique astigmatism, a vector analysis of the preoperative astigmatism and incision will yield the correct axis and magnitude of cylinder to be corrected.

As with any surgical procedure, there are potential complications associated with LRIs, but most are either temporary or correctable. The procedure is generally not associated with glare or starbursts, as may be seen with radial keratotomy or astigmatic keratotomy. The potential problems with LRIs include overcorrection, undercorrection, infection, perforation of the cornea, decreased corneal sensation, induced irregular astigmatism, and discomfort. For patients with significant remaining astigmatism, it may be necessary to retreat by redeepening or extending the LRI. For overcorrections, we recommend waiting and then later cleaning out the wound with a Sinskey hook and suturing it with 10-0 nylon if necessary. For smaller overcorrections, an excimer laser photoablation may be employed. We never recommend placing LRIs perpendicular to the original LRIs for consecutive cylinder, as this may induce irregular astigmatism. If the cornea is perforated, it may be self-sealing or a suture may be needed.

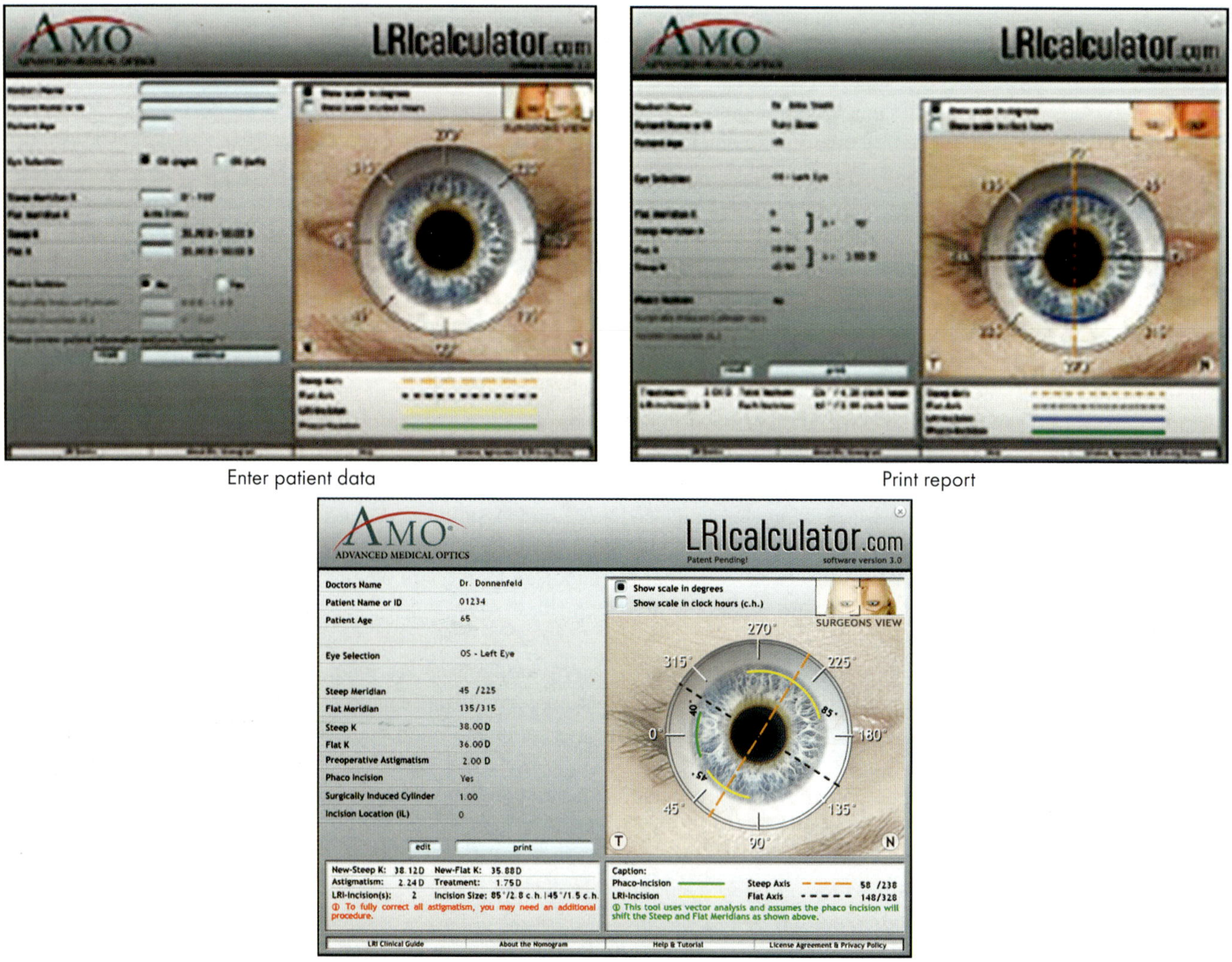

Fig. 18.3 The Donnenfeld Nomogram is available on the internet at: www.LRIcalculator.com

When a surgeon transitions from cataract to refractive IOL surgery, it is important to pay attention to detail. One of the most important concepts is to treat small residual refractive errors. Performing LRIs is a major step that many surgeons will take to become good refractive cataract surgeons.

Recent developments in femtosecond laser technology have shifted attention from manual LRI and astigmatic keratotomy procedures to femtosecond laser-guided procedures. These procedures offer a greater degree of precision and accuracy than manual methods.[24] The treatment of astigmatism during cataract surgery with femtosecond laser astigmatic incisions is an interesting development in femtosecond technology, which is positioned to revolutionize refractive cataract surgery.

Femtosecond lasers are photodisruptive lasers with extraordinarily short pulse durations of less than 800 femtoseconds (one femtosecond is 10^{-15} seconds). This extremely short pulse duration allows femtosecond lasers to cut tissue with considerably less energy than traditional lasers used in ophthalmic surgery. Per-pulse energies can be reduced approximately 1,000-fold, from around 1^{-10} millijoules for nanosecond lasers (neodymium-doped yttrium aluminum garnet [Nd:YAG]) to 1^{-10} microjoules for femtosecond lasers. These reductions in perpulse energy result in substantial reductions in collateral tissue damage, shifting from a few millimeters from the intended target for nanosecond lasers to just a few microns for femtosecond lasers. Similar to the Nd:YAG laser systems, femtosecond laser pulses pass through transparent tissues and can be focused at a predetermined depth.[25]

Femtosecond laser devices offer superior incision/cut accuracy and reproducibility, along with minimal collateral tissues effects, achieving levels that exceed those of highly sophisticated mechanical devices. Femtosecond lasers for ophthalmic surgery have been commercially available since 1999. The precision and extremely limited collateral effects of femtosecond pulses on the cornea have been established in more than 4 million procedures performed using femtosecond lasers for the creation of LASIK flaps and lamellar and penetrating keratoplasties.[26-30]

The US Food and Drug Administration has approved 510(k) clearance for the use of femtosecond lasers for clinical applications in cataract surgery. The most remarkable features of laser cataract surgery include the precision of the performed incisions with minimal collateral damage to surrounding tissues, accuracy of capsulotomy dimensions and diameters, enhancement of capsular edge strength, reduction of phaco power in lens fragmentation, and improved wound sealing and healing–all of which lead to better and more reproducible results compared with manual, mechanical surgeries.

A major clinical application of the femtosecond laser is in creating arcuate incisions that have precise and accurate arc length, depth, angular position, and optical zone. The femtosecond laser allows for precise and repeatable incisions, which are necessary for consistent results not normally achieved through manual methods.[31] In a case study, femtosecond laser arcuate keratotomy was used to correct astigmatism. The refraction of the patients' eyes was measured along with the corneal thickness. Then, the keratometric parameters were evaluated, and the femtosecond laser created the incisions, the length and position of which were calculated individually, per eye. OCT-controlled corneal pachymetry was performed directly in the area of the intended incisions and then programmed into the laser, which allows for extremely high levels of precision. The study results showed that with the femtosecond laser, astigmatism could be better corrected with an improved best-corrected visual acuity than was predicted in the preoperative values. Furthermore, the depth and location of the incisions were consistent with the surgical plan. Laser technology is able to create uniform corneal incisions precisely and predictably.[32]

Femtosecond lasers allow for more efficacious and safer surgical procedures. The corneal tissue does not absorb the laser wavelength. The photodisruption dissipates within 100 Ěm of the target, millimeters away from the cornea, which allows for a higher margin of safety because a sizeable distance is kept from Descemet's membrane, preventing perforation. Moreover, the laser can be programmed to create an ideal wound shape for enhanced sealing and healing of the incisions with reproducible induction of astigmatism that cannot be achieved with the use of manual keratomes. Femtosecond laser incisions provide superior reproducibility and reduced variability compared with conventional manual incisions.[33] In a study of intial results with the LenSx femtosecond laser, using 9 mm arcuate incisions and a 33 percent reduction of the Donnenfeld nomogram, a 70 percent reduction in astigmatism was achieved.[34] Femtosecond laser incisions can also be placed intrastromally in the sub-Bowman's layer of the cornea, which improves healing by sparing corneal epithelial damage. Further clinical investigation and nomogram development are currently underway to optimize this method, which would eliminate the need for corneal wound manipulation on the surface.

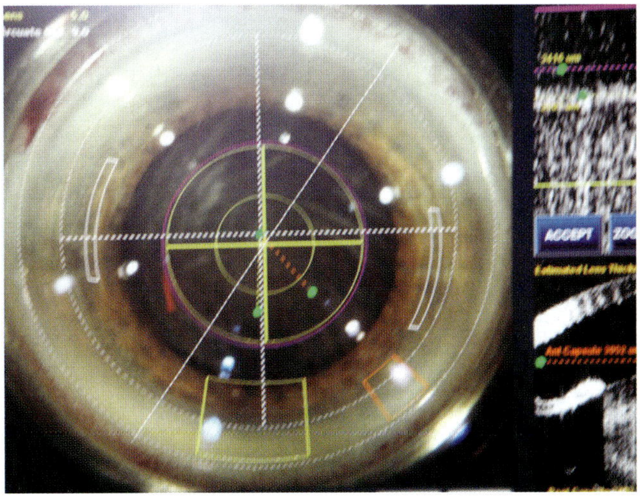

Fig. 18.4 Overlay of the arcuate incisions

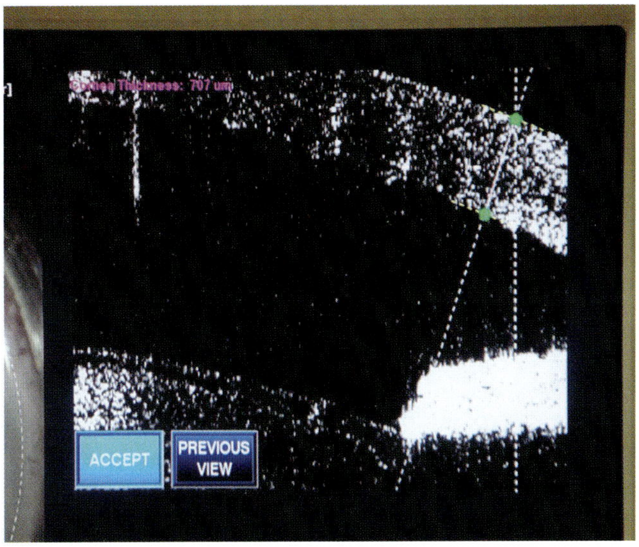

Fig. 18.5 OCT imaging of the cornea in the area of the arcuate incision

The surgical technique for performing femtosecond laser arcuate incisions first involves determining the length, posi-

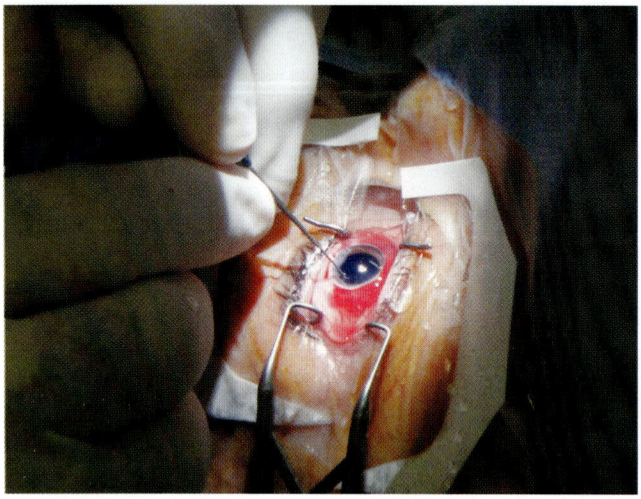

Fig. 18.6 Sinsky hook used to open the femtosecond arcuate incisions

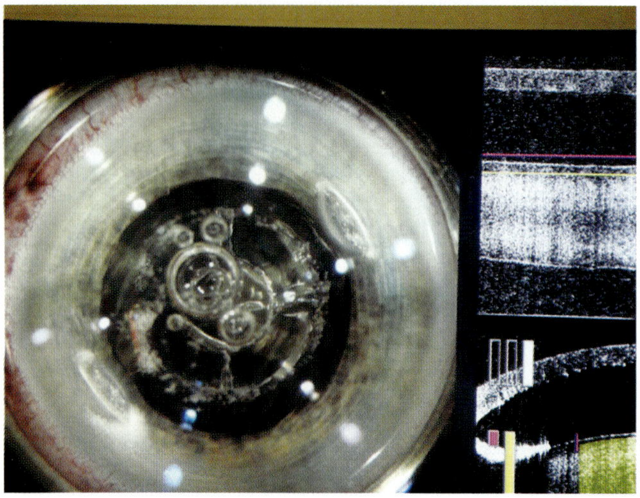

Fig. 18.7 Femtosecond arcuate incisions being performed

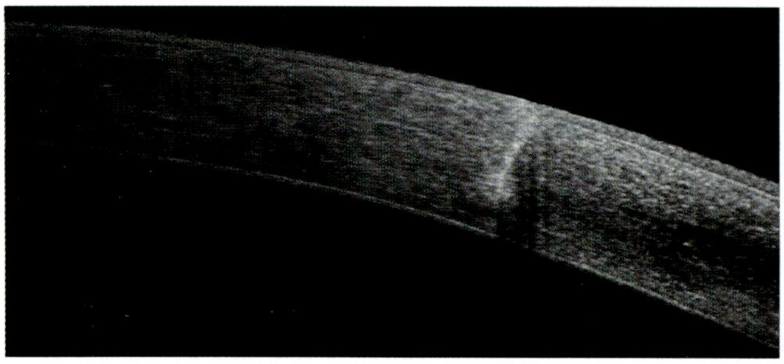

Fig. 18.8 OCT demonstrating the depth of the arcuate incision

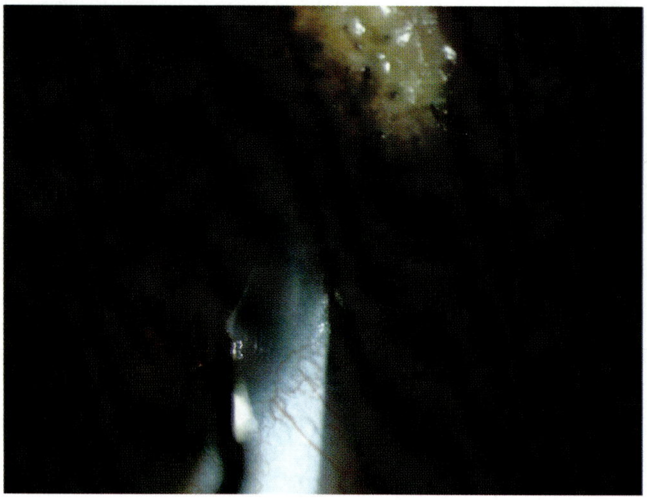

Fig. 18.9 Forcep used to open the femtosecond arcuate incison at the slit-lamp

tion, depth, and distance from the visual axis where the incisions will be created. We use a 33 percent reduction of the Donnenfeld nomogram and use the LRI calculator (www.lri-calculator.com) to determine the length and axis at which the incision should be placed (Fig. 18.3). We preset the depth of our incisions to 85 percent of the corneal pachymetry in the area of the incision. We have set our distance from the visual axis at 9 mm. This information is all downloaded onto the femtosecond laser (LenSx). We then begin the surgical procedure by docking the laser onto the cornea. An overlay of the incisions is then visible on the surgical screen (Fig. 18.4). OCT imaging of the cornea in the area of the arcuate incision is then visualized, and the depth is confirmed (Fig. 18.5). We first perform the capsulotomy, followed by the lens disruption, and then create our corneal incisions. Following the conclusion of the femtosecond laser treatment, the patient is brought to the operating microscope and the incisions are opened with a Sinskey hook (Fig. 18.6). Intraoperative aberrometry (Orange; WaveTec Vision, Aliso Viejo, CA) may be then performed to titrate the opening of the incisions. The incisions are symmetric and standardized at 9 mm from the visual axis (Fig. 18.7). OCT confirms the postoperative depth of the incisions (Fig. 18.8). If needed, the arcuate incisions may be opened postoperatively in the office at the slit lamp using forceps or a Sinskey hook and topical anesthetic (Fig. 18.9).

A major advantage of femtosecond arcuate incisions is that the refractive incisions may now be made before cataract surgery and then modified intraoperatively and/or postoperatively. The incisions are performed with the femtosecond laser but do not have significant effect until they are opened. To further refine our results, we have been performing intraoperative aberrometry (ORange) to titrate our results in the operating room. We remove the cataract, place the IOL, and open one of the femtosecond incisions with a Sinskey hook. The IOP is then raised to approximately 25 mm Hg. Next, we perform intraoperative aberrometry, which provides a very accurate reading of the existing astigmatism. The second femtosecond incision may then be opened partially or completely, based on the intraoperative aberrometry reading, which, if needed, can be taken again. For surgeons who do not have access to intraoperative aberrometry, patients can be examined with topography and refraction, performed the day after surgery. If needed, the remainder of the incision can be easily opened in the office to increase the effect of the incision and adjust the residual astigmatic refractive error.

In conclusion, the creation of femtosecond laser-assisted arcuate incisions is a novel technique that provides the precision of image-guided laser technology. Refractive incisions are now computer-controlled and do not rely on surgeon skill or experience. The use of a femtosecond laser system will provide faster, safer, easier, customizable, adjustable, and fully repeatable astigmatic incisions. Removing the inconsistencies in the astigmatic procedure will improve our understanding and accuracy of astigmatic incisions and should provide improved refractive results and patient satisfaction. The ability to perform intrastromal ablations cannot be achieved with manual incisions. This technology offers limitless potential but will require development with different nomograms and surgeons as well as expanded clinical experience.

REFERENCES

1. Pager CK, McClusky PJ, Retsas C. Cataract surgery in Australia: a profile of patient-centered outcomes. Clin Exp Ophthalmol. 2004;32:388-92.
2. Hawker MJ, Madge SN, Baddeley PA, Perry SR. Refractive expectations of patients having cataract surgery. J Cataract Refract Surg. 2005;31:1970-5.
3. Nichamin LD. Nomogram for limbal relaxing incisions. J Cataract Refract Surg. 2006;32(9):1048.
4. Ferrer-Blasco T, Montés-Micó R, Peixoto-de-Matos SC, González-Méijome JM, Cervino A. Prevalence of corneal astigmatism before cataract surgery. J Cataract Refract Surg. 2009;35(1):70-5.
5. Hill W. Expected effects of surgically induced astigmatism on AcrySof toric intraocular lens results. J Cataract Refract Surg. 2008;34(3):364-7.
6. Duffey RJ, Leaming D. US trends in refractive surgery: 2008 ISRS/AAO survey. International Society of Refractive Surgery & American Academy of Ophthalmology. Atlanta, Georgia. 8 November 2008. ISRS meeting.
7. Nichamin LD. Astigmatism control. Ophthalmol Clin North Am. 2006;19(4):485-93.
8. Wang L, Misra M, Koch DD. Peripheral corneal relaxing incisions combined with cataract surgery. J Cataract Refract Surg. 2003;29(4):712-22.
9. Budak K, Friedman NJ, Koch DD. Limbal relaxing incisions with cataract surgery. J Cataract Refract Surg. 1998;24(4):503-8.
10. Gills JP. Treating astigmatism at the time of cataract surgery. Curr Opinion Ophthalmol. 2002;13(1):2-6.
11. Oshika T, Shimazaki J, Yoshitomi F, et al. Arcuate keratometry to treat corneal astigmatism after cataract surgery: a prospective evaluation of predictability and effectiveness. Ophthalmology. 1998;105(11):2012-6.
12. Maloney WF, Grindle L, Sanders D, Pearcy D. Astigmatism control for the cataract surgeon: a comprehensive review of surgically tailored astigmatism reduction (STAR). J Cataract Refract Surg. 1989;15(1):45-54.
13. Price FW, Grene RB, Marks RG, Gonzales JS; for the ARC-T Study Group. Astigmatism reduction clinical trial: a multicenter prospective evaluation of the predictability of arcuate keratotomy. Arch Ophthalmol. 1995;113(3):277-82.
14. Devgan U. Corneal correction of astigmatism during cataract surgery. Cataract and Refractive Surgery Today. January 2007.pp41-4.
15. Tejedor J, Murube J. Choosing the location of corneal incision based on preexisting astigmatism in phacoemulsification. Am J Ophthalmol. 2005;139(5):767-76.
16. Kaufmann C, Peter J, Ooi K, Phipps S, Cooper P, Goggin M; for The Queen Elizabeth Astigmatism Study Group. Limbal relaxing incisions versus on-axis incisions to reduce corneal astigmatism at the time of cataract surgery. J Cataract Refract Surg. 2005;31(12):2261-5.
17. Muller-Jensen K, Fischer P, Siepe U. Limbal relaxing incisions to correct astigmatism in clear corneal cataract surgery. J Refract Surg. 1999;15(5):586-9.
18. Akura J, Matsuura K, Hatta S, Otsuka K, Kaneda S. A new concept for the correction of astigmatism: full-arc depth-dependent astigmatic keratotomy. Ophthalmology. 2000;107(1):95-104.
19. Faktorovich EG, Maloney RK, Price FW Jr. Effect of astigmatic keratotomy on spherical equivalent: results of the Astigmatism Reduction Clinical Trial. Am J Ophthalmol. 1999;127(3):260-9.
20. Price FW, Grene RB, Marks RG, Gonzales JS. Astigmatism reduction clinical trial: a multicenter prospective evaluation of the predictability of arcuate keratotomy. Evaluation of surgical nomogram predictability. ARC-T Study Group. Arch Ophthalmol. 1995;113(3):277-82. [Erratum in: Arch Ophthalmol. 1995;113(5):577.]
21. Oshika T, Shimazaki J, Yoshitomi F, et al. Arcuate keratotomy to treat corneal astigmatism after cataract surgery: a prospective evaluation of predictability and effectiveness. Ophthalmology. 1998;105(11):2012-6.
22. Bradley MJ, Coombs J, Olson RJ. Analysis of an approach to astigmatism correction during cataract surgery. Ophthalmologica. 2006;220(5):311-6.

23. AcrySof Toric SA60T4IOL [package insert]. Fort Worth, TX: Alcon Laboratories, Inc.; 2009.
24. Masket S, Sarayba M, Ignacio T, Fram N. Femtosecond laser-assisted cataract incisions: architectural stability and reproducibility. J Cataract Refract Surg. 2010;36(6):1048-9.
25. Vogel A, Noack J, Hüttman G, Paltauf B. Mechanisms of femtosecond laser nanosurgery of cells and tissues. Appl Phs B. 2005;81:1015-47.
26. Ratkay-Traub I, Ferincz IE, Juhasz T, Kurtz RM, Krueger RR. First clinical results with the femtosecond neodymium-glass laser in refractive surgery. J Refract Surg, 2003;19(2):94-103.
27. Nordan LT, Slade SG, Baker RN, Suarez C, Juhasz T, Kurtz RM. Femtosecond laser flap creation for laser in situ keratomileusis: six-month follow-up of initial U.S. clinical series. J Refract Surg. 2003;19:8-14.
28. Kezirian GM, Stonecipher KG. Comparison of the IntraLase femtosecond laser and mechanical keratomes for laser in situ keratomileusis. J Cataract Refract Surg. 2004; 30(4):804-11.
29. Montés-Micó R, Rodríquez-Galietero A, Alió JL. Femtosecond laser versus mechanical keratome LASIK for myopia. Ophthalmology. 2007;114:62-8.
30. Steinert RF, Ignacio TS, Sarayba MA. "Top hat"-shaped penetrating keratoplasty using femtosecond laser. Am J Ophthalmol. 2007;143(4):689-91.
31. Nichamin L. Femtosecond laser technology applied to lens-based surgery. Medscape Ophthalmology. June 22, 2010. http://www.medscape.com/viewarticle/723864. Accessed July 20, 2011.
32. Nagy Z, Takacs A, Filkom T, Sarayba M. Initial clinical evaluation of an intraocular femtosecond laser in cataract surgery. J Refract Surg. 2009;25(12):1053-60.
33. Abbey A, Ide T, Kymionis GD, Yoo SH. Femtosecond laser-assisted astigmatic keratotomy in naturally occurring high astigmatism. Br J Ophthalmol. 2009;93(12):1566-9.
34. Slade SG. Femtosecond laser arcuate incision astigmatism correction in cataract surgery. Presented at: the ASCRS Cornea Day; March 25, 2011; San Diego, CA.

19

An Update on Femtosecond Laser Refractive Surgery

Sharad Lakhotia (India)

Chapter Outline

- Technology
- Complication
- Astigmatism Correction
- Presbyopia Correction
- Intracor Procedure
- All Femtosecond Refractive Correction
- Complications
- Intralase FS laser
- Step-by-Step IntraLase Flap Creation
- Suction Loss within Raster Pattern
- Flap Complications

The growing role of femtosecond laser in refractive surgery is undisputable. The most common and important role of femtosecond laser in corneal refractive surgery is in making corneal flaps. Femtosecond is an ultrafast laser generating pulses as short as one quadrillionth of a second (10^{-15}). The world femto came from Danish word for number 15.

However all femtosecond refractive procedure are also available. Femtosecond laser has been used successfully in intrastormal astigmatic incision and presbyopia correction. Newer role of femtosecond cataract surgery is the most recent advancement in technology. More and more usage of femtosecond laser are expected in glaucoma, retina and cornea surgery.

INTRODUCTION

Femtosecond laser were first in limelight when bladeless flap creation in lasik was achieved by intralase femtosecond laser in USA.

Why Bladeless Flap is Better

Corneal flap created by femtosecond laser is better than microkerotome for the following reasons as seen in literature:

It shows better contrast sensitivity.

There is increased safety in comparison to flap related complication of microkerotome.

Faster unaided visual acuity postoperatively.

There is less induced aberration because of smooth flap.

There is less IOP variation during making of flap.

There is lesser incidence of dry eyes following lasik.

This is probably because femtosecond laser makes a thin flap with uniform geometry with greater predictability. This is clearly evident in optical coherence tomography, which reveals a very uniform flap.

TECHNOLOGY

All femtosecond lasers use a wave length of approximately 1053 nm. The main technical specifications are (i) Laser pulse repetition rate (ii) Spot size (iii)Pulse pattern.

Femtosecond laser produce photo disruption and optical break down. This creates plasma and a cavitation bubble is formed, which produces tissue cleavage. A shorter pulse (200–500fs) requires lower energy than longer pulse (500–1300fs) to reach threshold of photo disruption. The diameter and volume of laser spot is guided by numerical aperture of the lens. Higher numerical aperture use lower energy and increase depth and accuracy of lamellar cut.

If high energy photo disruption is used, the pulse need not be placed closed together as the bubble is large to produce tissue cleavage. However if low energy disruption is used than a greater number of pulse are placed in an overlapping pattern as the very small bubble does not induce tissue cleaving.

Thus an ideal device should have small spot size, a high repetition rate and low energy per pulse.

Two main patterns are raster and spiral.

In raster pattern pulse are applied in a liner pattern starting from the hinge and ending to the opposite edge. In spiral pattern, pulse begin centrally and proceed centrifugally. A

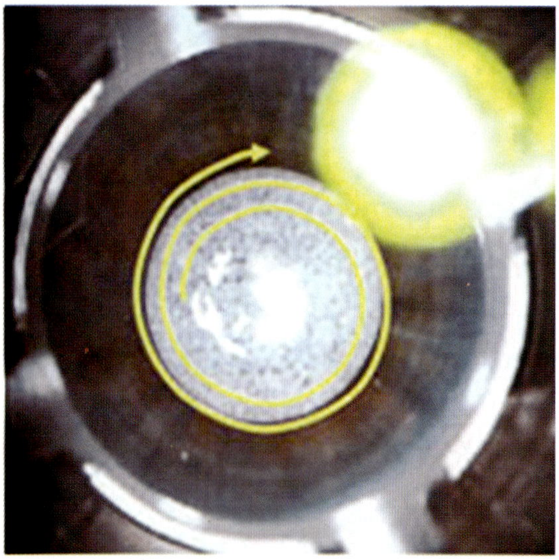

Fig. 19.1 Intralase spiral pattern

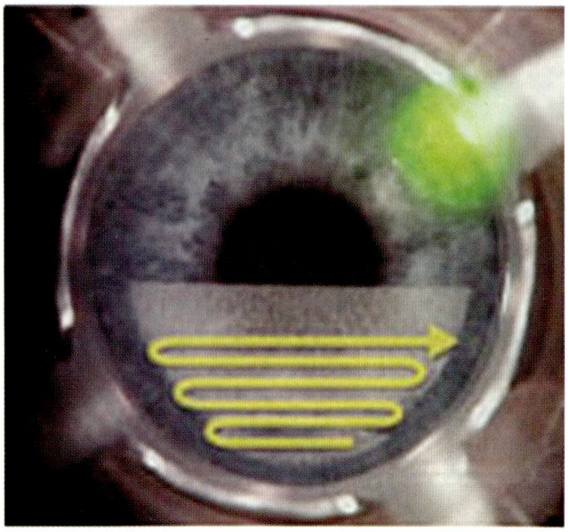

Fig. 19.2 Intralase raster pattern

centripetal approach is also available. Most devices use a raster pattern which ensures smooth stromal bed, e.g. intralase machine (AMO). However Visumax (Carl Zeiss Meditec AG, Jena, Germany) relies on spiral pattern.

There is a difference in suction ring and docking system amongst different machines. The devices which applanate the cornea induce higher pressure as is seen in intralase (AMO), Wavelight Ultraflap and LDV (Zeiss Ophthalmic Systems, Switzerland). Curved applanation docking surface induce lower pressure as seen in visumax and femtec (Technolas Perfect vision,Germany) lasers.

Commercially available femtosecond lasers with features:

IntraLase (AMO) achieved initial success when they first launched their femtosecond laser in USA in 2001. Since then, they have constantly upgraded in its fifth generation. Laser pulse rate has increased from 15-30-60-150–kHz allowing procedures to be faster. The side cut angle can be increased beyond 90 degrees thus allowing even an inverted bevel in edge. The increased geometry of cuts allows intra stromal ring placement and shaped corneal transplant.

The Visumax and Technolas laser used curved docking system, thus inducing lower pressure. Visumax, in addition, uses limbal suction mechanism instead of conjunctival suction mechanism. It has 500 kHz pulse and lower energy profile. It can thus also be used in femtosecond only refractive procedures like FLEx (femtosecond lenticular extraction) and SmILE (small incision lenticular extraction). Because the patient retain vision during the procedure, they actively co-operate throughout the procedure.

The LDV from Ziemer is small and portable femtosecond laser. It has very high pulse rate and low energy, causes very less bubble formation and ensures a smooth corneal bed. It has no vertical side cuts, so the flap making is similar to microkeratome with tapered flap edge. Because of tapered edge, the flap can be irregular if air is present. During application of pulse, the procedure is not visible and also an interface fluid is needed, which are some of its limitations. It can also perform intrastormal ring channels and lamellar keratoplasty. Femtec laser works similar to intralase in terms of pulse and energy. However it has a curve applanation docking system which induces less pressure. Correction of Presbyopia by Intracor is studied on this machine.

Alcon's new wavelight FS200 ultraflap laser is similar to intralase in technical specification. Specific cutting geometries are created by forming externalized channel to allow gas bubble to diffuse out of cornea. Thus less opaque bubble formation is expected and excimer laser can be done immediately in most of the cases.

COMPLICATION

Opaque Bubble Layer

During cutting by femtosecond laser some bubble escape into deeper stroma forming opaque bubble layer. They may take few hours to disappear and may also impair eye-tracking. It is less common in softer applanation devices which induces less pressure, lower energy and faster repetition rate of pulses. Prior to excimer laser treatment, density of OBL can be lessened by gently scraping the surface with a spatula.

Interface Haze

When very thin flaps of less than 100 microns are made, there can be interface haze as a possible complication.

Dense Interface Haze

Photophobia and mild pain can appear days after surgery and may persists for weeks. This is transient light sensitivity syndrome which was common in earlier versions because of high raster energy and lower numerical aperture optics. Another problem of earlier versions is rainbow glare. It is because of light scattering from perfect array of laser spots on the back surface of corneal flap. It is visual impact is not seen much in most cases.

ASTIGMATISM CORRECTION

In high degree of astigmatism as in post keretoplasty and also in high naturally occurring astigmatism, femtosecond laser astigmatic keretotomy is effective. In post keretoplasty astigmatism, its safe and more accurate than manual technique. Femtosecond laser can also create intrastromal-only astigmatic incisions which are safe and stable but may not be very effective. The technology for astigmatism is still evolving.

PRESBYOPIA CORRECTION

Intracor procedure increases depth of focus to combat presbyopia. Non invasive intralenticular femtosecond laser pulse can restore some accommodation in crystalline lens by increasing flexibility and sliding of lens fibers. The main fear to this approach is formaton of cataract and reduction in best corrected visual acuity. Studies shows development of pin point opacities at the site of laser interaction but there is no progressive cataract formation. The main positive feature of this technique is minimal invasiveness, negligible infection rate and possibility of performing a refractive lens exchange surgery, if it fails.

INTRACOR PROCEDURE

It is an intrastromal procedure done on Technolas Femtosecond work station. It applies energy inside the cornea without bringing it to the surface. There is no incision of epithelium, endothelium or Bowman's or Descemet's membrane and thus ensure better healing with minimal risk of infection. The pulses are placed on concentric intrastromal circles centered about visual axis and extended at least up to 100 microns from the surface. The concentric patterns of cut fibers shift the center of cornea slightly anteriorly and create a hyperprolate shape. At present myopia up to 3D and astigmatism up to 2D have been tried. However the results are not very accurate. Also its role in presbyopia has been emphasized, as it causes a biomechanical change in cornea that shifts center slightly forward creating a pattern of hypersphericity, thus allowing some near vision while retaining distance vision.

The surgeon can expand circle diameter or add radial intrastromal incision similar to those created in radial Keratotomy and is effective in biomechanically correcting small degree of myopia.

INTRACOR brings hope for correcting presbyopia with low refractive errors by biomechanical methods.

Visumax Femtosecond Laser (Carl Zeiss Meditec AG, Jena).

ALL FEMTOSECOND REFRACTIVE CORRECTION

ReLEx or Refractive Lenticule Extraction is a completely new technique for all femtosecond laser vision correction. Refractive Lenticule Extraction or ReLEx is performed with the Carl Zeiss Meditec VisuMax femtosecond laser. There are two types of ReLEx procedures–FLEx and SmILE.

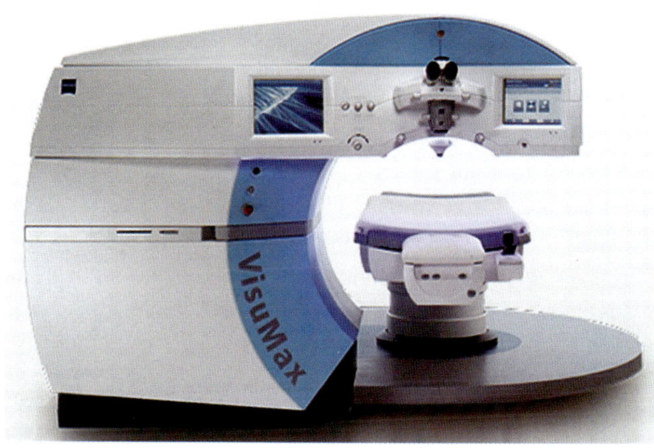

Fig. 19.3 VisuMax femtosecond laser

Femtosecond lenticule extraction or FLEx using VisuMax (Carl Zeiss Meditec, Jena Germany) have reported excellent results for total refractive correction.

Dr Walter Sekundo and Dr Marcus Blum were first surgeon to perform this procedure in August 2008. Over 1000 eyes have been operated at all centers by this procedure including one at Baroda (India) besides 3 other international sites. In FLEx treatment, Femtosecond laser creates two cuts, a refractive and a non-refractive as a single step. The first cut is made at the bottom of the refractive lenticle while the second one at its roof. Once the cuts are made, flap is lifted and refractive lenticule is removed. The flap is reposited in usual manner. It is important to make manipulation at correct plane between flap and lenticule and separate lenticule edge.

The lenticule can be of different diameters from 6 mm to 7.3 mm equivalent to optical zone of Eximer laser. The remaining stromal zone is usually 300 micron or more.

Walter Sekundo reported good results in 6 months follow up. Nine percent gained 2 lines of visual acuity, 43 percent gained one line 39 percent remain unchanged and 7 percent lost one line visual acuity. Seventy-four percent patients were 6/6 or better.

Dr Rupal Shah reported slow visual recovery at 1 week but changing scanning pattern produced good results. While scanning in spirals, they used to cleave posterior surface of the lenticule from center to periphery. The flap was then cleaved from periphery to center. The changed direction, i.e. posterior surface of lenticule was separated by scanning from periphery to center and anterior surface from center to periphery, led to dramatic improvement in visual recovery. Eighty-nine percent reached to preoperative BCVA after 1 week. With upgraded 500 Hz VisuMax, 95 percent reached or improved to pre-operation BCVA level at 1 week.

Some of the investigators are doing a modified version called small incision lenticular extraction or SmILE. This technique makes the procedure less invasive where by the entire lenticule can be extracted through a small incision (3-5 mm) without lifting up the flap.

Osama Ibrahim reported 90 percent of patients having better than 20/25 visual acuity after one month. He is of the opinion that results with this procedure are more stable than Lasik including high myopia patients. If the lenticule is dissected above than it is very easy to identify inferior surface by having enough resistance and empty spaces. Once dissected, it is very easy to lift flap by simple pull with forceps or spatula.

Journey from FLEx to SmILE is easy and has advantage of small incision thus providing all femtosecond laser solution to refractive problems.

Initial results with hyperopia suggest FLEx can be a good option. There was some regression as with Eximer laser but data at 1 month interval was great. In hyperopia, the flap is just 0.5 mm larger in diameter then the lenticule. The main point for consideration is that lenticule should not have sharp edge other wise it will result in regression. So the main challenge is to find an optimal form for lenticule edge.

COMPLICATIONS

They were minimal. Mild transient haze like reaction and microstria were reported which cleared. There were some problems in separating the refractive lenticule in 2 cases, of which one resolved after 6 months. However no patient showed signs of diffuse lamellar keratitis or transient light sensitivity syndrome.

Benefits of VisuMax System

While designing femtosecond laser a balance was kept between increasing pulse rate and focusing power and decreasing spot size and energy. The optics of VisuMax produce tightly focused beam with following effects.

i. The accuracy of pulse placement is optimized.
ii. Tissue dissection is carried out with low energy.
iii. Energy drop off from point of impact is steep thus causing less collateral tissue damage and hence faster recovery.

Some femtosecond lasers require scleral suction and flat applanation. High suction in such cases can cause discomfort and subconjunctival hemorrhages. In VisuMax the rise is intraocular pressure is low enough for patients to see and does not discomfort the patient. A curved contact glass and corneal suction allows IOP below 90 or 100 mg. In one study rise in IOP during flap creation was 84.9 ± 7.3 mm Hg for VisuMax, 180.6 ± 21.6 mmHg for intralase and 150.9 ± 17.2 mm Hg for LDV. Thus anxiety level of patient is less, patient can visualize fixation light and so better fixation is achieved during treatment.

Flap centration on corneal vertex is well achieved as it was intended for FLEx procedure. The vertex of cornea fit well in to vertex of contact glass of VisuMax femtosecond laser. The problem of opaque bubble layer is very thin as it dissipates very rapidly.

Intralase femtosecond have largest market share globally so we will discuss about Intralase technology and usage in bladeless flap making in detail.

Each pulse of focussed laser light lasts approximately 500 to 800 femtoseconds. In one second light travels 7.5 times around globe and in 100 femtosecond light travels across a human hair. As power is equivalent to Energy divided by Time, extremely high power is attained at a relatively low energy.

A sharply focused femtosecond laser pulse generates plasma. The plasma expands at a supersonic velocity displacing surrounding tissues and cools to form cavitation bubble. To get the effect of a surgical cutting instrument, thousands of individual laser pulse must be placed side by side to get incisional plane in cornea. These planes can be horizontal, vertical or oblique. The bubble is put in from deeper to superficial manner (insideout). The side cut of the flap can be created by stacking the pulses vertically or at an angle till they reach corneal surface. The hinge can be created in superior, nasal and temporal positions and hinge width can also be programmed.

The Intralase FS laser system uses sterile disposable sterile IntraLase FS patient interface consisting of suction ring assemblies and applanation lenses.

The normal ocular hazard distance is 54 cm. Fifty-eight centimetre and fifty-three centimetre for 15 kHz, 30 kHz, and 60 kHz lasers respectively.

Ideally temperature in the room should be between 18 °C and 24 °C. Relative humidity should be between 30 and 40 percent. The operative room should be preferably free from dust, particulate contamination and free of vibrations.

INTRALASE FS LASER

Laser console house laser source, control and power supply, electronics, cooling system and computer. The beam delivery device contains optics and laser components to precisely focus laser beam in the cornea. A green light is initiated by upper ward movement of applanation cone indicating eye contact.

When maximum contact pressure is reached, a red LED is visible in operating microscope and on vide display and audible tone reactivated. Joysticks control X, Y, Z movement of interface. Twisting joystick moves interface up or down.

Control panel contains surgical illumination knob, cone illumination knob and home button. The home button elevates the beam delivery device upward in 1.62 inch (10 sec) increments to predetermined point so patient interface can be removed safely. The loading deck assembly in located at the output of the Beam Delivery Device and accepts the applanation cone.

A flat panel color display, keyboard and trackball pointing device are mounted on a platform. It gives view of system status, patient logs, procedure data, use registry, live video display of microscope view. Laser resection parameters are entered, controlled and monitored through this.

Disposable patient interface consists of applanation cone, limbal suction ring and syringe. The suction ring fixes eye during the procedure.

The ring expands to receive the applanation cone and contracts to grip it firmly.

Docking is achieved by positioning Beam Delivery Device using joystick to couple the applanation cone in Suction ring assembly.

Soft docking is an advanced docking technique where an area of cornea is left unapplanated creating a meniscus.

This helps reduce incidence of opaque bubble layer. The intralase flaps are created in two parts resection—one is horizontal lamellar Cut and other is side Cut. A pocket is an optional resection which is used as a vent for bubble to evacuate.

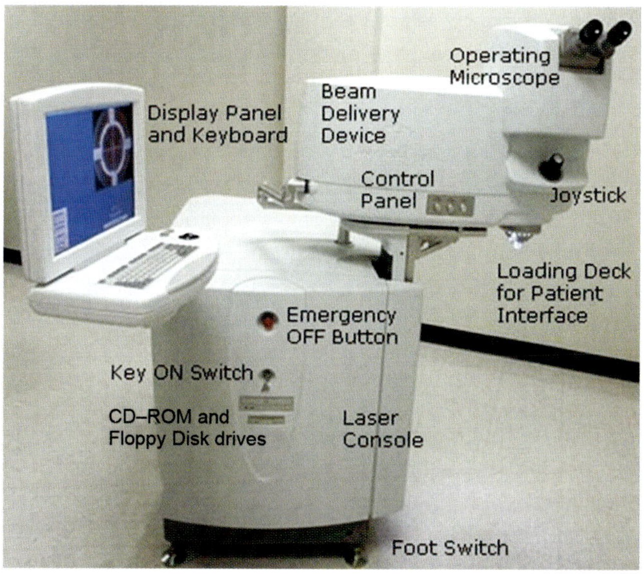

Fig. 19.4 IntraLase FS laser

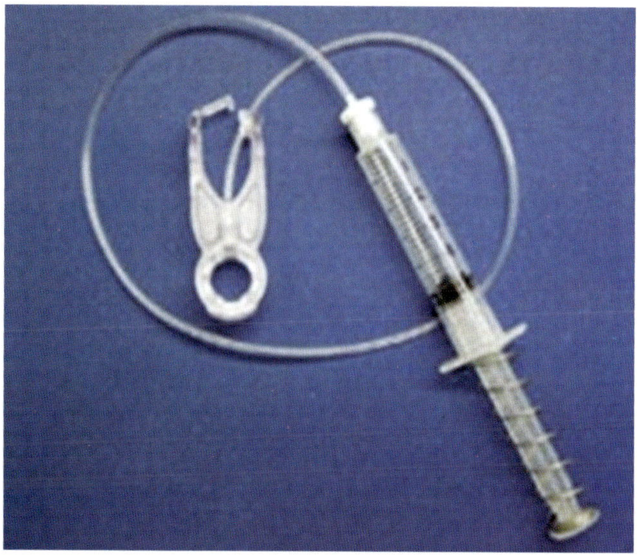

Fig. 19.5

Fig. 19.6 Applanation cone

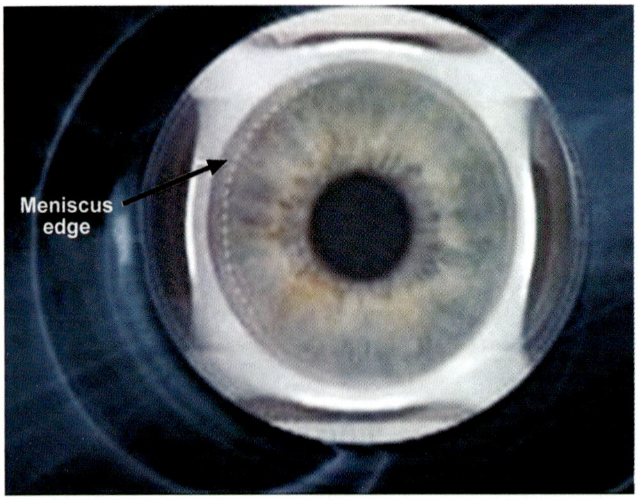

Fig. 19.7

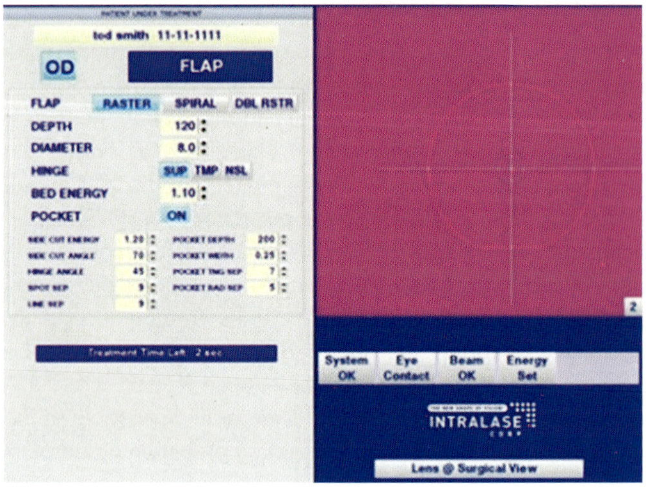

Fig. 19.8

For horizontal cut two patterns are available. A raster pattern where resection starts at hinge and a spiral pattern which starts at the center and moves peripherally.

STEP BY STEP INTRALASE FLAP CREATION

1. Patient interface contents are placed on sterile field.
2. The applanation cone is installed. The applanation glass is inspected for any scratches or defects prior to installation.
3. The magnification dial is set to position 2 which gives 15 mm field of view.
4. Suction ring assembly is connected to syringe. The syringe plunger is fully depressed. The suction ring is properly centered on the cornea and assistant slowly releases the plunger, allowing ring to firmly affix to the eye.
5. Select treat button. Once the patient is docked and green light is achieved, laser will automatically check the system. Once Treat is selected, one has 5 minutes to achieve green light.
6. The applanation cone is lowered and docked into suction ring using joystick.
7. Final adjustment to pattern centration may be done by clicking arrows to desired direction and to 'Adjust Params.'
8. The flap creation is initiated by stepping on Footswitch. The procedure is visible through microscope and video display.
9. On completion remove foot from footswitch. Release suction by depressing the plunger. Raise Beam Delivery Device. Release deck locking arm and remove applanation cone.
10. *Flap lift technique:* From the side cut, open a portion with a sinskey hook. The flap lifter or spatula is inserted through the opened portion of side cut sweeping 1/3rd of lamella starting at hinge. The same maneuver is repeated for the second 1/3rd of lamella and so on until the flap is dissected fully. The flap is then reflected up.

SUCTION LOSS WITHIN RASTER PATTERN

If a suction loss occurs, immediately one should disengage foot pedal. Select cancel on menubox. Retreat immediately after re-establishing suction and applanation, using same cone and same settings.

One can consider reducing planned diameter or increasing spot/line separation to decrease procedure time. If suc-

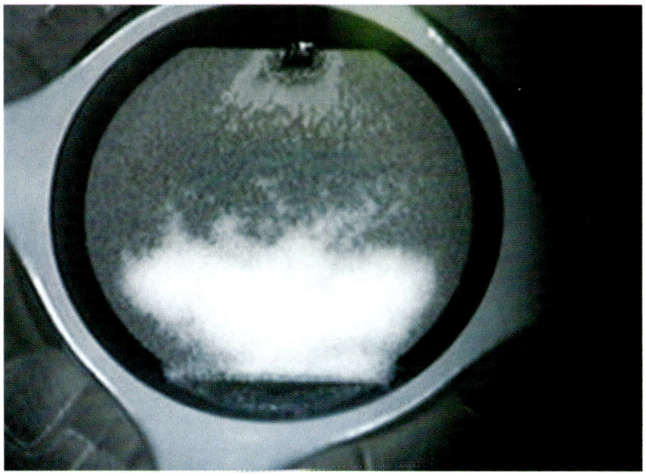

Fig. 19.9 Vertical gas breakthrough

tion loss occurs second time, the assistant is asked to pull on plunger adding vacuum during the procedure.

If it occurs third time, abort the procedure and post pone. Subsequent procedure on such patient should use a different depth of at least 40 micron from original depth.

If the suction loss occurs during side cut, the same procedure is adopted but on subsequent day then increase the attempted depth 10 micron for new side cut to ensure reaching depth of previous bed.

If Vertical Gas Breakthrough occurs, stop the procedure. Do not wait for side cut. Attempt to lift vertical Breakthrough may create buttonhole. In eyes with known previous trauma and surgery to Bowman's membrane, recut atleast 40 micron deeper.

Horizontal Gas Breakthrough appears as irregular or incomplete ablation pattern that appear watering. Adjusting the pocket setting can eliminate watering appearances. It is less likely to interface flap creation as compared to vertical gas breakthrough.

iFS Advanced Femtosecond Laser System (Upgrade)

The iFS laser user interface can be navigated through keyboard and touch screen monitor.

Elliptical flaps can be created. In a standard round ablation, if hinge encroaches on ablation zone, one has to either increase the diameter of the flap or decrease hinge angle. In elliptical flap option, one can simply increase the horizontal diameter peripherally and minimizing the vertical diameter, thus allowing larger hinge angle. The larger hinge angle increases flap stability and also cause less dry eyes by preserving corneal nerve.

Another advantage is that it can increase side cut angle from 30-90-150 degrees. Thus it is possible to have a bevel in side cut which provides better biomechanical strength compared to other angles.

Besides there is improved Accuracy Docking. Green light indicates contact of applanation lens and suction ring. Yellow light indicates stop downward Z travel and adjust meniscus by pinching suction ring. Red light indicates end of downward travel. Besides it prevents doctors having to turn 90 degrees during surgery with improved ergonomics.

FLAP COMPLICATIONS

The following are possible complications resulting from lamellar flap resection using femtosecond laser :

- Corneal edema
- Corneal pain
- Epithelial ingrowth
- Flap de-centration
- Incomplete flap creation
- Thin or thick flaps
- Flap tearing or incomplete lift-off
- Free cap
- Photophobia
- Corneal inflammation.

20

Femtosecond Lasers in Cataract Surgery: Designer Cataract Surgery

Arun C Gulani, Ming X Wang (USA)

Chapter Outline

- Gulani "Cataract Category Classification"

Millions of cataract surgeries are performed throughout the world every year and it is called a "Routine" procedure, I would like to propose and propel the thought that No Cataract surgery should be ROUTINE.

"Cataract surgery is an endeavor to help patients see what they have been missing and also an opportunity to plan for the best vision they can have, for the rest of their Life"- Gulani AC, MD.

With such an outlook, I would encourage ophthalmologists to classify every cataract patient into four categories and then individually plan to address all of their correctable visual issues using customized cataract surgery with Premium Lens Implants and/or staged Laser Vision Surgery.

GULANI "CATARACT CATEGORY CLASSIFICATION"

(Gulani AC. XXXII World Congress of Ophthalmology, Berlin, June, 2010):

1. Cataracts with Associated Refractive errors (i.e. Myopia, Hyperopia, astigmatism, presbyopia)
2. Cataracts with Previous Eye Surgery (i.e. RK, LASIK. etc)
3. Cataracts with Associated Pathology (i.e. Fuchs, Corneal scar, Keratoconus, etc.)
4. Enhancing Previous Cataract Surgery (i.e. Residual ref error, haloes, etc.)

Raise this bar further by applying one of the most sophisticated and accurate surgical technologies that makes technique, precision and outcomes reproducible.

Welcome the Femtosecond Laser technology in Cataract surgery in truly designing cataract surgery with tailored visual outcomes.

New generation patient populations have come to expect and believe that perfect vision is a possibility of cataract surgery and that too, at all distances without any glasses.

Add to this an ever-growing population of previous refractive surgery patients now transitioning to becoming cataract patients with the same mentality of perfect vision expectations and minimalistic surgical approaches, and we as eye surgeons need an armamentarium of techniques, combinations and technologies to help us live up to these expectations (Gulani AC: Multifocal IOL and Laser ASA Combination. Subspecialty meeting of AAO, Fl, Dec 2011)

Femtosecond laser has enjoyed wide range of application in anterior segment surgery in recent years and it is principally due to the specific requirement of the laser-tissue interaction and the unique solution that an ultrashort-pulsed laser such as femtosecond laser provides.

In order to achieve precise tissue separation in the cornea and lens, the laser needs to achieve sufficient power and yet needs to "stay cool" by not producing any collateral thermal damage. One of the ingenious ways to achieve this is to use a laser with an extremely short pulse duration, so that one can use very low energy (microjoules) and still be able to achieve high laser power (10^{7-9} watts) to accomplish corneal or lenticular tissue separation. Since power P (watts) = energy

E (joules) / pulse duration t (second), a femtosecond-pulsed ($10^{-13\text{-}15}$ seconds) laser with an energy level of 10^{-6} joules can indeed produce $10^{7\text{-}9}$ watts of laser power to cut the tisse. At such a low level of energy of 10^{-6} joules, there is minimal collateral thermal damage thereby one can achieve the principal goal of microscopic cornea and lens surgery, i.e., precise tissue cutting at any depth in any shape while laser causes no thermal damage to the target tissue or the tissue that it transmits through along the laser pathway.

Since all common types of femtosecond lasers all generate laser light in the infrared range of wavelength, they all have the ability to penetrate deep into a target tissue. Hence, to precisely control the depth of corneal and lenticular cut, one has to rely upon collimating a divergent laser beam and achieving precise laser focusing. The tight laser focusing generates sufficient number of photos per unit square area to produce instantaneous ionization of large number of electrons and generate a plasma wave which then separates the adjacent tissue. In recent years, there has been significant improvement in the technology of laser manufacturing to produce lasers with stable ultrashort laser pulses and in the technology of optical devices to achieve precise laser focusing of these short-pulsed laser beams. The successful development in these two areas are the key reasons for the rapid rise in popularity of femtosecond laser in anterior segment surgery in the last few years.

Symbiotic technologies that assist in patient selection, surgical technique planning and outcome evaluations have also advanced to plan and predict surgical endpoints with accuracy and consistency (Gulani AC, Pineda R, Ghaith A, ElKateb M: Corneal Topography from A To Z. International Forum of Ophthalmology, Cairo University, Egypt. Dec, 2010 (Figs 20.1 to 20.4).

We have described Femto Laser cataract surgery applications at all levels of the procedure in previous book chapters and summarize to say that the creation of femtosecond laser-assisted incisions including astigmatic keratotomies provide the precision of image-guided laser technology. Refractive incisions are now computer-controlled and do not rely on surgeon skill or experience. Capsulorhexis, nucleus cleavage and capsular bag preparation will allow variations just like we have with Femto LASIK flaps. The use of a femtosecond laser system will eventually provide faster, safer, easier, customizable, adjustable, and fully repeatable surgical techniques. This technology offers limitless potential but will require development with different nomograms and surgeons as well as expanded clinical experience.

There are five areas in which future improvement in femtosecond laser lens technology can further its application, widespread acceptance and surgical outcomes:

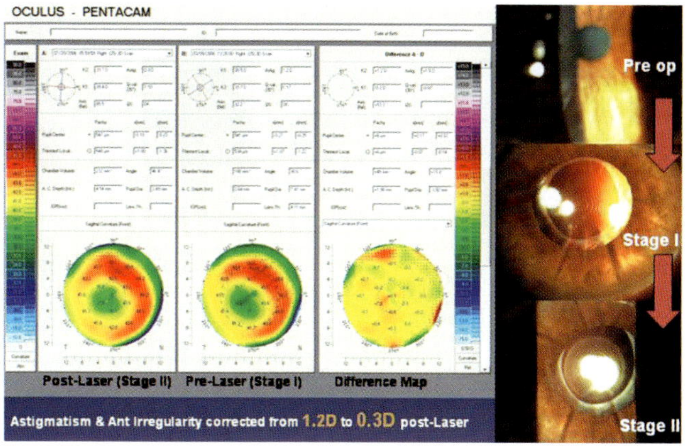

Fig. 20.1 GulaniRKReStor1

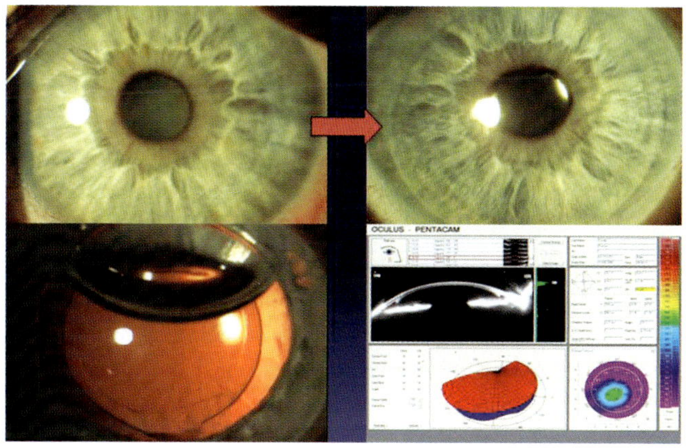

Fig. 20.2 Gulani DSAEKTORIC

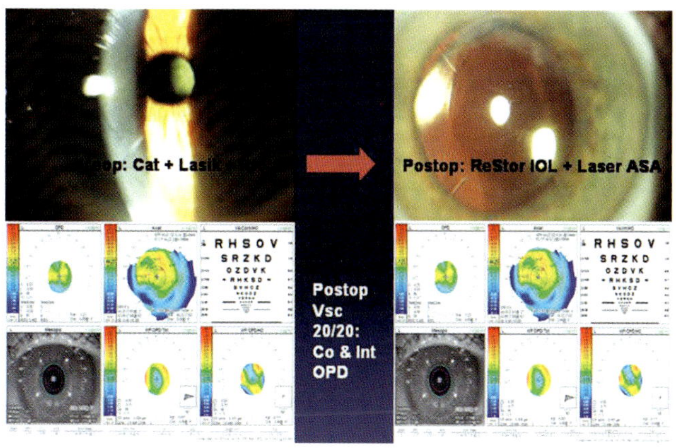

Fig. 20.3 GulaniReStorLasikRK

First, tissue separation via plasma wave generation does produce more inflammation compared with mechanical devices (microkeratome, as in the case of LASIK and chopper in the case of cataract surgery). When tissue is being torn apart

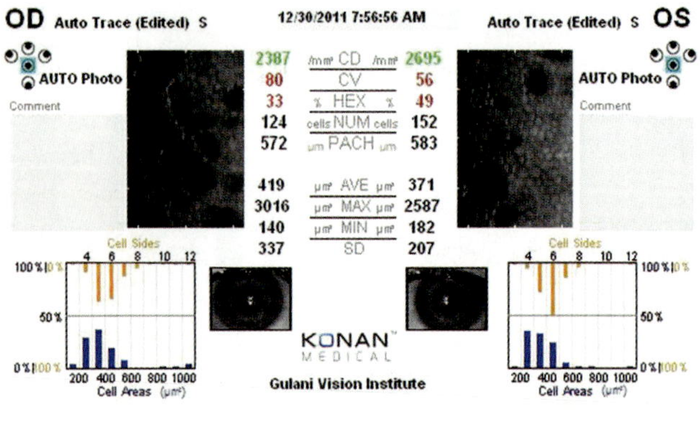

Fig. 20.4 Gulani cataract chapter

by gas waves such a plasma wave rather than clean mechanical blade cut, the tissue edges are rougher and there is more local trauma and inflammation in the tissue. This problem can be continuously reduced by a number of ways, including devising better laser tissue separation planes that follow more naturally the existing corneal or lenticular lamellar planes and developing better optics to achieve tighter laser focusing in a sub-micron space. Currently the plasma wave cavitation bubbles are on the order of 2-5 um and if we can reduce that to under one micron the smoothness of tissue cut will significantly improve and inflammation reduced.

Secondly, lenticular application of femtosecond laser such as that for cataract surgery is intrinsically more challenging than corneal application such as that for LASIK, due to the need to achieve a much deeper level of tissue penetration in the former. The deeper the laser goes, the larger the plasma wave cavitation bubble size and hence the lesser precise the tissue cut and also more collateral tissue damage. One can get a sense of this by comparing the interface smoothness of a 100 um versus 200 um corneal flaps in LASIK. Therefore femtosecond laser technologies need to continue improve to reduce the degradation of precision of focusing as a function of depth of laser-tissue interaction.

Thirdly, unique to lens surgery, the laser beam has to traverse a large volume of aqueous in the anterior chamber which partially absorbs the light. In fact, the most variable and inconsistent areas of laser-tissue interaction in the current generation of femtosecond laser cataract surgery technologies is the lack of consistency and reproducibility of capsulotomy cut. In about half of the surgeries, despite of maximally allowable laser energy setting, the tightest laser focusing volume and closest spot separation, there are still areas of "tags" of anterior capsule which are not cut by laser. Given the increased difficulty of visualization of capsulotomy edge when laser is used (due to the disturbances of cortex tissue), when compared with manual technique, and the significant implication of any incomplete cut or radial tear to the success of the entire cataract surgery itself, improvement in the precision and particularly completeness of capsulotomy with femtosecond laser represents one of the most important areas of improvement in femtosecond cataract technology in upcoming years.

Fourthly, current femtosecond laser cannot produce sufficient cut for very dense nucleus, due to the fact the lens nucleus partially blocks laser beam. This problem can be solved in upcoming years with designs of new femtosecond lasers with even shorter pulse durations (in sub-femtosecond) so the laser power (in watts) can be larger than the current level of $10^{7\text{-}8}$ watts. If laser power can truly one day exceed 10^9 watts, lens fragmentation should be much readily achieved even with 4+ nucleus density.

Finally, currently the incisions produced by femtosecond laser is still not satisfactory. We need to not only improve the consistency of tissue separation, but also and most importantly we need to achieve what laser should theoretically be able to uniquely achieve while a human hand could not, i.e., multiplane incision which does not require tissue hydration to seal the wound at the end of a case.

These above mentioned five areas represent the platform for improvement for greater adaptation of femtosecond cataract surgery technology among eye surgeons as we collectively enter the era of designer cataract surgery with multidimensional vision.

Removing the inconsistencies in the surgical procedure will improve our understanding and accuracy and should provide improved refractive results and patient satisfaction.

The ability to perform intrastromal corneal ablations with real time aberrometry for a truly combined keratolenticular refractive surgery is a dream of a very near future.

This technological advance may be the ship we have been waiting for to sail forward in our quest for supervision and leave behind memories of 'Routine" surgery to welcome the era for "Designer" cataract surgery.

ACKNOWLEDGMENT

Ms Aaisha Gulani

Research Assistant: Gulani Vision Institute

BIBLIOGRAPHY

1. Bansal J, Gulani AC. Excimer Laser Enhancements After IOL Surgery. SLACK Inc,Thorofare NJ, (2012-In press).
2. Colin J, et al, Ophthalmology, 2001;108:1409–14.
3. Colin J, Simonpoli S, J Fr Ophtalmo, 2005;28(2):205–17.

21 Femtosecond Laser for Astigmatism Correction During Cataract Surgery

Eric D Donnenfeld, Arun C Gulani (USA)

Chapter Outline

- Gulani "Cataract Category Classification"
- Femtosecond Laser Astigmatic Incisions in Cataracts

Cataract surgery is one of the most commonly performed procedures worldwide and has become a refractive procedure in which patients have increasingly high expectations for both functional visual improvement as well as spectacle independence.[1,2]

New generation of patients transitioning from being LASIK candidates to cataract patients have a new set of expectations and demands. As eye surgeons of this modern era we need to prepare, adapt and also consistently deliver to these expectations.

In the cataract category classification below, the most common "Residual" refractive error, i.e. Astigmatism, falls under category I.

GULANI "CATARACT CATEGORY CLASSIFICATION"

(Gulani AC. XXXII World Congress of Ophthalmology, Berlin, 2010:

1. Cataracts with Associated Refractive errors (i.e. Myopia, Hyperopia, Astigmatism, Presbyopia).
2. Cataracts with Previous Eye Surgery (i.e. Radial Keratotomy, LASIK, etc.).
3. Cataracts with Associated Pathology (i.e. Fuchs dystrophy, Corneal scar, Keratoconus, etc.).
4. Enhancing Previous Cataract Surgery (i.e. Residual ref error, haloes, etc.).

Astigmatism in cataract surgery has been approached with an array of preoperative, intra-operative and postoperative modalities in our continued quest for an emmetropic outcome which is a base requirement for patient satisfaction.

This desire is only fueled by the rapid acceptance of premium lens implants.

Excimer laser enhancements are a predictable and accurate way of correcting astigmatism as a postoperative application in premium lens implant cases.

Laser Vision Correction Enhancements to Premium IOL cases have been classified by Dr Gulani as follows:

1. Planned Laser combination surgeries.
2. Laser Enhancement for unexpected, visually significant refractive errors.
3. Laser surgery to correct complications.

Since most astigmatism associated with routine cataracts is actually in the low to moderate category *(Excimer Laser Enhancements After IOL Surgery: Bansal J, Gulani AC. SLACK Inc, Thorofare NJ, - In press)* and intraoperative astigmatic keratotomy is a convenient way to correct the same, what if we had technology that could produce consistent cuts and therefore predictable outcomes?

Technology today has answered that question in the form of femtosecond laser cataract surgery.

FEMTOSECOND LASER ASTIGMATIC INCISIONS IN CATARACTS

Management of astigmatism is the single most important aspect of the ability to provide good surgical outcomes, quality of vision, and satisfaction for patients requesting refractive cataract surgery and optimal postoperative uncorrected visual acuity. Astigmatism decreases visual acuity through meridional blur, in which one axis of the cornea is steeper than the other, causing distortion of the visual image. Residual astigmatism after cataract surgery of 0.50 diopters (D) or even less may result in glare, symptomatic blur, ghosting, and halos.[3] As a result, greater emphasis has been placed on treating corneal astigmatism at the time of cataract surgery with limbal relaxing incisions (LRIs). A recent study of 4,540 eyes of 2,415 patients showed that corneal astigmatism is present in the majority of patients undergoing cataract surgery, with at least 1.50 D measured in 22.2 percent of study eyes.[4] Approximately 38 percent of eyes undergoing cataract surgery have at least 1.00 D of pre-existing corneal astigmatism, and 72 percent of patients have 0.5 D or more.[5]

Astigmatism management is important regardless of the IOL that is used. Patients implanted with presbyopia-correcting IOLs, however, are significantly more sensitive to even minor refractive errors than those implanted with conventional monofocal IOLs. Refractive IOL surgeons must be willing and able to treat small amounts of postoperative astigmatism in order to achieve happy postoperative patients.

Astigmatic incisions are corneal incisions that are used to relax and flatten the steep axis of regular corneal astigmatism while steepening the flat axis. The procedure reduces corneal tension and allows the cornea to heal into a more spherical shape (Figs 21.1A and B). Incisions placed adjacent to the limbus are LRIs, and incisions placed further from the limbus and closer to the visual axis are astigmatic keratotomies. Because there are several advantages of LRIs over astigmatic keratotomies, LRIs are performed more frequently. These advantages include a reduced tendency to cause axial shift, less irregular astigmatism, a 1:1 coupling ratio, a decreased chance of visually significant scarring, and a reduced likelihood of perforation.

Unfortunately, the majority of cataract surgeons are uncomfortable performing manual incisional astigmatic incisions. In a survey of 233 surgeons, only 73 routinely performed LRIs.[6] In general, irregular astigmatism (Fig. 21.2) should not be treated with LRIs. There are several different options for treating astigmatism, including excimer laser photoablation, toric IOL implantation, and conductive keratoplasty. These options, however, may not be available to all cataract surgeons and patients. Correction of pre-existing corneal

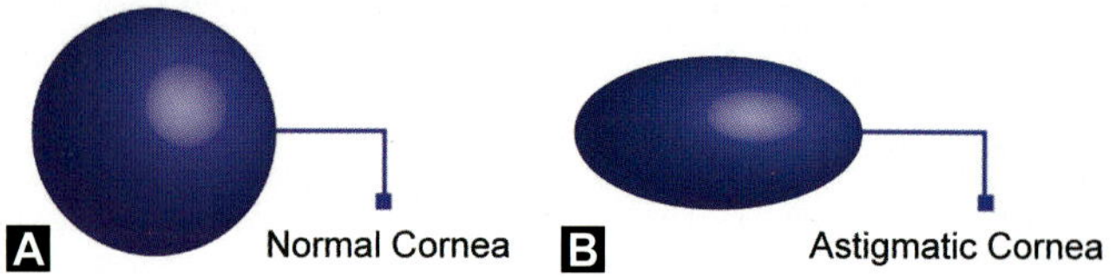

Figs 21.1A and B In a normal cornea (A), shaped like a basketball, both axes are equal; in an astigmatic cornea (B), shaped like a football, one axis is steeper than the other

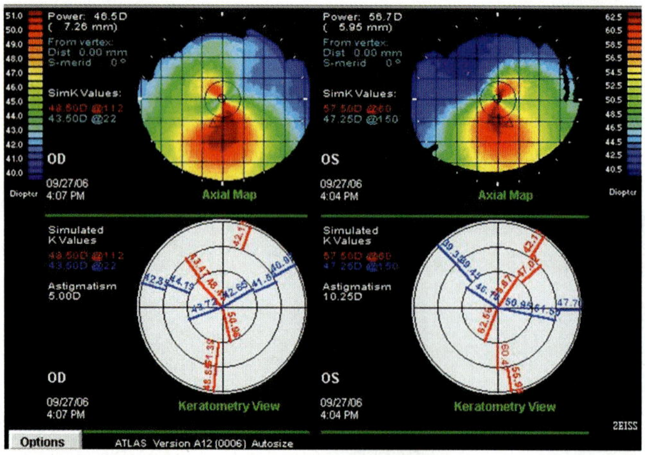

Fig. 21.2 Irregular astigmatism should generally not be treated with incisional surgery

astigmatism is not a covered benefit of Medicare and most other insurance plans. Therefore, the patient who is the beneficiary of a noncovered service has the option to pay for the procedure and the additional physician time and testing that accompanies it.

Once the required location of the incision has been ascertained, the incision length and depth are determined by the surgeon. Sometimes, a nomogram may be consulted. There are a number of LRI nomograms for correcting small amounts of cylinder, and many studies evaluating LRIs have been performed.[7-21] In one study on the effectiveness of LRIs, there was a 60 percent average reduction of cylinder,[22] with 79 percent of patients corrected to less than 1.00 D of cylinder and 59 percent corrected to less than 0.5 D of cylinder. The 60 percent reduction in cylinder compares favorably with the results achieved using toric IOLs, which result in a mean 58.4 percent reduction in cylinder.[23]

Many LRI nomograms are adjusted for age and cylinder axis, making them detailed and complex and giving the impression that the procedure is extremely precise and unforgiving. In my opinion, however, this simply is not the case. The LRI procedure has numerous chances for error, and the many variables that help determine how the incision is performed can each be measured incorrectly. The incision itself, if performed manually, is only as precise as the surgeon making the incision and the accuracy of the blade. Any of these errors can compound and result in suboptimal visual acuity. Manual LRIs are therefore as much an art as a science.

In addition, the phacoemulsification incision must be factored into the preoperative corneal astigmatism to determine the final postoperative residual cylinder. Using vector analysis, the surgeon should determine the proper axis to perform an astigmatic incision. All of these factors can be calculated online at www.lricalculator.com (Fig. 21.3). The online LRI calculator uses vector analysis to calculate where to make LRI incisions based on preoperative patient keratometry and the surgeon's induced astigmatism; it also uses the Donnenfeld and Nichamin nomogram and provides a visual map of the axis and length of incisions that should be performed. A printout of the LRI calculator can be brought to the operating room and used as a guide when marking the cornea and performing LRIs.

An LRI is performed on all patients who, based on topography readings and surgical incision, are likely to end up with visually significant cylinder. Surgeons who make their incisions superiorly need to be aware that additional against-the-rule cylinder will be induced. For a patient who has 0.5 D of against-the-rule cylinder, it would be appropriate to perform an LRI at 180° preoperatively. Conversely, for a patient who has preexisting 0.5 D of with-the-rule cylinder, this astigma-

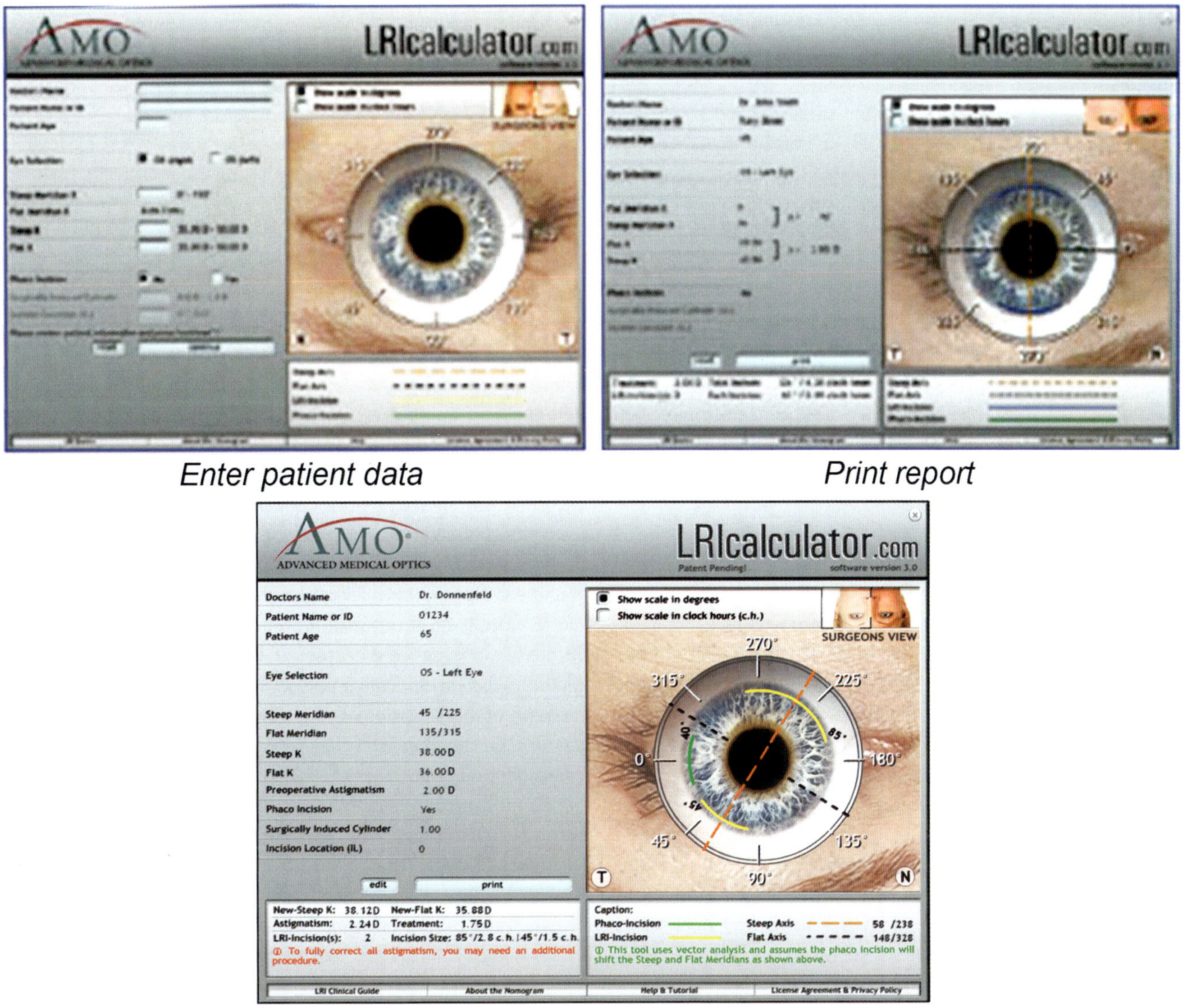

Fig. 21.3 The Donnenfeld Nomogram is available on the Internet at: www.lricalculator.com

tism will be corrected by the surgical technique of a superior incision. For oblique astigmatism, a vector analysis of the preoperative astigmatism and incision will yield the correct axis and magnitude of cylinder to be corrected.

As with any surgical procedure, there are potential complications associated with LRIs, but most are either temporary or correctable. The procedure is generally not associated with glare or starbursts, as may be seen with radial keratotomy or astigmatic keratotomy. The potential problems with LRIs include overcorrection, undercorrection, infection, perforation of the cornea, decreased corneal sensation, induced irregular astigmatism, and discomfort. For patients with significant remaining astigmatism, it may be necessary to retreat by redeepening or extending the LRI. For overcorrections, we recommend waiting and then later cleaning out the wound with a Sinskey hook and suturing it with 10-0 nylon if necessary. For smaller overcorrections, an excimer laser photoablation may be employed. We never recommend placing LRIs perpendicular to the original LRIs for consecutive cylinder, as this may induce irregular astigmatism. If the cornea is perforated, it may be self-sealing or a suture may be needed.

When a surgeon transitions from cataract to refractive IOL surgery, it is important to pay attention to detail. One of the most important concepts is to treat small residual refractive errors. Performing LRIs is a major step that many surgeons will take to become good refractive cataract surgeons.

Recent developments in femtosecond laser technology have shifted attention from manual LRI and astigmatic keratotomy procedures to femtosecond laser-guided procedures. These procedures offer a greater degree of precision and accuracy than manual methods.[24] The treatment of astigmatism during cataract surgery with femtosecond laser astigmatic incisions is an interesting development in femtosecond technology, which is positioned to revolutionize refractive cataract surgery.

Femtosecond lasers are photodisruptive lasers with extraordinarily short pulse durations of less than 800 femtoseconds (one femtosecond is 10^{-15} seconds). This extremely short pulse duration allows femtosecond lasers to cut tissue with considerably less energy than traditional lasers used in ophthalmic surgery. Per-pulse energies can be reduced approximately 1,000-fold, from around 1^{-10} millijoules for nanosecond lasers (neodymium-doped yttrium aluminum garnet [Nd:YAG]) to 1^{-10} microjoules for femtosecond lasers. These reductions in per-pulse energy result in substantial reductions in collateral tissue damage, shifting from a few millimeters from the intended target for nanosecond lasers to just a few microns for femtosecond lasers. Similar to the Nd:YAG laser systems, femtosecond laser pulses pass through transparent tissues and can be focused at a predetermined depth.[25]

Femtosecond laser devices offer superior incision/cut accuracy and reproducibility, along with minimal collateral tissues effects, achieving levels that exceed those of highly sophisticated mechanical devices. Femtosecond lasers for ophthalmic surgery have been commercially available since 1999. The precision and extremely limited collateral effects of femtosecond pulses on the cornea have been established in more than 4 million procedures performed using femtosecond lasers for the creation of LASIK flaps and lamellar and penetrating keratoplasties.[26-30]

The US Food and Drug Administration has approved 510(k) clearance for the use of femtosecond lasers for clinical applications in cataract surgery. The most remarkable features of laser cataract surgery include the precision of the performed incisions with minimal collateral damage to surrounding tissues, accuracy of capsulotomy dimensions and diameters, enhancement of capsular edge strength, reduction of phaco power in lens fragmentation, and improved wound sealing and healing—all of which lead to better and more reproducible results compared with manual, mechanical surgeries.

A major clinical application of the femtosecond laser is in creating arcuate incisions that have precise and accurate arc length, depth, angular position, and optical zone. The femtosecond laser allows for precise and repeatable incisions, which are necessary for consistent results not normally achieved through manual methods.[31] In a case study, femtosecond laser arcuate keratotomy was used to correct astigmatism. The refraction of the patients' eyes was measured along with the corneal thickness. Then the keratometric parameters were evaluated, and the femtosecond laser created the incisions, the length and position of which were calculated individually, per eye. OCT-controlled corneal pachymetry was performed directly in the area of the intended incisions and then programmed into the laser, which allows for extremely high levels of precision. The study results showed that with the femtosecond laser, astigmatism could be better corrected with an improved best-corrected visual acuity than was predicted in the preoperative values. Furthermore, the depth and location of the incisions were consistent with the surgical plan. Laser technology is able to create uniform corneal incisions precisely and predictably.[32]

Femtosecond lasers allow for more efficacious and safer surgical procedures. The corneal tissue does not absorb the laser wavelength. The photodisruption dissipates within 100 μm of the target, millimeters away from the cornea, which allows for a higher margin of safety because a sizeable distance is kept from Descemet's membrane, preventing perforation. Moreover, the laser can be programmed to create an ideal wound shape for enhanced sealing and healing of the incisions with reproducible induction of astigmatism that cannot be achieved with the use of manual keratomes. Femtosecond laser incisions provide superior reproducibility and reduced variability compared with conventional manual incisions.[33] In a study of initial results with the LenSx femtosecond laser, using 9 mm arcuate incisions and a 33 percent reduction of the Donnenfeld nomogram, a 70 percent reduction in astigmatism was achieved.[34] Femtosecond laser incisions can also be placed intrastromally in the sub-Bowman's layer of the cornea, which improves healing by sparing corneal epithelial damage. Further clinical investigation and nomogram development are currently underway to optimize this method, which would eliminate the need for corneal wound manipulation on the surface.

The surgical technique for performing femtosecond laser arcuate incisions first involves determining the length, position, depth, and distance from the visual axis where the incisions will be created. We use a 33 percent reduction of the Donnenfeld nomogram and use the LRI calculator (www.lricalculator.com) to determine the length and axis at which the incision should be placed (Fig. 21.3). We preset the depth of our incisions to 85 percent of the corneal pachymetry in the area of the incision. We have set our distance from the visual axis at 9 mm. This information is all downloaded onto the femtosecond laser (LenSx). We then begin the surgical procedure by docking the laser onto the cornea. An overlay of the incisions is then visible on the surgical screen (Fig. 21.4). OCT imaging of the cornea in the area of the arcuate incision is then visualized, and the depth is confirmed (Fig. 21.5). We first perform the capsulotomy, followed by the lens disruption, and then create our corneal incisions. Following the conclusion of the femtosecond laser treatment, the patient is brought to the operating microscope and the incisions are opened with a Sinskey hook (Fig. 21.6). Intraoperative aberrometry (Orange; WaveTec Vision, Aliso Viejo, CA) may then be performed to titrate the opening of the incisions. The incisions are symmetric and standardized at 9 mm from the visual axis (Fig. 21.7). OCT confirms the postoperative depth of the incisions (Fig. 21.8). If needed, the arcuate incisions may be opened postoperatively in the office at the slit lamp using forceps or a Sinskey hook and topical anesthetic (Fig. 21.9).

A major advantage of femtosecond arcuate incisions is that the refractive incisions may now be made before cataract surgery and then modified intraoperatively and/or postoperatively. The incisions are performed with the femtosecond laser but do not have significant effect until they are opened. To further refine our results, we have been performing intraoperative aberrometry (ORange) to titrate our results in the operating room. We remove the cataract, place the IOL, and open one of the femtosecond incisions with a Sinskey hook. The IOP is then raised to approximately 25 mm Hg. Next, we perform intraoperative aberrometry, which provides a very accurate reading of the existing astigmatism. The second femtosecond incision may then be opened partially or completely, based on the intraoperative aberrometry reading, which, if needed, can be taken again. For surgeons who do not have access to intraoperative aberrometry, patients can be examined with topography and refraction, performed the day after surgery. If needed, the remainder of the incision can be easily opened in the office to increase the effect of the incision and adjust the residual astigmatic refractive error.

In conclusion, the creation of femtosecond laser-assisted arcuate incisions is a novel technique that provides the pre-

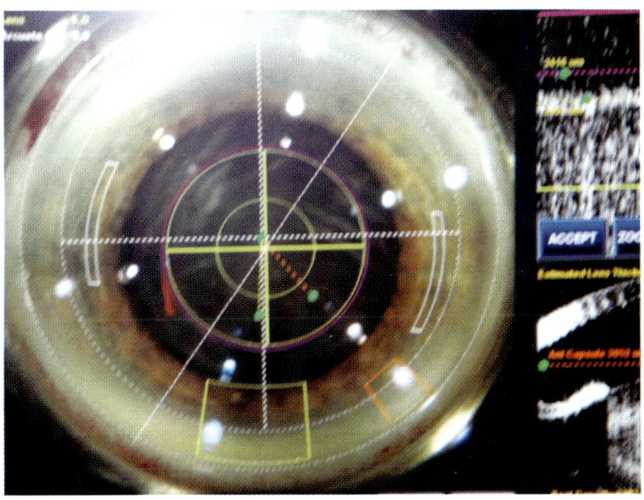

Fig. 21.4 Overlay of the arcuate incisions

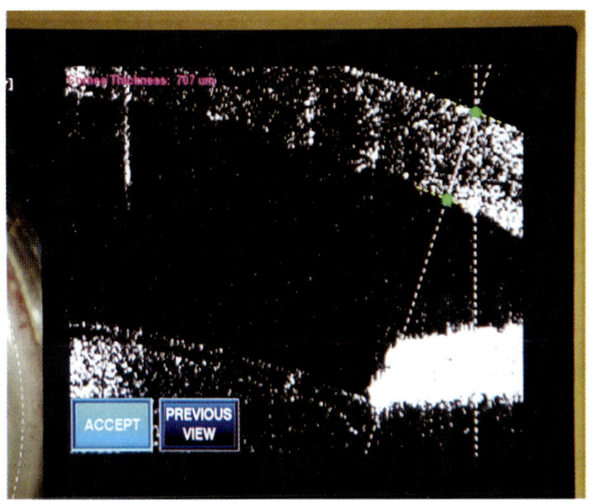

Fig. 21.5 OCT imaging of the cornea in the area of the arcuate incision

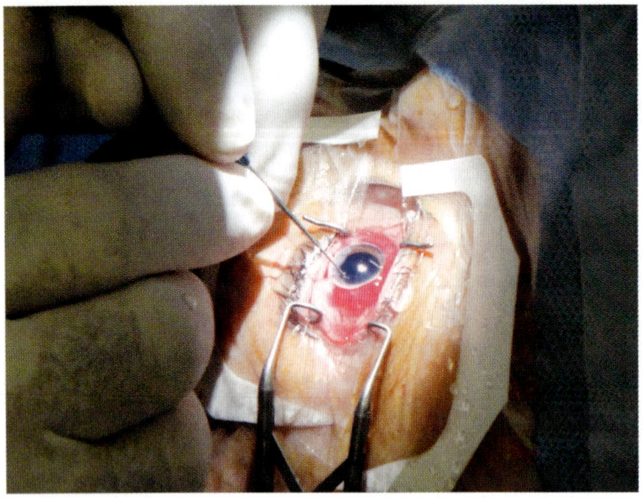

Fig. 21.6 Sinsky hook used to open the femtosecond arcuate incisions

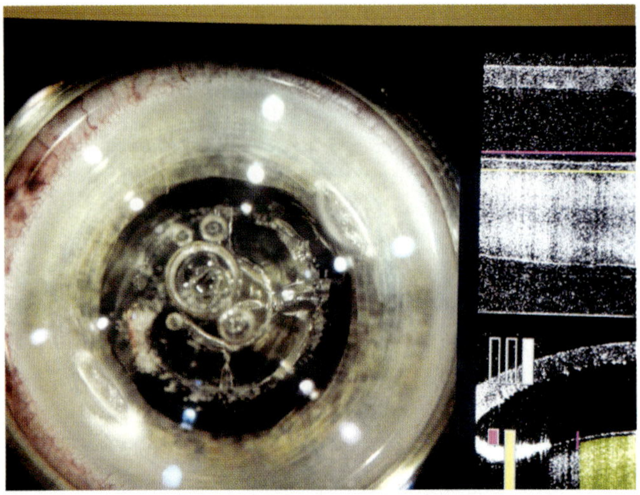

Fig. 21.7 Femtosecond arcuate incisions being performed

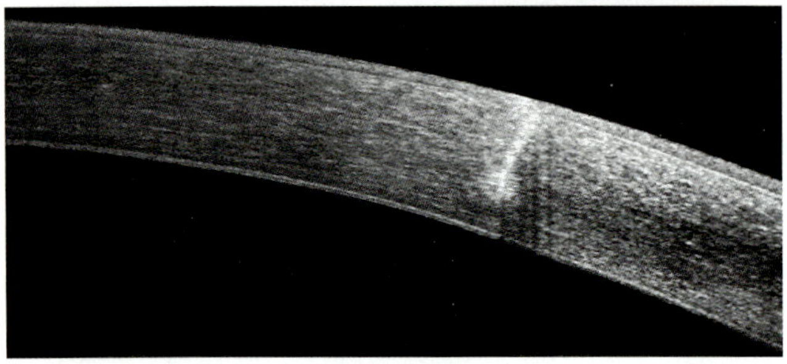

Fig. 21.8 OCT demonstrating the depth of the arcuate incision

cision of image-guided laser technology. Refractive incisions are now computer-controlled and do not rely on surgeon skill or experience. The use of a femtosecond laser system will provide faster, safer, easier, customizable, adjustable, and fully repeatable astigmatic incisions. Removing the inconsistencies in the astigmatic procedure will improve our understanding and accuracy of astigmatic incisions and should provide improved refractive results and patient satisfaction. The ability to perform intrastromal ablations cannot be achieved with manual incisions. This technology offers limitless potential but will require development with different nomograms and surgeons as well as expanded clinical experience.

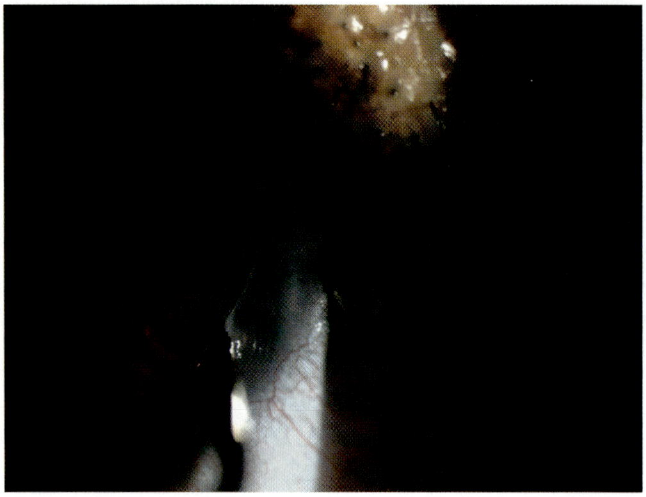

Fig. 21.9 Forceps used to open the femtosecond arcuate incison at the slit lamp

ACKNOWLEDGMENT

Ms Aaisha Gulani,
Research Assistant: Gulani Vision Institute

REFERENCES

1. Pager CK, McClusky PJ, Retsas C. Cataract surgery in Australia: a profile of patient-centered outcomes. Clin Exp Ophthalmol. 2004;32:388-92.
2. Hawker MJ, Madge SN, Baddeley PA, Perry SR. Refractive expectations of patients having cataract surgery. J Cataract Refract Surg. 2005;31:1970-5.
3. Nichamin LD. Nomogram for limbal relaxing incisions. J Cataract Refract Surg. 2006;32(9):1048.
4. Ferrer-Blasco T, Montés-Micó R, Peixoto-de-Matos SC, González-Méijome JM, Cerviño A. Prevalence of corneal astigmatism before cataract surgery. J Cataract Refract Surg. 2009;35(1):70-5.
5. Hill W. Expected effects of surgically induced astigmatism on AcrySof toric intraocular lens results. J Cataract Refract Surg. 2008;34(3):364-7.

6. Duffey RJ, Leaming D. US trends in refractive surgery: 2008 ISRS/AAO survey. International Society of Refractive Surgery and American Academy of Ophthalmology. Atlanta, Georgia. 8 November 2008. ISRS meeting.
7. Nichamin LD. Astigmatism control. Ophthalmol Clin North Am. 2006;19(4):485-93.
8. Wang L, Misra M, Koch DD. Peripheral corneal relaxing incisions combined with cataract surgery. J Cataract Refract Surg. 2003;29(4):712-22.
9. Budak K, Friedman NJ, Koch DD. Limbal relaxing incisions with cataract surgery. J Cataract Refract Surg. 1998;24(4):503-8.
10. Gills JP. Treating astigmatism at the time of cataract surgery. Curr Opinion Ophthalmol. 2002;13(1):2-6.
11. Oshika T, Shimazaki J, Yoshitomi F, et al. Arcuate keratometry to treat corneal astigmatism after cataract surgery: a prospective evaluation of predictability and effectiveness. Ophthalmology. 1998;105(11):2012-6.
12. Maloney WF, Grindle L, Sanders D, Pearcy D. Astigmatism control for the cataract surgeon: a comprehensive review of surgically tailored astigmatism reduction (STAR). J Cataract Refract Surg. 1989;15(1):45-54.
13. Price FW, Grene RB, Marks RG, Gonzales JS; for the ARC-T Study Group. Astigmatism reduction clinical trial: a multicenter prospective evaluation of the predictability of arcuate keratotomy. Arch Ophthalmol. 1995;113(3):277-82.
14. Devgan U. Corneal correction of astigmatism during cataract surgery. Cataract and Refractive Surgery Today. January 2007:41-4.
15. Tejedor J, Murube J. Choosing the location of corneal incision based on preexisting astigmatism in phacoemulsification. Am J Ophthalmol. 2005;139(5):767-76.
16. Kaufmann C, Peter J, Ooi K, Phipps S, Cooper P, Goggin M; for The Queen Elizabeth Astigmatism Study Group. Limbal relaxing incisions versus on-axis incisions to reduce corneal astigmatism at the time of cataract surgery. J Cataract Refract Surg. 2005;31(12):2261-5.
17. Muller-Jensen K, Fischer P, Siepe U. Limbal relaxing incisions to correct astigmatism in clear corneal cataract surgery. J Refract Surg. 1999;15(5):586-9.
18. Akura J, Matsuura K, Hatta S, Otsuka K, Kaneda S. A new concept for the correction of astigmatism: full-arc depth-dependent astigmatic keratotomy. Ophthalmology. 2000;107(1):95-104.
19. Faktorovich EG, Maloney RK, Price FW Jr. Effect of astigmatic keratotomy on spherical equivalent: results of the Astigmatism Reduction Clinical Trial. Am J Ophthalmol. 1999;127(3):260-9.
20. Price FW, Grene RB, Marks RG, Gonzales JS. Astigmatism reduction clinical trial: a multicenter prospective evaluation of the predictability of arcuate keratotomy. Evaluation of surgical nomogram predictability. ARC-T Study Group. Arch Ophthalmol. 1995;113(3):277-82. [Erratum in: Arch Ophthalmol. 1995;113(5):577.]
21. Oshika T, Shimazaki J, Yoshitomi F, et al. Arcuate keratotomy to treat corneal astigmatism after cataract surgery: a prospective evaluation of predictability and effectiveness. Ophthalmology. 1998;105(11):2012-6.
22. Bradley MJ, Coombs J, Olson RJ. Analysis of an approach to astigmatism correction during cataract surgery. Ophthalmologica. 2006;220(5):311-6.
23. AcrySof Toric SA60T4IOL [package insert]. Fort Worth, TX: Alcon Laboratories, Inc.; 2009.
24. Masket S, Sarayba M, Ignacio T, Fram N. Femtosecond laser-assisted cataract incisions: architectural stability and reproducibility. J Cataract Refract Surg. 2010;36(6):1048-9.
25. Vogel A, Noack J, Hüttman G, Paltauf B. Mechanisms of femtosecond laser nanosurgery of cells and tissues. Appl Phs B. 2005;81:1015-47.
26. Ratkay-Traub I, Ferincz IE, Juhasz T, Kurtz RM, Krueger RR. First clinical results with the femtosecond neodymium-glass laser in refractive surgery. J Refract Surg. 2003;19(2):94-103.
27. Nordan LT, Slade SG, Baker RN, Suarez C, Juhasz T, Kurtz RM. Femtosecond laser flap creation for laser in situ keratomileusis: six-month follow-up of initial US clinical series. J Refract Surg. 2003;19:8-14.
28. Kezirian GM, Stonecipher KG. Comparison of the IntraLase femtosecond laser and mechanical keratomes for laser in situ keratomileusis. J Cataract Refract Surg. 2004; 30(4):804-11.
29. Montés-Micó R, Rodríquez-Galietero A, Alió JL. Femtosecond laser versus mechanical keratome LASIK for myopia. Ophthalmology. 2007;114:62-8.
30. Steinert RF, Ignacio TS, Sarayba MA. "Top hat"-shaped penetrating keratoplasty using femtosecond laser. Am J Ophthalmol. 2007;143(4):689-91.
31. Nichamin L. Femtosecond laser technology applied to lens-based surgery. Medscape Ophthalmology. June 22, 2010. http://www.medscape.com/viewarticle/723864. Accessed July 20, 2011.
32. Nagy Z, Takacs A, Filkom T, Sarayba M. Initial clinical evaluation of an intraocular femtosecond laser in cataract surgery. J Refract Surg. 2009;25(12):1053-60.
33. Abbey A, Ide T, Kymionis GD, Yoo SH. Femtosecond laser-assisted astigmatic keratotomy in naturally occurring high astigmatism. Br J Ophthalmol. 2009;93(12):1566-9.
34. Slade SG. Femtosecond laser arcuate incision astigmatism correction in cataract surgery. Presented at: the ASCRS Cornea Day; March 25, 2011; San Diego, CA.

22

Femtosecond Laser in Ophthalmology—Past, Present and Future

Arun C Gulani, Ming X Wang (USA)

Chapter Outline

FEMTOSECOND LASER APPLICATIONS

As technology advances in parallel to our surgical finesse and outcome expectations, few industries have grown with the exponent power as femtosecond laser technologies in the field of ophthalmology.

Recent years have seen a rapid development of femtosecond laser technology in anterior segment surgery.

In this book chapter, we shall briefly look at the history and development of femtosecond laser technology over the past 15 years, summarize the present status and propose the future directions.

Femtosecond laser has found wide range of application in ophthalmology in recent years, particularly in corneal and lens surgery, and for a good reason. In contrast to excimer laser which works through a photoablative principle, femtosecond laser interacts with tissue through a photodisruptive process. Focusing laser energy in a very small spatial confinement in a very short period of time generates on the order of 10^{23} number of electrons. The sudden generation of such large number of electrons creates a plasma shock wave and its mechanical explosive force pushes apart the adjacent corneal or lens tissue, resulting in tissue separation. We can get a sense of the magnitude of such an explosive pushing-apart force if one images that there is a sudden generation of 1 billion people in a tiny room in a split of second, then the shear largeness of the mass of the tissue explosive expansion will push aside the four walls of the room with a tremendous force. This is the process of photodisruption through plasma shock wave generation. YAG laser capsulotomy works by this principle.

Why is it then we need femtosecond laser to perform these delicate corneal and lens surgery today? The need for femtosecond laser for anterior segment surgery comes from two considerations. First of all, though the power of the laser to separate the corneal tissue needs to be quite significant, the energy itself of the laser pulse (joule) can not be large at all. In fact, in order to harmlessly transmit through superficial corneal tissue and to reach corneal stroma and lens, a laser beam energy needs to be very small, i.e. on the order of 10^{-6} joules. At any energy level which is significantly higher than this, the laser will cause significant collateral thermal damage and burn of the corneal or lens tissue along the pathway of laser beam.

The second consideration is the tremendous power that is needed to separate corneal or lens tissue. The power needs to be huge, on the order of 10^9 watt. Now, how can we generate such a high laser power to separate corneal or lenticular tissue, but at the same time keeping the energy level of laser to such a low level? Here is where femtosecond laser comes in. An ultrashort laser such as femtosecond laser presents a unique solution to this problem. Let's look at this equation:

$$P = E/t \qquad \text{(Eq. 1).}$$

where P is the power of laser beam (in watts), E is the laser energy (in joules), and t is the laser pulse duration (in seconds). If we use a laser with an extremely short pulse duration t (on the order of 10^{-15}, or fe mtosecond), then, even with a laser at low energy level E of 10^{-6} joules, the power P that such a short-pulse laser generates will be of sufficient magnitude (10^{9}) to separate efficiently corneal or lens tissue.

To get a sense how short a femtosecond laser pulse is: light travels at a speed of 300,000 km/second and can wrap around the equator of the earth 7.5 times in one second. If one gives such a fast-traveling light 100 femtosecond to travel, light would just barely make one circle around one strand of hair!

It is also important to point out that, unlike excimer laser which is an ultraviolet laser (wavelength = 193 nm), femtosecond laser operates at infrared range (wavelength = 1060 nm). Hence, unlike excimer laser which has a very shallow level of tissue penetration, infrared laser such as femtosecond laser can in fact reach a deep penetration of tissue. We take advantage of the deep penetrating ability of YAG laser when we perform YAG posterior capsulotomy. The ability of femtosecond laser to penetrate deeply into the eye can also presents a risk. When we use femtosecond laser to perform corneal or lens surgery, we do not want the laser to reach too deep into the eye and cause unintended intraocular tissue damage. Therefore, when we use femtosecond laser to perform eye surgery, it is very important to realize that the focus of the laser, by which we achieve precise depth of laser-tissue interaction, is of paramount importance. When we use femtosecond laser for LASIK, it focuses at precisely 100 um at below corneal epithelial surface to create a corneal flap. In contrast, when femtosecond laser is used to perform cataract lens surgery, the focus needs to be much deeper (at more 4 mm below corneal epithelial surface). The deeper the level of laser focusing, the more imprecise the focusing itself is. As such, femtosecond laser-assisted lens surgery is much more technically challenging due to the increased difficulty in précising focusing on a deeper tissue level than corneal surgery such as LASIK. One can get a sense of such a degradation of precision of focusing if one makes two corneal flaps in LASIK, one at 100 um one at 200 um. One would notice that the interface surface irregularty with the 200 um cut is much more significant due to its deeper tissue penetration and increased lack of precision of laser focusing.

Femtosecond laser is an infrared laser (wavelength = 1040 nm) with a mode-locking system that generate ultrashort pulses (on the order of 100 femtosecond, or 10^{-13} seconds). The mechanism of laser-tissue interaction is through a massive instantenous ionization of electrons, which creates a plasma wave, and the rapid and forceful expansion of the plasma wave pushes adjacent tissue apart, achieving tissue separation. The laser power P (in watts), laser energy E (in joules) and pulse width t (in seconds) are related to each other through the following fomula:

$$P = E / t$$

With energy E on the order 10^{-6} joule (to avoid collateral thermal tissue damage), the pulse duration t needs to be very small (10^{-13} seconds), in order to produce sufficient amount of power P (10^{7} watts) need to achieve proper tissue separation.

Early research in mid 1990s of femtosecond laser for anterior segment research focused on its potential ability to achieve corneal refractive surgery through an intrastromal approach. The theory was that femtosecond laser could create a corneal intrastromal cavitation bubbles, which then collapse, resulting in flattening of central cornea, to achieve myopic refractive correction. By late 1990's, researchers realize that such an intrastromal approach would not work due to two reasons:

1. Cavitation bubbles can be varied in sizes, creating irregular corneal surface.
2. Intrastromal cavitation bubbles stay in the cornea afterwards, for weeks to even month, making a central myopic correction clinically not feasible since the bubbles accumulates in the central part of corneal stroma for a long time, blocking sight.

Having failed in such a total intrastromal approach, research in femtosecond laser changed direction in late 1990's and the laser was evaluated for its ability to make corneal flap for LASIK. 1999, FDA granted 510(k) approval, for Intralase (AMO) femtosecond laser for the application of flap-making for LASIK. 510(k) was subsequently granted for Femtec 20/10 Perfect Vision in 2004, and for VisuMax (Carl Zeiss Meditec) and Femto-LDV (Ziemer) in 2006, and Wavelight (Alcon) FS-200 in 2009.

Early generations of femtosecond lasers did not produce smooth corneal cut since they used low repetition rate (15 kHz) and high pulse energy levels (several microjoules). The newer generations of lasers reduced the pulse energy to sub-microjoule and nanojoule range, with a spot size of 1 um and an increased repetition rate to up to 1 MHz, achieving a much smoother corneal cut and a procedure time to as short as 12 seconds. In addition, increased flexibility of flap cutting geometry was achieved, giving surgeons improved ability to customize the flap.

Compared with microkeratome LASIK flap, femtosecond laser has the following advantages:

1. Being able to generate customized flap dimensions, including flap thickness, diameter, hinge position and angle and side cut angle.
2. Being able to customize roster lamellar cut energy. For post-RK eyes, the rostral lamellar cut energy an be increased to enable the flap to be lifted more readily. The rostral lamellar cut energy is optimized to avoid its being too small (which results in difficulty in lifting the flap) or too large [which results in increase risk of corneal stromal inflammation such as deep lamellar keratitis (DLK)].
3. Being able to produce consistent flap thickness and the degree of hydration of the stromal bed, thus reducing the enhancement rate and reducing the risk of too thin flap (risk of button-hole) and too thick flap (increasing the risk of ectasia).

In recent years, femtosecond laser application has been broadened beyond LASIK and used in the following areas:

1. Penetrating keratoplasty: Customized edge shape can be achieved, e.g. mushroom, top-hat or zig-zag, which increase the surface area of donor/host tissue contact.
2. Deep lamellar cut, to aid big-bubble DALK procedure.
3. Ring channel creation for Intacs segment for treating keratoconus.
4. Astigmatic keratotomy.
5. All-femtosecond laser (FLEx/SmILE) procedure (VisuMax) in which a second femtosecond laser cut is made to produce a central corneal intrastromal lenticule which can then be removed through a small size incision. The benefits of the small-incision technique presumably include: (a) achieving the benefit of using just one laser; (b) inducing lesser degree of dry eyes; (c) reducing the risk of ectasia, since it has been shown that the vertical side cut (rather than the horizontal lamellar cut) is the true culprit of causing biomechanical instability.
6. Possible application in endothelial transplantation.

In recent years, femtosecond laser has enjoyed rapid growth in the area of cataract surgery. Femtosecond laser has been shown to adequately perform corneal incision, astigmatic cut, capsulorhexis and lens fragmentation. Due to the increased interest in premium lens implantation, and the increased need to optimize the refractive outcome after cataract surgery, femtosecond laser may indeed find its place in cataract surgery with its more predictable and consistent tissue cutting capability such as centration, size and circularity of capsulorhexis which has been shown to be important for effective lens position and refractive outcome. The use of femtosecond in cataract surgery has been shown to reduce or even eliminate the use of phaco energy, thus reducing the amount of postoperative inflammation, trauma to intraocular tissues such as corneal endothelium and capsule and improving visual outcome.

Looking forward, femtosecond laser has potential to be used in a wide range of applications:

1. Intrastromal ring vertical cuts, to treat presbyopia (Intracor).
2. Direct lens modification of Crystalens for presbyopia.
3. Treating trabecular meshwork, for the treatment of glaucoma.
4. Possible application of scleral tunnel creation and implantation of scleral segment, for the treatment of presbyopia.
5. Presbyopic lentotomy in treatment of presbyopia performed with ultrashort FS-laser impulses.

Femtosecond laser technology has enjoyed a tremendous growth in anterior segment surgery in the past decade. It has led to novel approaches and raised the bar on surgical standards.

The intrinsic capability of this laser to go through transparent tissue (without causing collateral damage) and make intra-tissue cut in any shape or form renders this exciting technology a great future of expanding applications in the field for eye surgery as we continue our quest for "Supervision".

Acknowledgement: Aaisha Gulani.

Research Assistant: Gulani Vision Institute.

BIBLIOGRAPHY

1. Colin J, et al. Ophthalmology. 2001;108:1409-14.
2. Colin J, Simonpoli S. J Fr Ophtalmo. 2005;28(2):205-17.
3. Durrie DS, et al. Cataract Refract Surg. 2005;31:120-6.
4. Holzer MP, et al. Invest Ophthalmol Vis Sci. 2006;47:2828-31.
5. Rabsilber TM, et al. Klin Monatsbl Augenheilkd. 2005;222 (Suppl. 1):S24.
6. Ruckhofer J, Klin Monatsbl Augenheilkd. 2002;219(8):557-74.
7. Tunc Z, et al. J Fr Ophtalmol. 2003;26(8):824-30.

23

Presbyopia Treatment Using the IntraLase Femtosecond Laser

Francisco Sánchez León, Jaime Martiz (México)

Chapter Outline

- IntraLase Femtosecond Laser
- Intrastromal Rings

In 2008, at the NovaVision Center in Mexico City, we began an evaluation of presbyopia correction using the IntraLase FS femtosecond laser (Abbott Medical Optics, Santa Ana, CA) to create intrastromal rings. Our study shows we are able to reshape the tissue of the cornea and obtain the desired refractive effect (Figs 23.1 and 23.2). The exciting part of this project is that just as the IntraLase femtosecond technology improved safety and predictability of LASIK flap creation; it confirms it will also improve the safety, predictability and control of corneal presbyopia surgery.

The IntraLase FS can produce changes in the biomechanical forces of the cornea without disrupting descemet's or bowman's membrane and the epithelium and endothelium remain intact with no damage (Fig. 23.3). The procedure produces

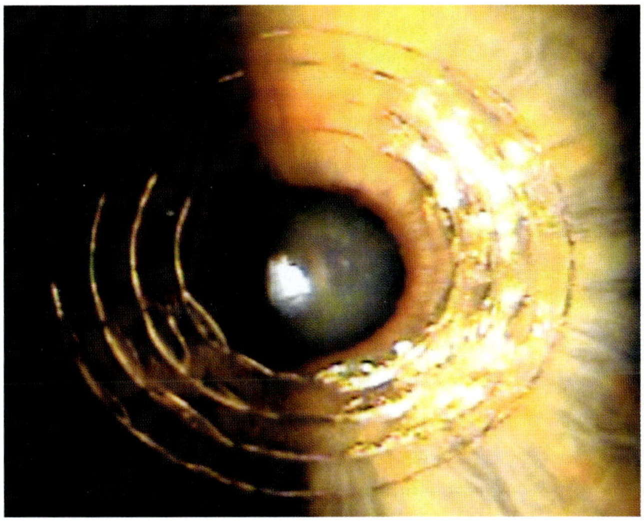

Fig. 23.1 Intrastromal rings in the cornea placed by the IntraLase femtosecond laser (Abbott Medical Optics). Postoperative 1 hour

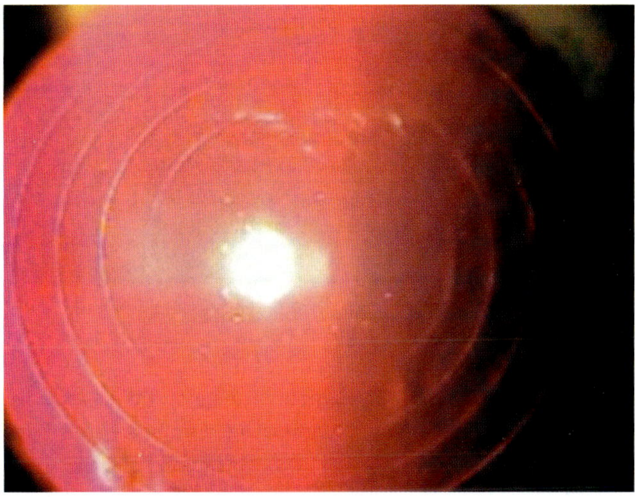

Fig. 23.2 Intrastromal rings in the cornea placed by the IntraLase femtosecond laser (Abbott Medical Optics). Postoperative 1 day

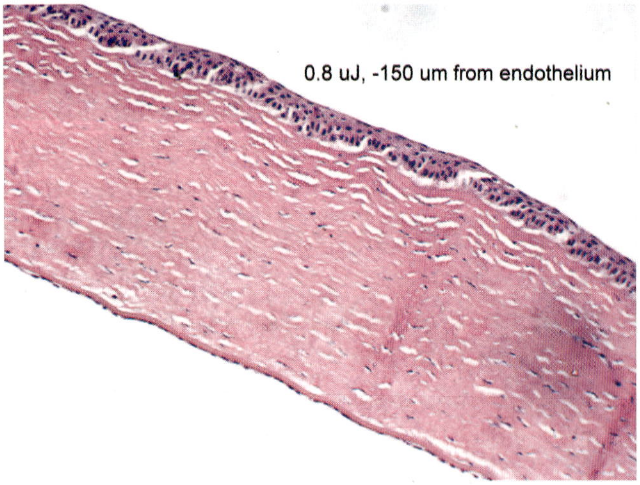

Fig. 23.3 Here is a cross-sectional image of a cornea showing areas of focal hyalinization (loss of organization) and occasional swirling of the collagen fibers in the stroma following intrastromal femtosecond laser ablation

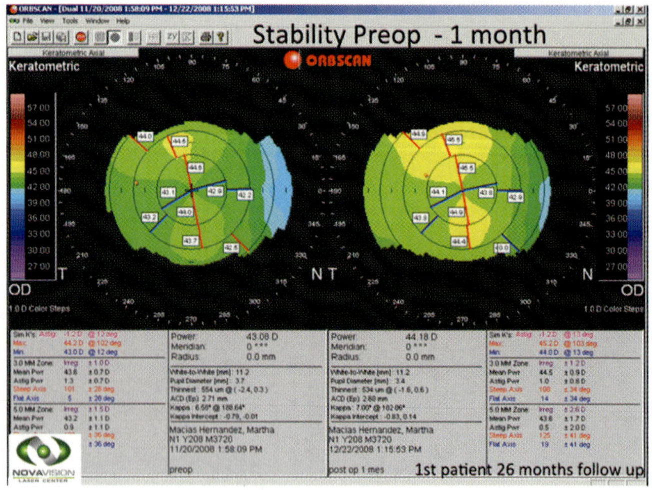

Fig. 23.4 ORBScan topography showing steepening of the central portion of the cornea. Preoperatively vs 26 months postoperatively

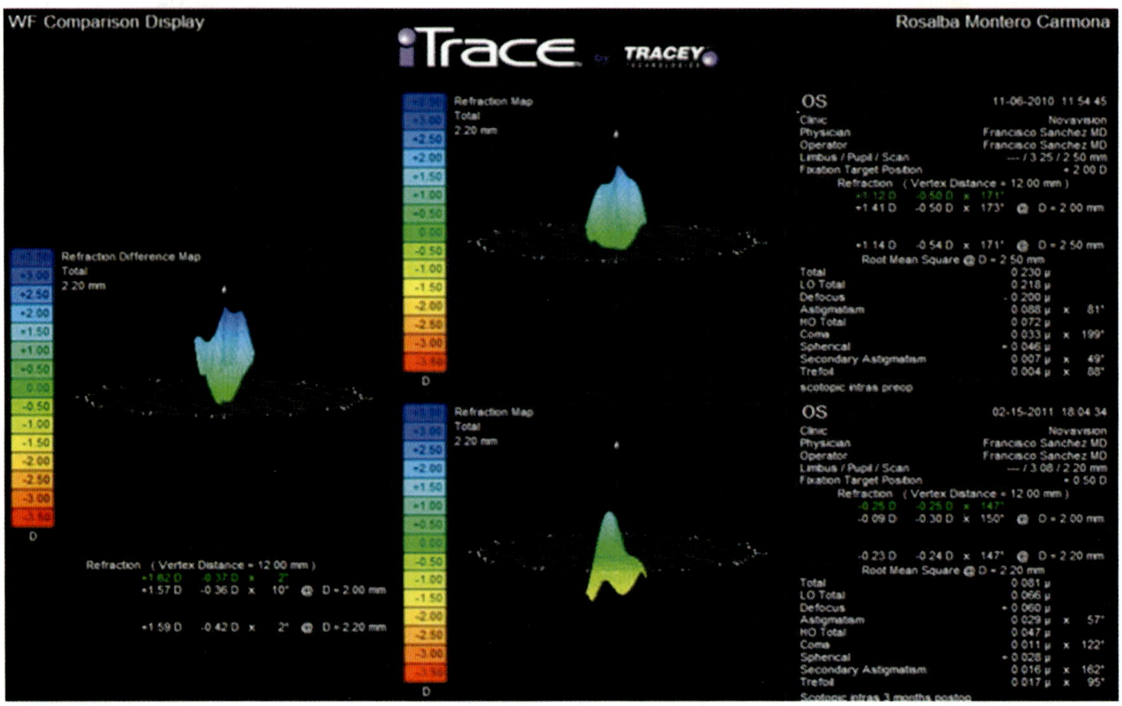

Fig. 23.5 Eyesys topography showing steepening of the central portion of the cornea. Preoperatively vs three month postoperatively

steepening of the central portion of the cornea to improve near vision and maintain preoperative distance vision (Figs 23.4 and 23.5).

The IntraLase femtosecond laser creates intrastromal rings that are uniform in size and depth (Fig. 23.6). The first step during the surgery is to mark the cornea on the Purkinjie's reflex and the treatment is centered based on that mark. (Figs 23.7 and 23.8) To ensure biomechanical effect, a soft applanation technique is used to create a flat patient interface and is applied to the eye to compress collagen fiber and guarantee intrastromal effect. One of the advantages of using a flat interface is this technique requires minimal energy and

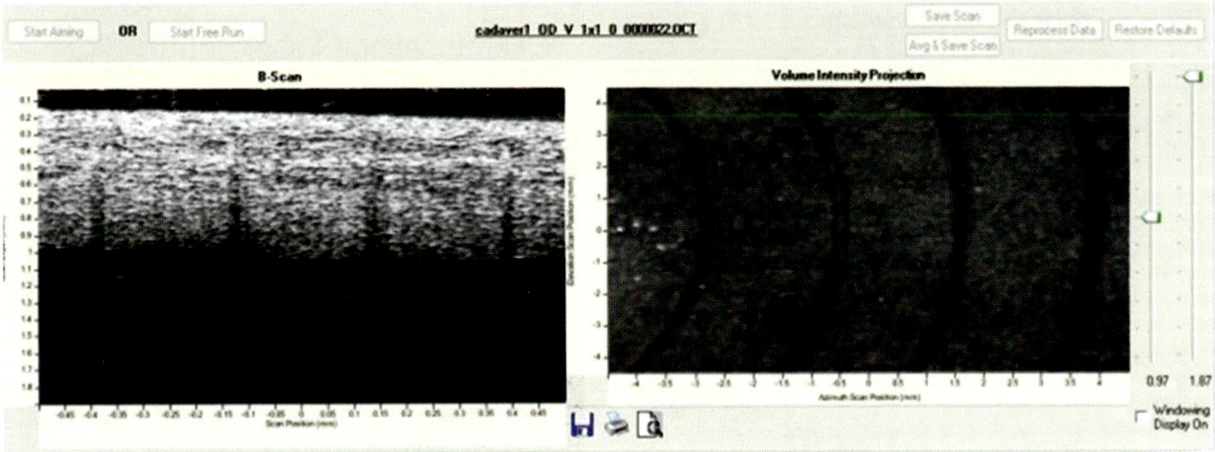

Fig. 23.6 Bioptigen OCT images of the femtosecond laser intracorneal rings that are uniform in size and depth

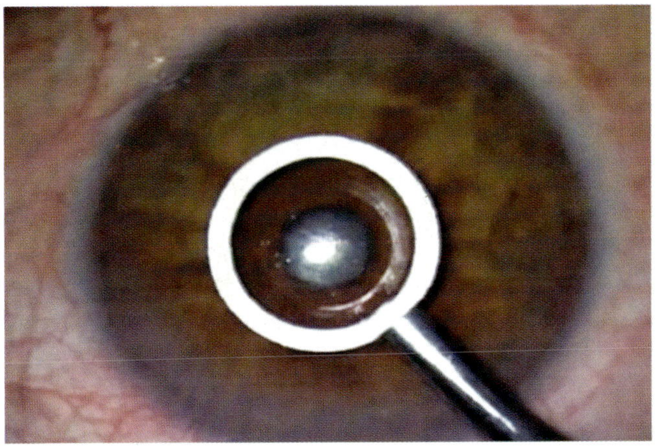

Fig. 23.7 Marking of the cornea on the first Purkinje's reflex

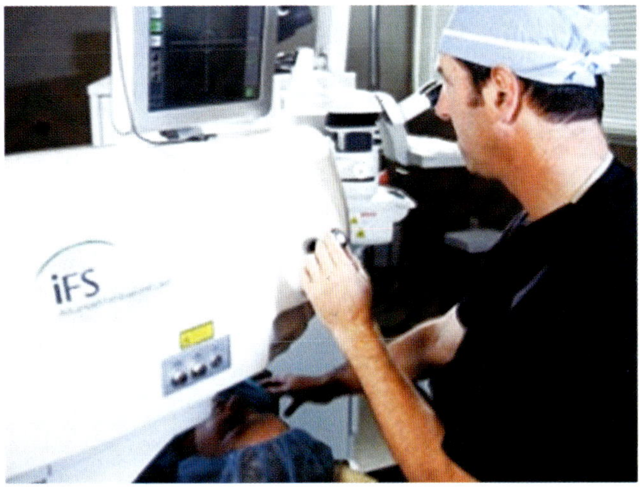

Fig. 23.9 Software is capable of allowing the surgeon to ensure centration as part of the last step before starting the procedure

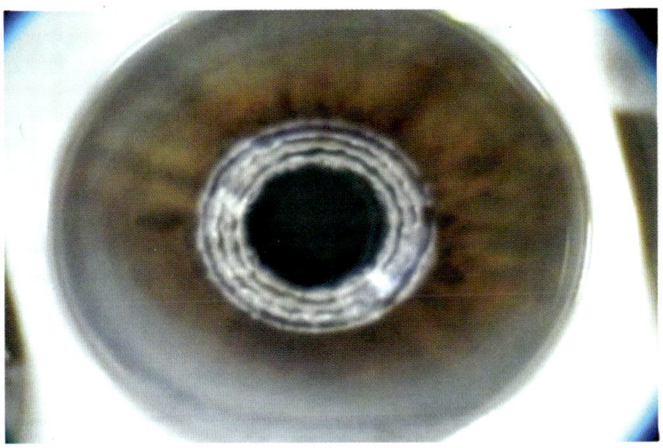

Fig. 23.8 Locate and verify the mark on the Purkinje's reflex to ensure centration

produces a straight incision. When performing the soft applanation technique, there is minimal or no distortion of the eye centration. However, the software is capable of assisting the surgeon in accurate centration as part of the last step before starting the procedure, if needed (Fig. 23.9).

A nomogram has been developed to utilize the technology which is customized to each patient according to age, refraction, keratometric values, corneal thickness and pupil diameter.

STUDY RESULTS

Encouraging early results at one day, one week, one month and three months were reported. There were 19 patients treated with a mean age of 52.58 years. The uncorrected near visual acuity was equal to or less than 20/80, minimal near addition of

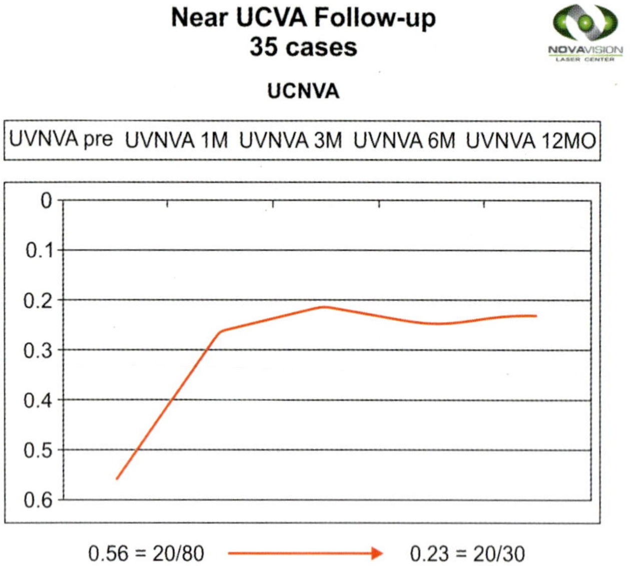

Fig. 23.10 A Visual Acuity LogMAR graph showing three months follow-up, the mean monocular uncorrected near visual acuity improved from 0.56 to 0.23 both photopic and mesopic conditions

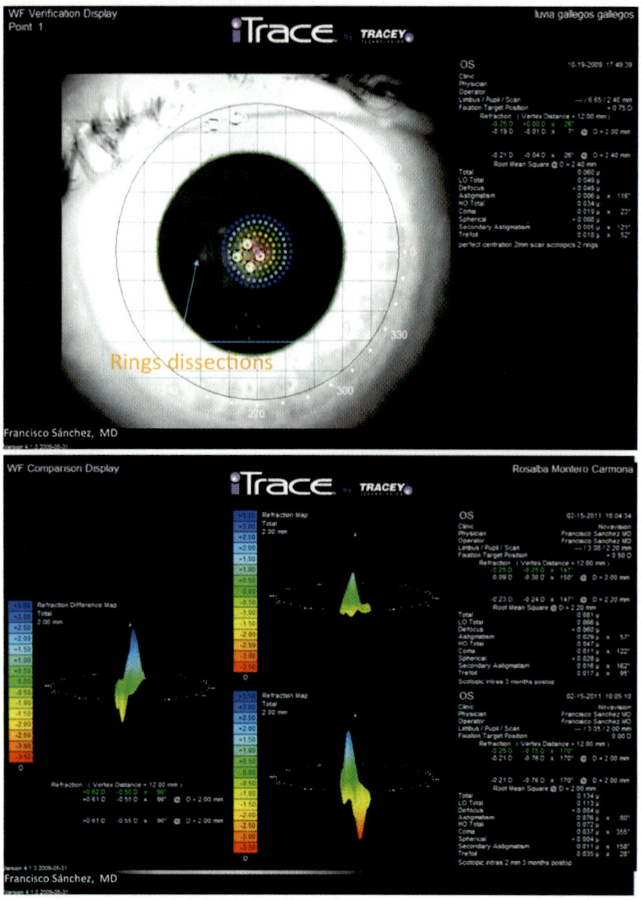

Fig. 23.11 Wavefront capture inside the intrastromal rings showing no change in cylinder

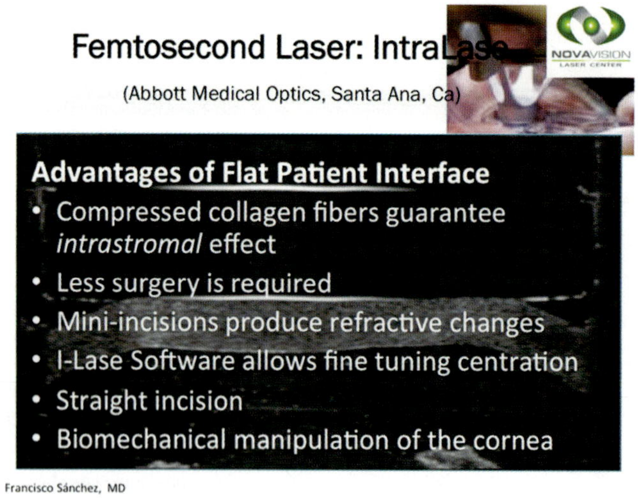

Fig. 23.12 Advantages of a flat patient interface

+1.50 diopters for best corrected near visual acuity (BCNVA) and the non-dominant eye was treated.

At three months follow-up, the mean monocular uncorrected near visual acuity improved from 20/80 (LogMAR 0.56) to 20/30 (LogMAR 0.23) in photopic and mesopic conditions (Fig. 23.10). The mean distance corrected near visual acuity (DCNVA) was 20/30 (J3) under photopic and mesopic conditions. The average best distance for near vision was 38 cm under mesopic and photopic conditions with and without correction. The mean intermediate visual acuity was J3 with and without correction for distance.

In addition, we observed no significant change in the monocular uncorrected distance visual acuity preoperatively 0.16 ± 0.19 diopters to postoperatively 0.14 ± 0.2 diopters. The mean refractive cylinder remained unchanged. (Fig. 23.11) None of the patients reported severe visual symptoms at one month or three months, as well as no night time glare or halos.

The early results of this ongoing study demonstrate good distance and near vision with minimal risk of infection with monocular intrastromal rings treatment for presbyopia created by the IntraLase femtosecond laser.

BIBLIOGRAPHY

1. Doccio F, Sacchi CA, Marshall J. Lasers Opthalmol. 1986;1:83.
2. Intrastromal correction of presbyopia using Femtosecond Laser. ASCRS. F Sanchez. M Martiz, 2011.
3. Juhasz T, Kurtz RM, et al. Optics & Photonics. 2002;13:24-9.
4. Martiz J, et al. Highlights of Ophthalmology, 9, The Art of laser in Ophthalmology. 2004;9:107.

Index

Page numbers followed by *f* refer to figure and *t* refer to table